High Resolution NMR

Theory and Chemical Applications

THIRD EDITION

High Resolution NMR

Theory and Chemical Applications

THIRD EDITION

Edwin D. Becker

National Institutes of Health
Bethesda, Maryland

Academic Press

San Diego   London   Boston   New York
Sydney   Tokyo   Toronto

Academic Press
A Harcourt Science and Technology Company
525 B Street, Suite 1900, San Diego, California 92101-4495, U.S.A.
http://www.apnet.com

Academic Press
24-28 Oval Road, London NW1 7DX, UK
http://www.hbuk.co.uk/ap/

Library of Congress Catalog Card Number: 99-62782

International Standard Book Number: 0-12-084662-4

PRINTED IN THE UNITED STATES OF AMERICA
99 00 01 02 03 04 MM 9 8 7 6 5 4 3 2 1

Contents

3   *Instrumentation and Techniques*

8   Relaxation

9   Pulse Sequences

10 *Two-Dimensional NMR*

11 *Density Matrix and Product Operator Formalisms*

13 *Elucidation of Molecular Structure and Macromolecular Conformation*

14 *NMR Imaging and Spatially Localized Spectroscopy*

Preface to the Third Edition

The preface to the first edition of this book begins with the statement: "Few techniques involving sophisticated instrumentation have made so rapid an impact on chemistry as has nuclear magnetic resonance." I went on to say that "dramatic instrumental developments have improved resolution and sensitivity by factors ~ 50 from the first commercial instruments. Today more than 1500 NMR spectrometers are in use, and the scientific literature abounds in reference to NMR data." That was in 1969! NMR had indeed made remarkable advances, but 30 years later, experimental NMR technology bears almost no resemblance to the then "sophisticated" instruments that were almost exclusively based on slow passage, continuous wave techniques. Fourier transform methods were in their infancy, with utilization severely limited by the use of dedicated computers with 4–8 kilobytes (not megabytes) of memory. New conceptual approaches to NMR, together with remarkable advances in materials, computers, electronics, and magnet and probe technology, have brought NMR to its current level, with no indication of an end in sight.

In 1969 there were few books on NMR; indeed, I wrote *High Resolution NMR* as a textbook that would provide a sound theoretical approach but at a level that would be more acceptable to graduate students and other chemists than the more rigorous books then available. Now there are many excellent books on NMR, covering a wide range from elegant treatises on NMR theory to comprehensive treatments of modern techniques and applications to the elucidation of the structure of both small molecules and macromolecules. Why then a third edition?

I think there is still a gap between several very good introductory books that take a rather empirical approach to NMR and books written for more sophisticated users. In particular, product operator and density matrix formalisms are virtually ignored in the more elementary books but are treated as almost self-evident in the more advanced works. My experience is that most chemists are frightened by the terminology, yet realize that some understanding of these approaches is needed to understand many contemporary applications of NMR.

In this edition I have tried to retain the largely "conversational" approach of previous editions, with the assumption that the reader has no prior background in NMR but does have some understanding of basic quantum mechanics. I have given careful attention to developing and interrelating four approaches to NMR theory—steady-state energy levels, the rotating vector picture, the density matrix, and the product operator formalism. Each has its advantages in treating different areas, and I hope that I have been able to use the more familiar methods to clarify the concepts and language of the density matrix and product operator formalisms. Within the space available, I have tried to introduce and explain the major 2D and 3D NMR methods and to indicate in general terms how they are used in the elucidation of the structure of small molecules and proteins. I have also included introductions to high resolution NMR studies in solids and to NMR imaging.

I have benefited from reading many of the classic and contemporary books on NMR, and at the end of each chapter I have suggested many of these as sources for further reading. In an introductory book, I believe it is usually more important to point the reader to other books and reference works than to cite original literature.

I am deeply indebted to many colleagues for their help and encouragement. In particular, Ad Bax (National Institutes of Health) educated me on many aspects of 2D NMR and encouraged me to prepare the third edition of this book, as did the late Regitze Vold (University of California, San Diego) and Attila Szabo (NIH). Several people kindly read specific chapters and provided corrections and suggestions for improvement, including Ad Bax, Marius Clore, Robert Tycko (all at NIH), Angel de Dios (Georgetown University), Thomas C. Farrar (University of Wisconsin), Eugene Mazzola (Food and Drug Administration), and Kyou-Hoon Han (Korea Research Institute of Bioscience and Biotechnology). Herman J. C. Yeh (NIH) generously obtained a number of spectra to illustrate specific topics, as did Joseph Barchi (NIH) and Daron Freedberg (FDA). Dennis Torchia, Robert Tycko, and Marius Clore (all at NIH) gave me original spectra and figures. I thank all of these individuals as well as a number of scientists and publishers who agreed to the reproduction of published figures, as given in specific references.

Edwin D. Becker

CHAPTER 1

Introduction

1.1 **Origins and Early History of NMR**
1.2 **High Resolution NMR: An Overview**
1.3 **Additional Reading and Resources**

The first nuclear magnetic resonance (NMR) was detected early in 1938 in a molecular beam, and the first studies of NMR in bulk materials were carried out about 8 years later. Over the following decades, NMR has grown from an interesting and important study of a physical phenomenon to an indispensable technique in a very wide variety of fields. In organic chemistry NMR is arguably one of the two most important tools for the elucidation of molecular structure. In structural biology NMR rivals x-ray crystallography in providing precise three-dimensional structures for proteins and other macromolecules, but NMR goes beyond x-ray crystallography in furnishing information on internal mobility and overall molecular motion in both large and small molecules.

NMR has become one of the best methods for obtaining anatomical images of human subjects and animals (under the common name *magnetic resonance imaging*, MRI) and for exploring physiological processes. Materials science uses NMR spectroscopy and imaging to describe the structure, motion, and electronic properties of heterogeneous and technologically important substances. NMR is widely used in the food industry to measure moisture content and to assess the quality of certain foodstuffs. NMR is used to measure the flow of liquids in pipes in industrial processes and to observe the flow of blood in human beings. NMR is used in the exploration for petroleum, and it has even been used to search for submarines during wartime.

There is a vast literature on NMR and many books that describe all of these applications (and others). For example, the eight-volume *Encyclopedia of Nuclear Magnetic Resonance* describes a wide variety of NMR studies.[1] Underlying all of these diverse applications, however, is a basic theory that treats the behavior of nuclear magnets in a magnetic field. In this book we cannot hope to provide a comprehensive treatment of NMR theory or application. Our intention, rather, is to explore the basic physical phenomena, primarily from the viewpoint of the chemist, to illustrate ways in which NMR spectroscopy can be applied in chemistry and structural biology and to point out very briefly how NMR imaging is applied in materials science and biomedicine. Although we use mathematics as needed, we try to provide an intuitive understanding insofar as possible, sometimes at the expense of mathematical rigor.

This chapter begins with some historical background to put NMR into perspective and concludes with a survey of the topics that we shall examine in the remainder of the book.

1.1 ORIGINS AND EARLY HISTORY OF NMR

Many atomic nuclei behave as though they are spinning, and as a result of this spin each nucleus possesses an angular momentum ($\mathbf{p}$) and a magnetic moment ($\boldsymbol{\mu}$). These two nuclear properties were first observed indirectly in the very small splittings of certain spectral lines (hyperfine structure) in the visible and ultraviolet spectra of atoms. In 1924 Pauli[2] suggested that this hyperfine structure resulted from the interaction of magnetic moments of nuclei with the already recognized magnetic moments of electrons in the atoms. Analysis of the hyperfine structure permitted the determination of the angular momentum and approximate magnetic moments of many nuclei. The concept of nuclear spin was strengthened by the discovery (through heat capacity measurements) of *ortho* and *para* hydrogen[3]—molecules that differ only in having the spins of the two constituent nuclei oriented in the same and opposite directions, respectively.

In the early 1920s Stern and Gerlach[4,5] had shown that a beam of atoms sent through an inhomogeneous magnetic field is split into two discrete beams because of magnetic effects arising from the quantized orbital angular momentum of the electrons. As the atoms move through the inhomogeneous field, the interaction between the magnetic moment of the atom and the magnetic field gradient causes the beam to deviate in either a positive or negative direction, depending on the quantum state of the electrons in the atom. During the 1930s, refinements of the Stern–Gerlach technique permitted the observation of much smaller effects from nuclear magnetic moments, and the laboratory of I. I. Rabi at Columbia University became a major center for such studies. With the development of

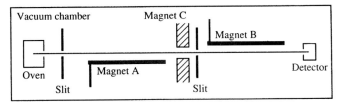

FIGURE 1.1   Schematic representation of the apparatus for molecular beam studies. Magnets A and B were electric wires that produce inhomogeneous magnetic fields. Magnet C was also an electric wire in initial experiments but was replaced by a magnet producing a homogeneous magnetic field for the resonance experiments. From *Encyclopedia of Nuclear Magnetic Resonance,* D. M. Grant and R. K. Harris, Eds. Copyright 1996 John Wiley & Sons Limited. Reproduced with permission.

improved techniques that employed three successive inhomogeneous magnets, as illustrated in Fig. 1.1, Rabi's group was able to measure the signs and magnitudes of the magnetic moments of hydrogen and deuterium to an accuracy of about 5%.[6]

In 1938 Rabi and his colleagues made a major improvement in beam experiments by substituting a very homogeneous magnet for the middle inhomogeneous magnet of Fig. 1.1 and by applying a radio frequency (rf) electromagnetic field to the molecules as they passed through the homogeneous field. With all molecules thus experiencing the same static magnetic field B, quantum theory showed that nuclear magnetic moments would interact with the field to give quantized energy levels separated by energy ΔE,

$$\Delta E = \pm (\mu/p)B \qquad (1.1)$$

with the orientation of the spin differing in the two levels. Electromagnetic energy of a sharply defined frequency, $\nu = \Delta E/h$ (where h is Planck's constant), would then be absorbed by the nuclear spin system and cause a small but measurable deflection of the beam. The plot in Fig. 1.2 shows such a deflection and was the first observation of nuclear magnetic resonance.[7] This work earned Rabi a Nobel Prize in 1944.

With the resonance method, far higher precision in measurement of magnetic moments was possible, but such studies could be performed only in molecular beams under very high vacuum. The idea of observing NMR in *bulk* materials— solids, liquids, or even gases at normal pressure—was also considered. However, as we see in Chapter 2, the NMR signal in such circumstances is expected to be weak, and its observation would be difficult. In fact, in 1936 C. J. Gorter had attempted to detect resonance absorption by measuring the very slight increase in temperature of the sample on absorption of radio frequency energy. Although his experiment was unsuccessful (probably because of a poor choice of samples and experimental conditions), Gorter published the negative results[8] and discussed them with Rabi. In fact, Gorter's visit to Rabi's laboratory in 1937 played a major role in Rabi's decision to attempt to measure NMR in a molecular beam.[9]

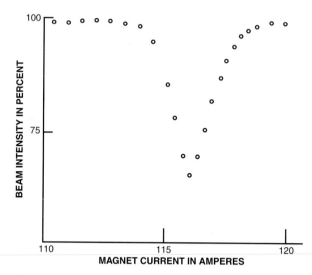

FIGURE 1.2   Plot of detector current resulting from refocused molecular beam as a function of the value of the field of magnet C. One ampere corresponds to 18.4 gauss. A radio frequency of 3.518 MHz was applied. From Rabi *et al.*[7]

It was not until 1945 that nuclear magnetic resonance was actually found in bulk materials. In that year Edward M. Purcell, Henry C. Torrey, and Robert V. Pound, three physicists who were completing wartime work on radar at the Massachusetts Institute of Technology (MIT), decided to look for resonance absorption from nuclear magnetic moments, using radio frequency techniques developed during World War II. Aware of Rabi's work in molecular beams and of Gorter's failed attempt in 1936 (and a second unsuccessful effort in 1942 using a different method), they nevertheless believed that a measurable signal could be obtained. They located a suitable magnet at nearby Harvard University and constructed a simple radio frequency spectrometer based on radar techniques. On December 15, 1945, they observed the first NMR signal from a large sample of paraffin wax.[10]

Meanwhile, several months earlier at Stanford University, Felix Bloch, a well-known and highly respected physicist, had formed a collaboration with W. W. Hansen, an electronics expert, and recruited a graduate student, Martin Packard, to attempt to observe the effects of nuclear magnetism by a very different method. In early January 1946, they observed a signal from the hydrogen nuclei in a sample of liquid water by a method that Bloch called *nuclear induction*[11] (which we describe in detail in Chapter 2). Within a few months, Bloch and Purcell recognized that they were observing two different aspects of the same phenomenon, and Bloch agreed that the method should be called *nuclear magnetic resonance* to conform with Rabi's terminology. Ironically, it is Bloch's experimental approach

I. I. Rabi   Felix Bloch   Edward M. Purcell
Photos © The Nobel Foundation.

to nuclear induction that is now used in virtually all NMR spectrometers. Bloch
and Purcell shared the 1952 Nobel Prize in Physics, because the two discoveries
at MIT–Harvard and at Stanford were regarded as independent and essentially
concurrent.

When we now speak of nuclear magnetic resonance, we are discussing the kind
of NMR discovered by Bloch and Purcell, that is, nuclear magnetic resonance in
bulk materials. The early work in NMR was concentrated on the elucidation of
the basic phenomena (much of which we cover in Chapters 2 and 8) and on the
accurate determination of nuclear magnetic moments, which were of interest in
elucidating aspects of the structure of the atomic nucleus.

NMR attracted little attention from chemists until, in 1950, it was discovered
that the precise resonance frequency of a nucleus depends on the state of its
chemical environment. Proctor and Yu[12] discovered accidentally that the two nitrogen nuclei in NH_4NO_3 had different resonance frequencies, and similar results
were reported independently for fluorine and hydrogen.[13–15] In 1951, Arnold,
Dharmatti, and Packard found separate resonance lines for chemically different
protons in the same molecule.[16] The discovery of this so-called *chemical shift* set
the stage for the use of NMR as a probe into the structure of molecules.

1.2 HIGH RESOLUTION NMR: AN OVERVIEW

It is found that chemical shifts are very small, and in order to observe such effects
one must study the material under suitable conditions. In solids, where intermolecular motion is highly restricted, internuclear interactions cause such a great
broadening of resonance lines that chemical shift differences are masked (as we
discuss in detail in Chapters 2 and 7). In liquids, on the other hand, rapid molecular tumbling causes these interactions to average to zero, and sharp lines are
observed. Thus, in the early days of NMR studies, there came to be a distinction

between *broad line* NMR and *high resolution,* or narrow line, NMR. Broad line NMR has played an important role in developing the theory of NMR, in elucidating structural features in many solids, and in measuring magnetic properties in superconductors. However, we shall deal almost exclusively with high resolution NMR, in which chemical shifts can be distinguished. The instrumental requirements for observing high resolution NMR spectra demand extremely homogeneous magnets of high field strength and very stable electronics, as we point out in Chapter 3. Although initially restricted to liquids and gases with rapid molecular tumbling, high resolution NMR was later extended to solids by means of clever techniques that we describe in Chapter 7.

An NMR spectrum is obtained by placing a sample in a homogeneous magnetic field and applying electromagnetic energy at suitable frequencies to conform to Eq. 1.1. In Chapter 2 we examine in detail just how NMR spectra arise, and in Chapter 3 we delve into the procedures by which NMR is studied. Before we do so, however, it may be helpful to see by a few examples the type of information that can be obtained from an NMR spectrum.

Basically there are three quantities that can be measured in a high resolution NMR spectrum: (1) frequencies of the resonance lines, (2) their areas, and (3) widths or shapes of the lines. Figure 1.3 shows the spectrum of a simple compound, diacetone alcohol. This spectrum arises only from the resonance of the hydrogen nuclei (^{1}H, or protons) in the molecule, because (as we see in Chapter 2) we are able to set conditions to obtain a spectrum from only one type of

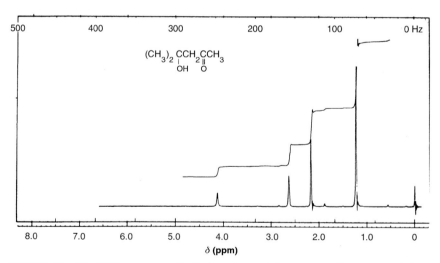

FIGURE 1.3 ^{1}H NMR spectrum of 4–hydroxy–4–methyl–2–pentanone (diacetone alcohol) at 60 MHz. Assignments of lines: δ/ppm = 1.23, $(CH_3)_2$; 2.16, $CH_3C{=}O$; 2.62, CH_2; 4.12, OH. See Chapter 4 for definition of the δ scale.

nuclide at a time. Each of the spectral lines can be assigned to one of the functional groups in the sample, as indicated in the figure. The step function shown along with the spectrum is an integral, with the height of each step proportional to the area under the corresponding spectral line. The chemical shift, on the scale shown at the bottom, is given the symbol δ and, as we see in Chapter 4, is expressed in parts per million (ppm) relative to a reference nucleus, usually the proton in tetramethylsilane (TMS), which gives the line at zero in this spectrum.

Several important features are illustrated in Fig. 1.3. First, the chemical shift is clearly demonstrated, for the resonance frequencies depend on the chemical environment (as we study in detail in Chapter 4). Second, the areas under the lines are different and, as we shall see when we examine the theory in Chapter 2, the area of each line is proportional to the number of nuclei contributing to it. Third, the widths of the lines are different; in particular, the line due to the OH is considerably broader than the others. (We examine the reasons for different line widths in Chapters 2 and 8.)

The spectrum in Fig. 1.3 is particularly simple. A more typical spectrum—that of a natural product, ferrugone—is given in Fig. 1.4. This spectrum consists of single lines well separated from each other, as were the lines in Fig. 1.3, and of simple multiplets. (The inset shows the multiplets on an expanded abscissa scale.) The splitting of single lines into multiplets arises from interactions between the nuclei called *spin–spin coupling*, a phenomenon first recognized and explained by Herbert Gutowsky and coworkers in 1951. This is an important type of information obtainable from an NMR spectrum. In Chapter 5 we inquire into the origin

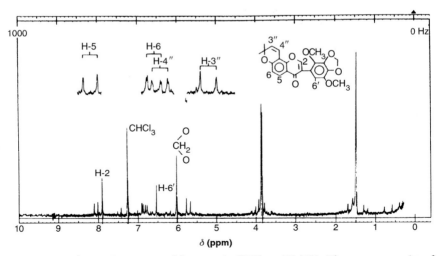

FIGURE 1.4 ^{1}H NMR spectrum of ferrugone in CDCl$_3$ at 100 MHz. The spectrum consists of single lines and multiplets, with assignments indicated. Courtesy of R. J. Highet (National Institutes of Health).

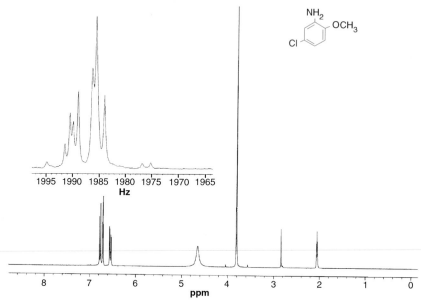

FIGURE 1.5 ^{1}H NMR spectrum of 5-chloro-*ortho*-anisidine in a mixed solvent of CDCl$_3$ and acetone-d_6 at 300 MHz. Assignments of lines; δ/ppm = 3.8, CH$_3$; 4.65, NH$_2$; 6.5–6.7, aromatic protons; 2.1, acetone-d_5; 2.8, water impurity. The inset shows the aromatic region on an expanded scale. Eleven lines with no regularity in spacing are clearly observed over a range of about 20 Hz (0.067 ppm.) Spectrum courtesy of Herman J. C. Yeh (National Institutes of Health).

of spin coupling and note that we can obtain information of chemical value from measurements of the coupling.

Figure 1.5 shows the spectrum of a simple molecule, 5-chloro-*ortho*-anisidine, but the three aromatic protons give rise to a spectrum at δ = 6.5–6.7 ppm that shows little regularity in spacing or intensity distribution. In Chapter 6 we find that such spectra can be explained by a very simple application of basic quantum mechanical concepts to the interaction of nuclear spins with each other and with an applied magnetic field. We shall find that when the magnetic field (and the corresponding observation frequency) is increased by a sufficient amount, the apparently irregular spacings of lines often give way to more readily discerned simple multiplets (called *first-order* spectra) of the sort shown in Fig. 1.4. However, as indicated in Fig. 1.5, even at the moderately high field of a 300 MHz spectrometer, complex spectral patterns may appear. Moreover, as we see in Chapter 6, other features in the spectrum are independent of field, so that even at the highest magnetic field obtainable, spectra are occasionally observed that are far from the first-order approximation.

In Chapter 7 we look at certain effects that are seen only in solids and other ordered phases where molecules are fixed in position relative to each other.

Herbert S. Gutowsky   John S. Waugh   Alexander Pines

As noted earlier, NMR lines from solids are very broad, but as early as 1958 Raymond Andrew and Irving Lowe independently showed that rapid spinning of the sample under certain conditions can reduce the line width. During the period 1965–1975 a number of clever new methods were developed (primarily in the laboratories of John Waugh, Alex Pines, Peter Mansfield, and Robert Vaughn) to employ both pulsed and continuous wave NMR methods to obtain narrow lines in solids. We shall describe the basic ideas behind these techniques and give some illustrations of their application.

The ways in which nuclei *relax*, or return to their equilibrium state after some perturbation, will be taken up briefly in Chapter 2 and discussed more fully in Chapter 8. A great deal of useful information on molecular structure and dynamics can be extracted from an analysis of nuclear magnetic relaxation processes.

Most modern NMR studies depend on the application of sequences of rf pulses and the analysis of the observed data by methods based on the Fourier

Richard R. Ernst   Ray Freeman   Erwin L. Hahn

transform (FT). The use of very short rf pulses to excite NMR in simple systems (usually giving only one NMR line) was introduced by Erwin Hahn in 1949, but application to molecules with high resolution multiline spectra came only after a seminal publication by Richard Ernst and Weston Anderson in 1966.[17] FT NMR has now become *the* method for observing high resolution NMR spectra. In Chapter 2 we describe the concepts of the use of rf pulses and FT NMR, and we discuss many experimental aspects of these methods in Chapters 3 and 9. We shall see that with suitable rf pulse sequences nuclear magnetizations can be manipulated at will. In fact, the "spin gymnastics" that the magnetizations undergo provide us with a versatile means of examining molecular structure.

In 1971, Jean Jeener conceived of a new way of applying pulse sequences and displaying the results in terms of two separate frequency scales. The concept was soon developed into the very important method of *two-dimensional NMR*,

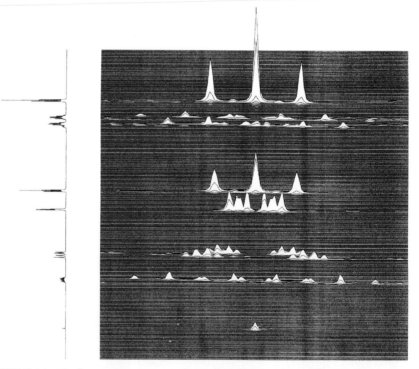

FIGURE 1.6   The ^1H NMR spectrum of allylbutylether ($CH_2{=}CHCH_2OCH_2$ $CH_2CH_2CH_3$) in $CDCl_3$ at 300 MHz, shown at the left, is separated into its components by a two-dimensional NMR experiment. The multiplets arising from spin–spin coupling interactions are spread into a plane on the basis of the ^1H chemical shifts, thus better displaying overlapping peaks and facilitating their assignment. Details of this experiment are given in Chapter 10. Spectrum courtesy of Herman J. C. Yeh (National Institutes of Health).

Kurt Wüthrich   Ad Bax   Paul C. Lauterbur

particularly in the laboratory of Richard Ernst (who received the Nobel Prize in Chemistry in 1991 for his major contributions to FT NMR in both one and two frequency dimensions) and the laboratory of Ray Freeman. Suitable pulse sequences can provide spectra spread in a plane instead of along one axis, as illustrated in Fig. 1.6. As we see in Chapter 10, such 2D spectra contain resonance peaks that are very informative in elucidating the structure of molecules.

A complete understanding of the processes involved in 2D NMR requires a more powerful theoretical underpinning than used in most of the book, so Chapter 11 is devoted to an introduction to the *density matrix* and *product operator* formalisms. These methods are not familiar to many chemists, but they are simple outgrowths of ordinary quantum mechanics. We examine the basic ideas and apply this theory in Chapters 11 and 12 to describe some of the most frequently used 1D and 2D NMR experiments.

In Chapter 13, we summarize the ways in which NMR methods can be applied to determine the structures of moderate size (usually organic) molecules. We then describe briefly the use of NMR methods in conjunction with computational energy minimization procedures to determine the three-dimensional structures of macromolecules. With the systematic approach put forth by Kurt Wüthrich and his coworkers beginning in the 1980s, along with more sophisticated pulse techniques developed by Ad Bax and others, it is now possible to determine structures of proteins with precision rivaling that of x-ray crystallography.

Finally, Chapter 14 demonstrates the way in which a discovery by Paul Lauterbur in 1973 has been developed into a widely applied NMR method for obtaining two-dimensional and three-dimensional *images* of macroscopic objects and for permitting high resolution NMR spectra to be obtained in localized volumes of such objects. These methods combine many of the techniques discussed in earlier chapters and are of great importance in studying both anatomy and function in living biological systems (including humans) and in examining the structure of heterogeneous solids.

1.3 ADDITIONAL READING AND RESOURCES

This book is intended to provide only an introduction to the vast field of high resolution NMR, and many aspects are treated only superficially. To facilitate further study, we provide at the end of each chapter a few suggestions for additional reading—occasionally to original literature, but primarily to some of the many good books in the field and to databases where useful NMR information may be obtained.

The *Encyclopedia of Nuclear Magnetic Resonance*[1] is the most comprehensive work in the field, with hundreds of authoritative articles on specialized topics. Most of the material is at a rather advanced level, but many articles can be understood with the background provided in this book. Each article in the *Encyclopedia* provides many references to original literature. We shall refer to a number of specific articles in later chapters.

Two other much shorter books, which give excellent definitions and summaries of a wide range of NMR topics are *A Dictionary of Concepts in NMR* by S. W. Homans[18] and *A Handbook of Nuclear Magnetic Resonance* by Ray Freeman.[19]

Several serial publications provide continuing updates in the field of high resolution NMR. *Concepts in NMR* gives largely pedagogical presentations of specific topics. *Advances in Magnetic and Optical Resonance, Progress in NMR Spectroscopy, NMR: Basic Principles and Progress,* and *Annual Reports on NMR Spectroscopy* include a variety of articles ranging from explanatory presentation to comprehensive reviews of individual areas. *Specialist Periodical Reports: NMR* gives thorough annual reviews of NMR literature in selected areas.

The history of NMR, summarized very briefly in Section 1.1, is covered most comprehensively in the *Encyclopedia of NMR*, the first volume of which contains a long overview article and 200 primarily autobiographical articles by leaders in the field from its inception to recent years. Shorter articles on specific historical topics are given in volume 28 of *Progress in NMR Spectroscopy*.

The Theory of NMR

In this chapter we discuss the basic physics underlying NMR. A few fundamental concepts in both classical and quantum mechanics are assumed. Standard texts in these areas should be consulted if additional background is needed.

2.1 NUCLEAR SPIN AND MAGNETIC MOMENT

We know from basic quantum mechanics that angular momentum is always quantized in half-integral or integral multiples of $\hbar$, where $\hbar$ is Planck's constant divided by 2π. For the electron spin, the multiple (or spin quantum number) is $\frac{1}{2}$, but the value for nuclear spin differs from one nuclide to another as a result of interactions among the protons and neutrons in the nucleus. If we use the symbol I to denote this nuclear spin quantum number (or, more commonly, just *nuclear spin*), we can write for the maximum observable component of angular momentum

$$p = I\hbar = Ih/2\pi \qquad (2.1)$$

We can classify nuclei, then, according to their nuclear spins. There are a number of nuclei that have $I = 0$ and hence possess no angular momentum. This class of nuclei includes all those that have both an even atomic number and an even mass number; for example, the isotopes ^{12}C, ^{16}O, and ^{32}S. These nuclei, as we shall see, cannot experience magnetic resonance under any circumstances. The spins of a few of the more common nuclei are:

$$I = \tfrac{1}{2}: {}^1H, {}^3H, {}^{13}C, {}^{15}N, {}^{19}F, {}^{31}P$$

$$I = 1: {}^2H(D), {}^{14}N$$

$$I > 1: {}^{10}B, {}^{11}B, {}^{17}O, {}^{23}Na, {}^{27}Al, {}^{35}Cl, {}^{59}Co$$

The nuclei that have $I > \tfrac{1}{2}$ have a nonspherical nuclear charge distribution and hence an electric quadrupole moment Q. We shall consider the effect of the quadrupole moment later. Our present concern is with all nuclei that have $I \neq 0$, because each of these possesses a magnetic dipole moment, or a *magnetic moment* μ. We can think of this moment qualitatively as arising from the spinning motion of a charged particle. This is an oversimplified picture, but it nevertheless gives qualitatively the correct results that (1) nuclei that have a spin have a magnetic moment and (2) the magnetic moment is collinear with the angular momentum vector. We can express these facts by writing.

$$\boldsymbol{\mu} = \gamma \mathbf{p} \tag{2.2}$$

(We use the customary boldface type to denote vector quantities.) The constant of proportionality γ is called the *magnetogyric ratio* and is different for different nuclei, because it reflects nuclear properties not accounted for by the simple picture of a spinning charged particle. (Sometimes γ is termed the gyromagnetic ratio.) While p is a simple multiple of $\hbar$, μ and hence γ are not and must be determined experimentally for each nuclide (usually by an NMR method). Properties of most common nuclides that result from nuclear spin are given in Appendix A.

2.2 THEORETICAL DESCRIPTIONS OF NMR

The theory of NMR can be approached in several ways, each of which has both advantages and disadvantages. We will emphasize several apparently different theoretical strategies in explaining various aspects of NMR. Actually, these approaches are entirely consistent but utilize differing degrees of approximation to give simple pictures and intuitively appealing explanations that are correct within their limits of validity.

Transitions between Stationary State Energy Levels

We have seen that NMR is a quantum phenomenon, and to some extent we can develop a theoretical framework by using the same sort of treatment used for other branches of spectroscopy: We first find the eigenvalues (energy levels) of the quantum mechanical Hamiltonian operator that describes the nuclear spin system and then use time-dependent perturbation theory to predict the probability of transitions among these energy levels. This procedure provides the frequencies and relative intensities that characterize an NMR spectrum. As we see in the following section and in more detail in Chapter 6, this approach works very well in explaining many basic NMR phenomena, and it accounts quantitatively for many spectra illustrated in Chapter 1. However, because this approach is based on the time-*in*dependent Schrödinger equation, it cannot account for the time-dependent phenomena that occur as nuclear spins respond to perturbations such as application of short radio frequency pulses.

Classical Mechanical Treatment

By using the time-dependent Schrödinger equation, we can follow the behavior of a nuclear spin in the presence of an applied magnetic field, and we show in Section 2.6 that the results are consistent with an appealing vector picture. Moreover, we shall show that the summation of nuclear magnetic moments over the whole ensemble of molecules that constitutes a real sample leads to a macroscopic magnetization that can be treated according to simple laws of classical mechanics. Many NMR phenomena can be quite well understood in terms of such simple classical treatments of magnetization vectors; indeed, in the first full paper on NMR, Bloch used classical mechanics to describe many important features of the technique. We shall employ this vector picture extensively, particularly in Chapters 2, 9, and 10.

Yet, it is clear that the classical mechanics approach is inadequate simply because it ignores quantum effects. To some extent, these features can be arbitrarily grafted onto a classical picture, but as we shall see, many of the newer and now most important NMR studies cannot be understood this way.

The Density Matrix

Fortunately there is a simple mathematical formalism that gives us the best of the quantum and classical approaches. By recasting the time-dependent Schrödinger equation into a form using a so-called density operator, physicists have long been able to follow the development of a quantum system with time. This formalism preserves all quantum features but permits a natural explanation of time-dependent phenomena, so that effects of phase and coherence can be understood. For NMR it turns out to be rather easy to set up the necessary theory in terms of a matrix—the density matrix—and to use simple matrix manipulations to follow the time course of the nuclear spin system.

Why, then, do we not ignore the preceding approaches and go directly to the density matrix? The answer is that the density matrix method, while conceptually simple, becomes very tedious and gives very little physical insight into processes that occur. We will come back to the density matrix in Chapter 11, but only after we have developed a better physical feeling for what happens in NMR experiments.

Product Operators

In Chapter 11 we shall also introduce the product operator formalism, in which the basic ideas of the density matrix are expressed in a simpler algebraic form that resembles the spin operators characteristic of the steady-state quantum mechanical approach. Although there are some limitations in this method, it is the general approach used to describe modern multidimensional NMR experiments.

2.3 STEADY-STATE QUANTUM MECHANICAL DESCRIPTION

As with other branches of spectroscopy, an explanation of many aspects of NMR requires the use of quantum mechanics. Fortunately, the particular equations needed are simple and can be solved exactly.

The Hamiltonian Operator

We are interested in what happens when a magnetic moment $\boldsymbol{\mu}$ interacts with an applied magnetic field $\mathbf{B}_0$—an interaction commonly called the *Zeeman* interaction. Classically, the energy of this system varies, as illustrated in Fig. 2.1*a*, with the cosine of the angle between $\boldsymbol{\mu}$ and $\mathbf{B}_0$, with the lowest energy when they are aligned. In quantum theory, the Zeeman appears in the Hamiltonian operator $\mathscr{H}$,

$$\mathscr{H} = -\boldsymbol{\mu} \cdot \mathbf{B}_0 \tag{2.3}$$

By substituting from Eqs. 2.1 and 2.2, we have

$$\mathscr{H} = -\gamma \hbar \mathbf{B}_0 \cdot \mathbf{I} \tag{2.4}$$

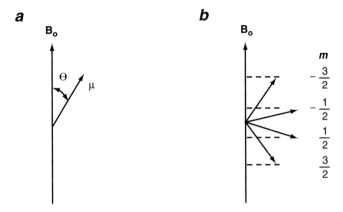

FIGURE 2.1   (*a*) Relative orientations of magnetic moment μ and magnetic field B_0. (*b*) Quantized projection of μ on B_0

Here **I** is interpreted as a quantum mechanical operator. From the general properties of spin angular momentum in quantum mechanics, it is known that the solution of this Hamiltonian gives energy levels in which

$$E_m = -\gamma \hbar m B_0 \tag{2.5}$$

The quantum number m may assume the values

$$-I, -I + 1, \ldots , I - 1, I$$

There are thus $2I + 1$ energy levels, each of which may be thought of as arising from an orientation of μ with respect to B_0 such that its projection on B_0 is quantized (See Fig. 2.1*b*). Equation 2.5 shows that the energy separation between the states is linearly dependent on the magnetic field strength.

Spin Wave Functions

For the particularly important case of $I = \frac{1}{2}$ there are just two energy levels and two wave functions, which are called α and β. It will be worthwhile to explore the properties of the functions α and β in more detail. α and β are orthogonal and normalized eigenfunctions of the quantum mechanical operators I_z and I^2 but not of I_x and I_y. Specifically, as shown in standard quantum mechanics texts, the effects of these operators may be expressed as

$$
\begin{array}{lll}
I_z\alpha = \frac{1}{2}\alpha & I_x\alpha = \frac{1}{2}\beta & I_y\alpha = \frac{1}{2}i\beta \\[2mm]
I_z\beta = -\frac{1}{2}\beta & I_x\beta = \frac{1}{2}\alpha & I_y\beta = -\frac{1}{2}i\alpha
\end{array}
\tag{2.6}
$$

For many purpose, it is simpler to use the so-called *raising* and *lowering* operators, I^+ and I^-, respectively. These operators are defined as

$$I^+ = I_x + iI_y \qquad I^- = I_x - iI_y \tag{2.7}$$

The terminology comes from the fact that I^+ applied to any spin wave function causes its quantum number to increase by one unit, while I^- causes a decrease of one unit. Thus, for α and β

$$I^+\beta = \alpha \qquad I^-\alpha = \beta$$

Alternatively, we can write these relationships in the quantum mechanical bra-ket notation that emphasizes the results of integration of wave functions or expression as matrix elements:

$$\langle \alpha,\alpha \rangle = \langle \beta,\beta \rangle = 1 \qquad \langle \alpha,\beta \rangle = \langle \beta,\alpha \rangle = 0$$

$$\langle \alpha| I_z |\alpha \rangle = \tfrac{1}{2} \qquad \langle \alpha| I_x |\beta \rangle = \tfrac{1}{2} \qquad \langle \beta| I_x| \alpha \rangle = \tfrac{1}{2} \tag{2.9}$$

$$\langle \beta| I_z |\beta \rangle = -\tfrac{1}{2} \qquad \langle \alpha| I_y |\beta \rangle = -\tfrac{1}{2}i \qquad \langle \beta| I_y |\alpha \rangle = \tfrac{1}{2}i$$

Finally, the same information can be given in the concise form of Pauli matrices:

$$I_x = \frac{1}{2}\begin{bmatrix} 0 & 1 \\ 1 & 0 \end{bmatrix} \qquad I_y = \frac{1}{2}\begin{bmatrix} 0 & -i \\ i & 0 \end{bmatrix} \qquad I_z = \frac{1}{2}\begin{bmatrix} 1 & 0 \\ 0 & -1 \end{bmatrix} \tag{2.10}$$

It is very important to grasp these relationships and to become familiar with them, because they are fundamental to an understanding of the quantum aspects of NMR. We shall make use of these relationships repeatedly in this and later chapters.

Spectral Transitions

NMR arises from transitions between energy levels, just as in other branches of spectroscopy, with transitions induced by the absorption of energy from an applied electromagnetic field with a magnetic amplitude of $2B_1$. The interaction between this oscillating field and the nuclear magnetic moment is treated as a time-dependent perturbation:

$$\mathcal{H}' = 2\mu_x B_1 \cos 2\pi\nu t \tag{2.11}$$

From the well-established results of time-dependent perturbation theory, we find that the probability of transition per unit time between levels m and m' is

$$P_{mm'} = \gamma^2 B_1^2 |(m|I_x| m')|^2 \delta(\nu_{mm'} - \nu) \tag{2.12}$$

$2B_1$ is the magnitude of the radio frequency field applied in the x direction, with $\mathbf{B}_0$ in the z direction; $(m|I_x|m')$ is the quantum mechanical matrix element of the

x component of the nuclear spin operator and is zero unless $m = m' \pm 1$; and $\delta(\nu_{mm'} - \nu)$ is the Dirac delta function, which is zero unless $\nu_{mm'} = \nu$. The frequency corresponding to the energy difference between states m and m', $\nu_{mm'}$, is given by the Bohr relation

$$\nu_{mm'} = \frac{\Delta E_{mm'}}{h} = \frac{\gamma B_0 |m' - m|}{2\pi} \tag{2.13}$$

Several important points are contained in Eq. 2.12. First, the transition probability increases quadratically with both γ and B_1. [Whereas the energy absorbed varies as $B_1{}^2$, the observed NMR signal, which is proportional to an induced *voltage* in a coil (see Chapter 3), varies linearly with B_1.] Second, the matrix element furnishes the selection rule $\Delta m = \pm 1$, so that transitions are permitted only between adjacent energy levels and thus give only a single line at a frequency

$$\nu = \frac{\gamma}{2\pi} B_0 \tag{2.14}$$

Third, the resonance condition is expressed in the delta function. Actually, the delta function would predict an infinitely sharp line, which is unrealistic; therefore, it is replaced by a line shape function $g(\nu)$, which has the property that

$$\int_0^\infty g(\nu)\, d\nu = 1 \tag{2.15}$$

[In practice, $g(\nu)$ often turns out to be Lorentzian or approximately Lorentzian in shape.] Equation 2.12 becomes, then,

$$P_{mm'} = \gamma^2 B_1{}^2 |(m|I_x|m')|^2 g(\nu) \tag{2.16}$$

For nuclei with $I = \frac{1}{2}$ there is only one transition, so Eq. 2.16 becomes

$$P_{mm'} = \frac{1}{4}\gamma^2 B_1{}^2 g(\nu) \tag{2.17}$$

2.4 EFFECT OF THE BOLTZMANN DISTRIBUTION

The tendency of nuclei to align with the magnetic field and thus to drop into the lowest energy level is opposed by thermal motions, which tend to equalize the populations in the $2I + 1$ levels. The resultant equilibrium distribution is the usual compromise predicted by the Boltzmann equation. For simplicity we consider only nuclei with $I = \frac{1}{2}$, so that we need include only two energy levels, the lower corresponding to $m = +\frac{1}{2}$ (state α) and the upper to $m = -\frac{1}{2}$ (state β). For the $I = \frac{1}{2}$ system the Boltzmann equation is

$$\frac{n_\beta}{n_\alpha} = e^{-\Delta E/kT} \tag{2.18}$$

By substitution of the values of E from Eq. 2.5, and by introducing Eqs. 2.1 and 2.2, we find that this becomes

$$\frac{n_\beta}{n_\alpha} = e^{-\gamma\hbar B_0/kT} = e^{-2\mu B_0/kT} \tag{2.19}$$

Even for the largest magnetic fields available for NMR, the energy levels are separated only by *milli*calories, and the argument in the exponential is very small except at extremely low temperature. Hence, the "high temperature" approximation $e^{-x} \approx 1 - x$ may be employed to show that the fractional excess population in the lower level is

$$\frac{n_\alpha - n_\beta}{n_\alpha} = \frac{2\mu B_0}{kT} \tag{2.20}$$

For ^{1}H, which has a large magnetic moment, in a field of 7 tesla this fractional excess is only about 5×10^{-5} at room temperature.

The near equality of population in the two levels is an important factor in determining the intensity of the NMR signal. According to the Einstein formulation, the radiative transition probability between two levels is given by

$$P_+ \propto B_+\rho(\nu)n_\alpha$$
$$P_- \propto B_-\rho(\nu)n_\beta + A_-n_\beta \tag{2.21}$$

P_+ and P_- are the probabilities for absorption and emission, respectively; B_+ and B_- are the coefficients of absorption and of *induced* emission, respectively; A_- is the coefficient of *spontaneous* emission; and $\rho(\nu)$ is the density of radiation at the frequency that induces the transition. Einstein showed that $B_+ = B_-$, while $A_- \propto \nu^3 B_-$. As a result of this strong frequency dependence, spontaneous emission (fluorescence), which usually dominates in the visible region of the spectrum, is an extremely improbable process in the rf region and may be disregarded. Thus the *net* probability of absorption of rf energy, which is proportional to the strength of the NMR signal, is

$$P \propto B\rho(\nu)[n_\alpha - n_\beta] \tag{2.22}$$

The small value of $(n_\alpha - n_\beta)$ accounts in large part for the insensitivity of NMR relative to other spectroscopic methods (see Chapter 3).

2.5 SPIN–LATTICE RELAXATION

It is important now to consider the manner in which the Boltzmann distribution is established. Again we shall for simplicity treat only the case $I = \frac{1}{2}$. If initially the sample containing the nuclear spin system is outside a polarizing magnetic field, the difference in energy between the two levels is zero, and the populations n_α and n_β must be equal. When the sample is placed into the magnetic field, the Boltzmann distribution is not established instantaneously, because the spin system must first undergo an internal energy adjustment to conform to the new constraint of a magnetic field. Because spontaneous emission is negligible, the redistribution of population must come about from an nonradiative interaction of the nuclei with their surroundings (the "lattice"), which supplies the connection between the spin system and the external world in which the temperature is T. As we shall see, this is a first-order rate process characterized by a rate constant R_1, or a lifetime T_1, called the *spin–lattice relaxation time*.

The origin of this process may be seen in the following: Let $n = (n_\alpha - n_\beta)$ be the difference in population; let $n_0 = (n_\alpha + n_\beta)$; let W_+ be the probability for a nucleus to undergo a transition from the lower to the upper level as a result of an interaction with the environment, and let W_- be the analogous probability for the downward transition. Unlike the radiative transition probabilities, W_+ and W_- are not equal; in fact, at equilibrium, where the number of upward and downward transitions are equal,

$$W_+ n_\alpha = W_- n_\beta \tag{2.23}$$

From Eqs. 2.19 and 2.23,

$$\frac{W_+}{W_-} = \left(\frac{n_\beta}{n_\alpha}\right)_{eq} = e^{-2\mu B_0/kT} \tag{2.24}$$

Using the same approximations as in Section 2.4 for $2\mu B_0/kT \ll 1$, and defining W as the mean of W_+ and W_-, we may write

$$\frac{W_+}{W} = \frac{(n_\beta)_{eq}}{n_0/2} = 1 - \frac{\mu B_0}{kT}$$
$$\frac{W_-}{W} = \frac{(n_\alpha)_{eq}}{n_0/2} = 1 + \frac{\mu B_0}{kT} \tag{2.25}$$

The total rate of change of n is

$$\frac{dn}{dt} = \frac{dn_\alpha}{dt} - \frac{dn_\beta}{dt} = 2\frac{dn_\alpha}{dt} \tag{2.26}$$

But by definition of W_+ and W_-,

$$\frac{dn_\alpha}{dt} = n_\beta W_- - n_a W_+ \tag{2.27}$$

So, from Eqs. 2.25 to 2.27,

$$\frac{dn}{dt} = -2W\left(n - n_0\frac{\mu B_0}{kT}\right) \tag{2.28}$$

This rate equation describes a first-order decay process, characterized by a rate constant $2W$, which may be defined as R_1 or $1/T_1$. The quantity $n_0 \mu B_0 / kT$ is the value of n at equilibrium or n_{eq}. With this notation, Eq. 2.28 becomes

$$\frac{dn}{dt} = -R_1(n - n_{eq}) \tag{2.29}$$

Thus R_1 or $1/T_1$ serves as a measure of the rate with which the spin system comes into equilibrium with its environment.

The magnitude of T_1 is highly dependent on the type of nucleus and on factors such as the physical state of the sample and the temperature. For liquids T_1 is usually between 10^{-2} and 100, but in some cases may be in the microsecond range. In solids T_1 may be much longer—sometimes days. The mechanisms of spin–lattice relaxation and some chemical applications will be taken up in Chapter 8.

Saturation

In the presence of an rf field the fundamental rate equation for spin–lattice relaxation (Eq. 2.29) must be modified by including a term like that in Eq. 2.22, which expresses the fact that the rf field causes net upward transitions proportional to the difference in population n. If the rate at which rf energy is absorbed by the spin system is sufficiently large relative to the relaxation rate, the populations can become equal (i.e., $n = 0$), a situation called *saturation*.

Partial saturation can be an impediment in continuous wave NMR studies (as described in Section 2.9), because the reduced magnitude of the magnetization causes a reduced NMR signal, but it is of little consequence in modern pulse experiments, in which the full magnetization is manipulated, as we shall see, in a time that is normally much less than T_1. In some instances in which there is no component of magnetization along the z axis parallel to $\mathbf{B}_0$, but there is a component in the xy plane, the term "saturation" is occasionally (but inappropriately) used.

Other Non-Boltzmann Distributions

Saturation leads to equalization in the populations of the energy levels, contrary to the Boltzmann distribution. On the other hand, a number of NMR techniques can be employed to increase the population difference well beyond that given by the Boltzmann distribution. In some instances it is convenient to retain the formalism of the Boltzmann relation by defining a *spin temperature* T_S that satisfies Eq. 2.19 for a given ratio n_β/n_α. For times much less than T_1 it is meaningful to have $T_S \neq T$, the macroscopic temperature of the sample, because the spin system and lattice do not interact in this time frame. Viewed in this way, saturation corresponds to $T_S = \infty$. In fact, it is quite feasible to obtain a population distribution in which $n_\beta/n_\alpha > 1$, which means that T_S is negative (i.e., "above" ∞).

There are a number of ways in which non-Boltzmann distributions can be obtained in NMR. Two involve chemical reactions—CIDNP (chemically induced dynamic nuclear polarization) and PASADENA (para-hydrogen and synthesis allow dramatically enhanced nuclear alignment). CIDNP refers to transient positive and negative signals that are greatly enhanced relative to those obtained from molecules in normal Boltzmann equilibrium. It arises from interactions between electrons in certain free radical–mediated chemical reactions and their subsequent interactions with nuclear spins.[20] PASADENA also gives greatly enhanced signals that originate in chemical reactions with para-hydrogen. The signals result from the presence of solely antiparallel proton spins in the hydrogen molecule.[21]

Another important method is the production of *hyperpolarized* ^{129}Xe or ^{3}He by interactions in the vapor phase with alkali metal atoms in excited electronic states that are obtained by *optical pumping* with circularly polarized laser radiation.[22] Because the lifetimes of these nuclei are very long (T_1 for ^{129}Xe can be several hours), it is possible to produce a very large population in the β state and negligible population in the α state, so that the polarization and resulting NMR signal are about 10^5 times as large as obtained with a Boltzmann distribution at normal temperatures and magnetic field strengths.

Discussion of the these three methods is outside the scope of this book, but in later chapters we consider other methods for producing much less dramatic non-Boltzmann distributions. By using rf irradiation to alter spin populations, the *nuclear Overhauser effect* results in signal enhancement (Chapters 8 and 10). Several techniques use pulse sequences to transfer polarization from nuclei with large γ to nuclei with small γ in solids (Chapter 7) and liquids (Chapters 9 and 12), hence to provide significant signal enhancement.

Line Widths

We pointed out in Section 2.3 that an NMR line is not infinitely sharp, and we assumed some function $g(\nu)$ as the line shape. The existence of spin–lattice relaxation implies that the line must have a width at least as great as can be estimated from the uncertainty principle:

$$\Delta E \cdot \Delta t \approx \hbar \tag{2.30}$$

Because the average lifetime of the upper state cannot exceed T_1, this energy level must be broadened to the extent of h/T_1, and thus the half-width of the NMR line resulting from this transition must be *at least* of the order of $1/T_1$.

There are, however, several other effects that can increase line widths substantially over the value expected from spin–lattice relaxation. Phenomenologically, these can be taken into account by means of a second relaxation time, T_2. The distinctions between T_1 and T_2 and the resultant effects on NMR spectra can be understood more easily in terms of the classical mechanical picture that we develop in Section 2.7, so we defer further discussion. However, we note here that $T_2 \le T_1$, and the line width, which is proportional to $1/T_2$, may be much greater than indicated before for certain samples, such as solids. For ^1H NMR in small molecules in the liquid phase, T_1 and T_2 are often of the order of magnitude of seconds, so that line widths of 1 Hz or less are common.

2.6 PRECESSION OF NUCLEAR MAGNETIC MOMENTS

Often it is convenient to have a pictorial representation of the behavior of nuclear magnetic moments. It is possible to use purely classical mechanics to develop a vector picture of the motion of a nuclear moment that is useful and to some extent correct. However, the classical approach to a system that is clearly quantum mechanical in nature is unsatisfying and can lead to misunderstandings. Fortunately, an equally appealing picture can be obtained directly from simple wave mechanical considerations.

We consider, for simplicity, a spin $\frac{1}{2}$ nucleus, with eigenfunctions α and β, as described in Section 2.3. Note that α and β represent stationary eigenstates, but this does not imply that a nuclear spin must reside only in one of them. A general, time-dependent wave function can be constructed as a linear combination (or *coherent superposition*) of α and β as follows:

$$\Psi = c_\alpha \alpha e^{-iE_\alpha t/\hbar} + c_\beta \beta e^{-iE_\beta t/\hbar}$$
$$= c_\alpha \alpha e^{i\gamma B_0 t/2} + c_\beta \beta e^{-i\gamma B_0 t/2} \tag{2.31}$$

where c_α and c_β give the relative portions of α and β and are chosen to produce an overall normalized wave function.

To obtain a picture of the way in which the magnetic moment behaves, let us consider the expectation values of the three components of μ, or alternatively, in accord with Eq. 2.1, the expectation values of the x, y, and z components. of $\mathbf{I}$:

$$\langle I_z \rangle = \langle \Psi^* | I_z | \Psi \rangle \tag{2.32}$$

$$= c_\alpha^2 \langle \alpha | I_z | \alpha \rangle + c_\beta^2 \langle \beta | I_z | \beta \rangle$$
$$+ c_\alpha c_\beta \langle \alpha | I_z | \beta \rangle e^{-i\gamma B_0 t} + c_\alpha c_\beta \langle \alpha | I_z | \beta \rangle e^{i\gamma B_0 t} \tag{2.33}$$

$$= c_\alpha^2 (1/2) + c_\beta^2 (-1/2) \tag{2.34}$$

$$= 1/2 [c_\alpha^2 - c_\beta^2] \tag{2.35}$$

In Eq. 2.33 the time dependence in the exponentials cancels out when the complex conjugates are multiplied, and the two cross terms give zero because of the orthogonality of α and β. The evaluation of the remaining quantities follows from Eq. 2.9. The result is that the z component of the spin (and of the magnetic moment) has a fixed, time-independent value, as could have been predicted from the fact that α and β are eigenfunctions of I_z.

The expectation values of I_x and I_y are obtained in a similar manner but give very different results:

$$
\begin{aligned}
\langle I_x \rangle &= c_\alpha{}^2 \langle \alpha | I_x | \alpha \rangle + c_\beta{}^2 \langle \beta | I_x | \beta \rangle + c_\alpha c_\beta \langle \alpha | I_x | \beta \rangle e^{-i\gamma B_0 t} \\
&\quad + c_\alpha c_\beta \langle \beta | I_x | \alpha \rangle e^{i\gamma B_0 t} \\
&= c_\alpha{}^2 \langle \alpha | \tfrac{1}{2} | \beta \rangle + c_\beta{}^2 \langle \beta | \tfrac{1}{2} | \alpha \rangle + c_\alpha c_\beta \langle \alpha | \tfrac{1}{2} | \alpha \rangle e^{-i\gamma B_0 t} \\
&\quad + c_\alpha c_\beta \langle \beta | \tfrac{1}{2} | \beta \rangle e^{i\gamma B_0 t} \\
&= 0 + 0 + \tfrac{1}{2} c_\alpha c_\beta [e^{i\gamma B_0 t} + e^{-i\gamma B_0 t}] \\
&= c_\alpha c_\beta \cos \gamma B_0 t
\end{aligned}
\tag{2.36}
$$

Likewise, for I_y,

$$
\langle I_y \rangle = c_\alpha c_\beta \sin \gamma B_0 t
\tag{2.37}
$$

Thus, this completely wave mechanical development leads to a picture in which $\boldsymbol{\mu}$ is a vector with fixed z component but x and y components that vary sinusoidally, 90° out of phase—i.e., executing a circular motion, as shown in Fig. 2.2. Overall, then, the motion is a precession about the magnetic field axis and is similar to that which would have been obtained from a classical mechanical treatment

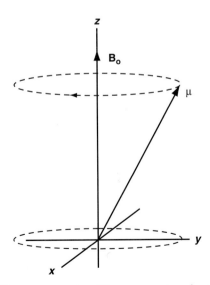

FIGURE 2.2 Vector representation of Larmor precession of magnetic moment $\boldsymbol{\mu}$.

of a spinning top in the Earth's gravitational field. However, the present treatment is quantum mechanical and shows clearly the relation to the quantum state occupations c_α and c_β. For a nucleus solely in one of the eigenstates there is only a z component of $\mathbf{\mu}$, and the general state in which α and β are mixed gives rise to a *coherent* motion at the angular frequency

$$\omega = \gamma B_0 \tag{2.38}$$

as given in Eqs. 2.36 and 2.37. The angular frequency ω (or its equivalent $\nu = \omega/2\pi$) is called the *Larmor frequency* or the *precession frequency*.

Macroscopic Magnetization

In practice, we never deal with a single nucleus, but with an ensemble of identical nuclei. Even in the smallest sample we can imagine there will be a significant fraction of Avogadro's number of nuclei. It is possible, then, to define a macroscopic magnetization $\mathbf{M}$ as the vector sum of the individual magnetic moments and to treat this macroscopic quantity by classical mechanics. As shown in Fig. 2.3, an ensemble of identical nuclei precessing about $\mathbf{B}_0$ (taken, as usual, along the z axis) have random phase in the x and y directions when they are at equilibrium, so that the resultant macroscopic magnetization $\mathbf{M}$ is oriented along the z direction, parallel to $\mathbf{B}_0$. As indicated in Eq. 2.35, $M_0 = |\mathbf{M}|$ is proportional to $c_\alpha^2 - c_\beta^2$, the difference between the populations of the two levels, and at equilibrium the population ratio is given by the Boltzmann distribution

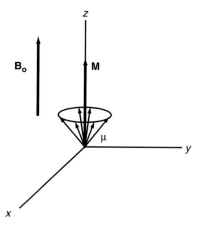

FIGURE 2.3　Precession of an ensemble of magnetic moments at random phase and their resultant vector sum $\mathbf{M}$, which represents the macroscopic nuclear magnetization.

$$\frac{n_\beta}{n_\alpha} = \frac{c_\beta^2}{c_\alpha^2} = e^{-\frac{E_\beta - E_\alpha}{kT}} \tag{2.39}$$

M_0 is responsible for a small static nuclear paramagnetic susceptibility, but the nuclear susceptibility is virtually undetectable at room temperature in the presence of the diamagnetic susceptibility from paired electrons in the molecule, which is about four orders of magnitude larger. Although the static nuclear susceptibility has been detected for hydrogen at very low temperatures, this type of measurement is impractical.

NMR provides a different and far superior means to measure M_0 by tipping **M** away from the z axis into the xy plane, where it precesses at the Larmor frequency and can be selectively detected without interference from the static electron susceptibility. We can examine this process in classical terms.

2.7 CLASSICAL MECHANICAL DESCRIPTION OF NMR

Figure 2.3 indicates that although **M** is static, it arises from a very large number of individual magnetic moments $\boldsymbol{\mu}_i$ that are precessing at the Larmor frequency. If we impose a magnetic field $\mathbf{B}_1$ that rotates about the z axis at the same frequency and in the same direction as the nuclear moments are precessing, there is an in-phase interaction between $\mathbf{B}_1$ and each $\boldsymbol{\mu}_i$, with the result that the ensemble of individual moments, hence **M** itself, is tipped. Specifically, classical mechanics shows that the interaction between a given $\boldsymbol{\mu}$ and $\mathbf{B}_1$ generates a torque **L**, which causes a change in the angular momentum **p** of the nuclear spin according to the following relations:

$$\frac{d\boldsymbol{p}}{dt} = \boldsymbol{L} = \boldsymbol{\mu} \times \boldsymbol{B}_1 \tag{2.40}$$

By multiplying both sides of this equation by γ and using Eq. 2.2, we obtain

$$\frac{d\boldsymbol{\mu}}{dt} = \gamma\boldsymbol{\mu} \times \boldsymbol{B}_1 \tag{2.41}$$

Because the length of $\boldsymbol{\mu}$ is constant, Eq. 2.41 describes the rate at which the direction of $\boldsymbol{\mu}$ changes. If this change in direction is just a rotation with an angular momentum and direction given by a vector $\boldsymbol{\omega}_1$, the motion is described classically by

$$\frac{d\boldsymbol{\mu}}{dt} = \boldsymbol{\omega}_1 \times \boldsymbol{\mu} = -\boldsymbol{\mu} \times \boldsymbol{\omega}_1 \tag{2.42}$$

Comparing Eqs. 2.41 and 2.42, we see that

$$\boldsymbol{\omega}_1 = -\gamma\mathbf{B}_1 \tag{2.43}$$

The negative sign in Eq. 2.43 indicates that the vector $\boldsymbol{\omega}$ that describes the rotation is directed opposite to $\mathbf{B}_1$, a point that we take up further in Section 2.11.

We assumed a $\mathbf{B}_1$ that rotates about z. Although such a rotating field can be generated and is used in some NMR experiments, it is easier to generate a radio frequency field polarized along one axis (say x') from a suitably placed coil, as we see in Chapter 3. However, a linearly polarized field of magnitude $2B_1$ can be thought of as the superposition of two counterrotating fields, each of fixed magnitude B_1 and rotating at ω_{rf} in the xy plane. One of these rotating fields is just what we want, while the other can be ignored for our present purposes because it rotates in the direction opposite to the nuclear precession, hence does not remain in phase with the nuclear moments to cause any interaction.

As we see in Chapter 3, there are several ways of performing NMR experiments, but the most common is to use a radio frequency pulse, that is, to turn $\mathbf{B}_1$ on for only a limited period T_p, which is usually called the *pulse width*. At the end of this period, $\mathbf{M}$ has moved in the yz plane through an angle

$$\theta = \omega_1 T_p = \gamma B_1 T_p \text{ radians} \tag{2.44}$$

as illustrated in Fig. 2.4. We discuss experimental parameters in detail in Chapter 3, but a typical value of B_1 is 10 gauss (or 1 millitesla), which causes $\mathbf{M}$ for ^{1}H to precess at a rate of about 40 kilohertz (kHz). In about 6 μs (a typical pulse width) $\mathbf{M}$ thus moves through an angle of 90° ($\pi/2$ radians).

Now what happens to $\mathbf{M}$? Because it is no longer collinear with $\mathbf{B}_0$, it experiences a torque from $\mathbf{B}_0$, just as it did from $\mathbf{B}_1$, and it precesses about

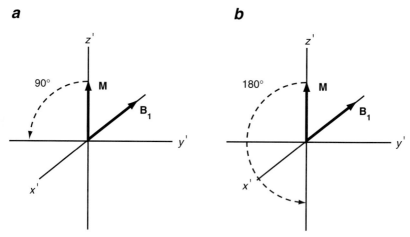

FIGURE 2.4   Rotation of magnetization $\mathbf{M}$ about rf field $\mathbf{B}_1$. Flip angles of 90° and 180° are illustrated.

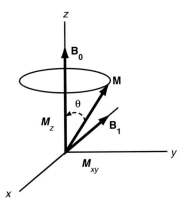

FIGURE 2.5   Precession of magnetization **M** about magnetic field **B**$_0$.

B$_0$ at a frequency

$$\boldsymbol{\omega}_0 = -\gamma \mathbf{B}_0 \tag{2.45}$$

as illustrated in Fig. 2.5. The electrical coil along the x axis, which was used to generate the exciting field **B**$_1$, can now serve as a detector of the precession motion of **M**, because as **M** precesses, its component in the xy plane M_{xy} induces, by Faraday's law, an electrical current in this coil at the precession frequency of **M**. The concept of measuring this signal was the basis for Felix Bloch's invention of nuclear induction, as we discussed in Chapter 1, and this technique underlies virtually all current NMR measurements.

Equation 2.45, often called the Larmor equation, is the basic mathematical expression for NMR. The precession (or Larmor) frequency is directly proportional to the applied magnetic field and is also proportional to γ (or μ/I), which varies from one nuclide to another. One significant feature of the Larmor equation is that the angle θ does not appear. Hence the magnetization precesses at a frequency governed by its own characteristic properties and that of the magnetic field. On the other hand, the energy of this spin system does depend on θ, because

$$E = -\boldsymbol{\mu} \cdot \mathbf{B}_0 = -\mu B_0 \cos \theta \tag{2.46}$$

The value of B_0 in common NMR instruments ranges from about 1 to 18 tesla, putting ν_0 in the range of a few MHz to more than 800 MHz, depending on the magnitude of the magnetic moment of the nuclide, as illustrated in Fig. 2.6. For example, at 7 T, Larmor frequencies for common nuclides range from about 10 MHz for ^{57}Fe to 300 MHz for ^{1}H and 320 MHz for ^{3}H. Because of this wide range of values, we can select a frequency so as to observe one kind of nuclide at a time.

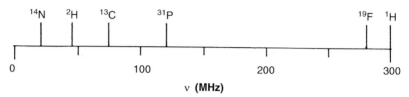

FIGURE 2.6   Larmor frequencies of several nuclides in a magnetic field of approximately 7 teslas.

The Bloch Equations

In Bloch's original treatment of NMR,[23] he postulated a set of phenomenological equations that accounted successfully for the behavior of the macroscopic magnetization **M** in the presence of an rf field. These relations are based on Eq. 2.41, where **M** replaces **μ**, and **B** is any magnetic field—static (**B**$_0$) or rotating (**B**$_1$). By expanding the vector cross product, we can write a separate equation for the time derivative of each component of M:

$$dM_x/dt = \gamma(M_y B_z - M_z B_y)$$
$$dM_y/dt = \gamma(-M_x B_z + M_z B_x) \qquad (2.47)$$
$$dM_z/dt = \gamma(M_x B_y - M_y B_x)$$

By the usual convention, the component $B_z = B_0$, the static field, while B_x and B_y represent the rotating rf field, as expressed in Eq. 2.41, when it is applied, and are normally zero otherwise. To account for relaxation, Bloch assumed that M_z would decay to its equilibrium value of M_0 by a first-order process characterized by a time T_1 (called the *longitudinal* relaxation time, because it covers relaxation along the static field), while M_x and M_y would decay to their equilibrium value of zero with a first-order time constant T_2 (the *transverse* relaxation time). Overall, then the Bloch equations become

$$dM_x/dt = \gamma(M_y B_0 + M_z B_1 \sin \omega t) - M_x/T_2$$
$$dM_y/dt = \gamma(-M_x B_0 + M_z B_1 \cos \omega t) - M_y/T_2 \qquad (2.48)$$
$$dM_z/dt = \gamma(M_x B_1 \sin \omega t + M_y B_1 \cos \omega t) - (M_z - M_0)/T_1$$

The Bloch equations can be solved analytically under certain limiting conditions, as described in Section 2.9.

The longitudinal relaxation time T_1 measures the restoration of M_z to its equilibrium value M_0, which is the same process that we discussed from the perspective of populations in energy levels in Section 2.5, spin–lattice relaxation. The factors that are important in determining the rate at which energy from the spin system flows to its surroundings will be discussed in Chapter 8.

The transverse relaxation time T_2 may be equal to T_1 or may be significantly shorter than T_1. It is clear from Fig. 2.4 that if the macroscopic magnetization **M** is restored to its equilibrium position along the z axis, there is no remaining xy component; hence T_2 cannot be longer than T_1. However, if we recall that **M** is really the sum of many components from individual nuclei in various parts of the sample, then we can recognize at least two ways in which M_{xy} could decrease to zero well before spin–lattice relaxation causes M_z to reach its equilibrium value. One mechanism involves random exchange of energy between pairs of spins, all of which are precessing at nominally the same frequency. If some spins thus precess slightly faster than average and others slightly slower, the spin ensemble dephases, as illustrated in Fig. 2.7, and the resultant M_{xy} decreases without any exchange of energy with the lattice. Thus T_2 becomes shorter than T_1. An apparently similar (but actually rather different) result occurs if nuclei in different parts of the sample experience slightly different values of the magnet field B_0. As we point out in Chapter 3, great efforts are made to provide a highly homogeneous magnetic field for NMR studies, but there are always some imperfections, so that packets of spins in one part of the field precess more rapidly than those in other parts of the field. Thus, the packets tend to get out of phase and cause M_{xy} to decrease.

In terms of the Bloch equations as given in the preceding text, both of these processes result in decrease in T_2, but only the first is considered as contributing to a fundamental determination of T_2 as a molecular parameter. The second process is an artifact of the experimental measurement (inhomogeneity in B_0). The term T_2^* is used to denote the *effective* T_2, that is, to encompass both processes:

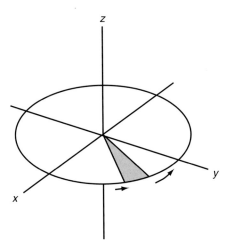

FIGURE 2.7　Dephasing of an ensemble of magnetization vectors as they precess in the xy plane.

$$1/T_2^\star = 1/T_2 + \gamma\Delta B_0 \tag{2.49}$$

We pointed out in Section 2.5 that the minimum line width as given by the uncertainty principle is proportional to $1/T_1$. We can now be more precise. Because $T_2^\star \leq T_2 \leq T_1$, we can express the "true" line width in terms of T_2 and the observed line width in terms of $T_2^\star$. If the line shape function $g(\nu)$ in Eq. 2.16 is Lorentzian, it can be shown that the observed width of the line at half-maximum is given by

$$\nu_{1/2} = 1/\pi T_2^\star \tag{2.50}$$

In Section 2.9 we show one way of distinguishing between T_2 and $T_2^\star$.

2.8 MAGNETIZATION IN THE ROTATING FRAME

Both the mathematical treatments and pictorial representations of many physical phenomena can be simplified by transforming to a coordinate system that moves in some way, rather than one which is fixed in the laboratory. We shall find it convenient to define a coordinate system, or frame of reference, that rotates about $\mathbf{B}_0$ at a rate ν_{rf} revolutions/second, or ω_{rf} radians/second, where the frequencies are those of the applied field $\mathbf{B}_1$. The axes in the rotating frame are designated x', y', and z', the last being coincident, of course, with the z axis in the laboratory frame of reference.

Suppose an observer is in a frame rotating at frequency ν_{rf}. An ensemble of nuclear magnetic moments actually precessing in the fixed laboratory frame of reference at the angular frequency ω_{rf} appears to be static (not precessing) in the rotating frame. Thus, to the observer in the rotating frame it appears that there is no magnetic field acting on this magnetization, because if there were, a torque would cause precession according to Eq. 2.45. In general, if the actual (laboratory frame) precession frequency is $\omega \neq \omega_{rf}$, $\mathbf{M}$ appears to precess in the rotating frame at $\omega - \omega_{rf}$, and thus there appears to be acting on it a magnetic field along the z' axis of

$$\mathbf{B}_{apparent} = (1/\gamma)(\omega - \omega_{rf}) \tag{2.51}$$

In addition to this apparent field along z', we must take into account the rf field, $\mathbf{B}_1$. The component of $\mathbf{B}_1$ rotating in the same direction as the nuclei precess can be taken to lie along the x' axis, while the counterrotating component (which rotates at $-2\omega_{rf}$ in the rotating frame) can safely be ignored. Thus, the overall, effective magnetic field acting on $\mathbf{M}$ in the rotating frame is given by the vector sum

$$\mathbf{B}_{eff} = \mathbf{B}_1 + (1/\gamma)(\omega - \omega_{rf}) \tag{2.52}$$

as depicted in Figure 2.8.

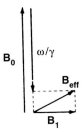

FIGURE 2.8   Formation of the effective rf field $\mathbf{B}_{eff}$ in a frame rotating at ω radians/second. $\mathbf{B}_{eff}$ is the vector sum of applied rf field $\mathbf{B}_1$ along x' and the residual field along z' resulting from $\mathbf{B}_0$ and the "fictitious" field that represents the effect of the rotating frame.

2.9   METHODS OF OBTAINING NMR SPECTRA

We found in Section 2.7 that the fundamental process for obtaining an rf NMR signal requires tipping the magnetization $\mathbf{M}$ away from the z axis to create a precessing component in the xy plane. We now inquire a little more closely into the means by which a set of magnetizations of differing Larmor frequency (e.g., a set of nuclei of different chemical shifts) can be so tipped.

Early high resolution NMR studies were restricted to methods for *sequential* excitation of nuclear magnetizations, initially by applying a fixed radio frequency ν_0 and varying the magnetic field until the Larmor relation (Eq. 2.45) is obeyed, in turn, for each nuclear moment. Indeed, as pointed out in Chapter 1, this is the method by which NMR in bulk materials was discovered. Later sequential excitation methods held the static magnetic field B_0 fixed as the applied frequency was varied. Currently, however, most high resolution NMR spectra are obtained by application of rf pulses, as mentioned previously. Before discussing pulse excitation in detail, we describe some sequential excitation methods, because they are sometimes used, and the methods themselves provide some fundamentally important concepts.

Adiabatic Passage

To understand many NMR phenomena we must recognize that the rate at which resonance is approached can be quite important. A very slow change of magnetic field or frequency is called *adiabatic*, and the adiabatic theorem tells us that if the rate of change is slow enough that

$$dB_0/dt \ll \gamma B_1^2 \tag{2.53}$$

$\mathbf{M}$ remains aligned with $\mathbf{B}_{eff}$. Far from resonance, $\mathbf{B}_{eff} \approx \mathbf{B}_0$ (see Eq. 2.52), and $\mathbf{M}$ is, at equilibrium, aligned with $\mathbf{B}_0$. If the value of either B_0 or ω is varied under conditions of adiabatic passage, **M** slowly tips with $\mathbf{B}_{\text{eff}}$ until at resonance it is aligned along $\mathbf{B}_1$. If the sweep is stopped, then **M** precesses until transverse relaxation reduces its value to zero. On the other hand, if the sweep is continued, $\mathbf{B}_{\text{eff}}$ changes sign, and **M** can be inverted. We shall see later that short rf pulses usually provide a more convenient and flexible means for inversion of **M**, but adiabatic passage is useful in some instances.

Slow Passage

Because application of $\mathbf{B}_1$ for a significant time can cause partial saturation and reduction in the magnitude of **M**, an adiabatic sweep must be fast enough to avoid saturation ($\gg 1/T_2$). In order to conform simultaneously to the adiabatic theorem, B_1 must then be large, and distortions of sharp lines result. Hence, high resolution spectra cannot normally be obtained with such *adiabatic fast passage*. Instead, the value of B_1 is kept very small (of the order of a few microgauss), the scan rate is made slow enough to avoid distortions of lines, and ideally the scan conforms to *slow passage* conditions in which **M** is tipped by only a few degrees from its equilibrium position along z and returns to z after passage through the line. In practice, however, scan rates are such that the magnetization cannot quite follow $\mathbf{B}_{\text{eff}}$, and after resonance some magnetization is left in the xy plane. The result is that the resonance line, instead of having a symmetric Lorentzian shape, shows *ringing* or "wiggles" after the line (Fig. 2.9), as M_{xy} continues to precess at one frequency and interferes with $\mathbf{B}_1$ as it moves to a different frequency in the scan. Many older NMR spectra obtained by frequency sweep or field sweep (both termed *continuous wave*, or *cw* excitation) show ringing, as do several examples in this book.

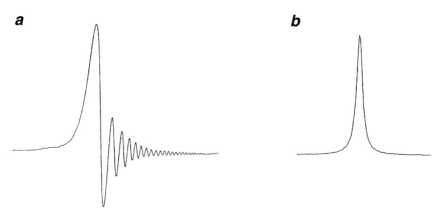

FIGURE 2.9   (*a*) Ringing that occurs after a scan through resonance. (*b*) Ideal slow passage Lorentzian line shape.

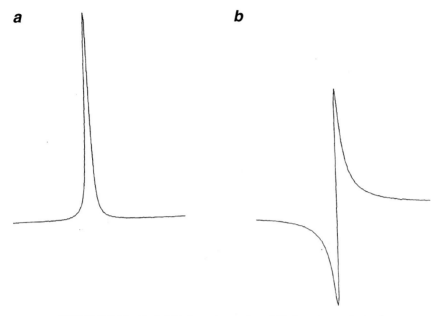

FIGURE 2.10   Typical (*a*) absorption mode and (*b*) dispersion mode signals.

Absorption and Dispersion

The Bloch equations can be solved analytically under the condition of slow passage, for which the time derivatives of Eq. 2.48 are assumed to be zero to create a steady state. The nuclear induction can be shown to consist of two components, *absorption*, which is 90° out of phase with $\mathbf{B}_1$ and has a Lorentzian line shape, and *dispersion*, which is in phase with $\mathbf{B}_1$. The shapes of these signals are shown in Fig. 2.10. By appropriate electronic means (see Section 3.3), we can select either of these two signals, usually the absorption mode.

Radio Frequency Pulses

In general, the motion of $\mathbf{M}$ in the rotating frame follows from the classical torque exerted on it by $\mathbf{B}_{\text{eff}}$. The effect of an rf pulse is then to tip $\mathbf{M}$ away from the z' axis and to generate a component in the $x'y'$ plane. As viewed from the laboratory frame of reference, this component precesses in the xy plane and induces an electrical signal at frequency ω in a coil placed in this plane. As the nuclear moments that make up $\mathbf{M}$ precess, they lose phase coherence as a result of interactions among them and magnetic field inhomogeneity effects, as described in Section 2.7. Thus M_{xy} decreases toward its equilibrium value of zero, and the signal decays (a *free induction decay*, or FID) with a time constant T_2^*. If the pulse width T_p is selected so that

$$\gamma B_{eff} T_p = \pi/2 \qquad (2.54)$$

M "flips" through 90°, and the pulse is called a 90° or $\pi/2$ pulse. Pulse widths for other flip angles are determined in an analogous manner.

It is important to note from Eq. 2.52 that the magnitude and direction of $\mathbf{B}_{eff}$ are dependent on the resonance offset $\Omega = (\omega - \omega_{rf})$, so that nuclei with different precession frequencies respond somewhat differently to the pulse. As illustrated in Fig. 2.11, a nominal 90° pulse on resonance results in nuclear magnetizations fanning out as a function of resonance offset, so that the projections in the $x'y'$ plane lie in different directions, thus causing phase distortions in their signals (mixtures of absorption and dispersion modes). Moreover, because the projections differ in magnitude, the intensities of the signals are also distorted. For a nominal 90° pulse on resonance the amplitude of M_{xy} is insensitive to resonance offset up to about $\Omega = \gamma B_1$, as the increased magnitude of $\mathbf{B}_{eff}$ relative to $\mathbf{B}_1$ partially compensates for its failure to lie precisely along $\mathbf{B}_1$. On the other hand, there are always at least small phase errors because any finite value of B_1 implies a pulse of finite duration, during which precession of off-resonance magnetizations occurs. For $\Omega \leqslant 0.5\gamma B_1$, the phase error ϕ can be approximated by

$$\phi = \tfrac{1}{2}\Omega T_p \qquad (2.55)$$

Equation 2.55 is useful for correcting phases, especially for two-dimensional NMR experiments, as discussed in later chapters.

If $B_1 \gg B_{apparent}$, then clearly $\mathbf{B}_{eff} \approx \mathbf{B}_1$ in both magnitude and direction for all nuclei, and gross distortions are avoided. For smaller values of B_1 there is considerable compensation in magnitude for a nominal 90° pulse, as we have seen, but the situation is quite different for a 180° pulse. However, even for smaller values of B_1, it is possible to create *composite pulses* that can minimize the distortions, as we shall see in Chapter 9.

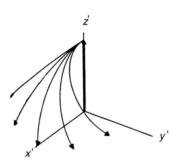

FIGURE 2.11   Magnetization vectors responding to a nominal 90° pulse. The vector representing magnetization precisely at resonance follows the desired trajectory to the x' axis, and those representing off-resonance magnetizations fan out and do not quite reach the xy plane at the end of the pulse period.

Pulse Sequences

With an understanding of the response of the nuclear magnetization to an rf pulse, we can further manipulate the spin system by applying a *sequence* of pulses. The magnetization is rotated by the first pulse, allowed to precess for some period, and rotated further by subsequent pulses. Chapters 9 and 10 are devoted largely to the application of a variety of pulse sequences and the resultant "spin gymnastics" that the nuclear magnetizations experience. Here we consider only two simple pulse sequences to illustrate the versatility and to lay the foundation for later discussion.

Suppose we wish to measure T_1 for a nuclear spin that gives only a single NMR line. There are a number of ways in which this can be done, but one of the simplest and most accurate is the *inversion-recovery* method, in which a 180° pulse inverts the magnetization, and its recovery by spin–lattice relaxation is monitored by application of a 90° pulse. Figure 2.12 shows that an initial 180° pulse flips **M**

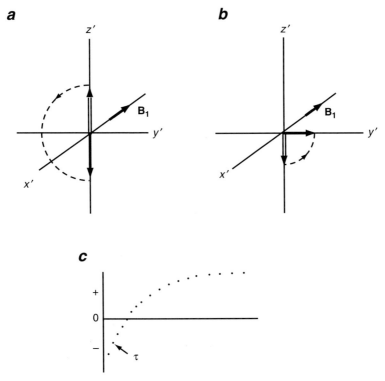

FIGURE 2.12 Determination of T_1 by the inversion-recovery method. (*a*) **M** is inverted by a 180° pulse at time 0. (*b*) After a time interval τ, during which spin–lattice relaxation occurs, a 90° pulse rotates the remaining magnetization to the y' (or $-y'$) axis. (*c*) The initial height of the FID or the height of a given line in the Fourier-transformed spectrum (see Chapter 3) is plotted as a function of τ. Note that each point results from a separate 180°, τ, 90° pulse sequence.

to the negative z axis, where it recovers to equilibrium by longitudinal relaxation. (Because no xy component is generated, transverse relaxation is irrelevent). After a relaxation period τ, a 90° pulse along the x' axis moves the remaining portion of **M** to the y' axis—either positive or negative y' axis depending on the extent of recovery of **M** toward equilibrium. If the spin system is now allowed to relax completely and the sequence repeated for another value of τ, a relaxation plot may be constructed as shown and easily analyzed to give the value of T_1.

Suppose now that we wish to measure T_2. It is easy to determine T_2^* from the free induction decay after a 90° pulse or from the width of the NMR line, and if magnetic field inhomogeneities are negligible, this gives the correct value of T_2. In many instances, however, inhomogeneity cannot be neglected, and this approach fails. Fortunately a simple pulse sequence—90°, τ, 180°—permits us to measure T_2 without interference from magnetic field inhomogeneity. The rationale of the method is shown in Fig. 2.13, which depicts the behavior of the magnetization in the rotating frame. In (a) **M** is shown being tipped through 90° by application of **B**$_1$ along the positive x' axis. The total magnetization **M** can be thought of as the vector sum of individual macroscopic magnetizations **m**$_i$ arising from nuclei in different parts of the sample and hence experiencing slightly

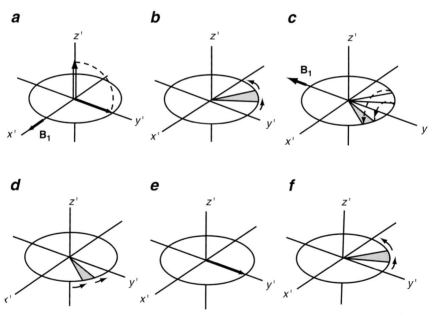

FIGURE 2.13   Illustration of the formation of a spin echo. (a) A 90° pulse rotates the macroscopic magnetization **M** to the $x'y'$ plane. (b) The individual **m**$_i$ dephase as they precess. (c) A 180° pulse rotates the **m**$_i$ as indicated. (d) The **m**$_i$ again precess, with the faster **m**$_i$ now beginning to catch up with the slower **m**$_i$. (e) All **m**$_i$ come into phase. (f) The **m**$_i$ again dephase as step (b) is repeated.

different values of the applied field, which is never perfectly homogeneous. There is thus a range of precession frequencies as indicated in (b) as the $\mathbf{m}_i$ begin to fan out. At a time τ after the 90° pulse, a 180° pulse is applied along the y' axis, as shown in (c). The effect of this pulse is to rotate each $\mathbf{m}_i$ by 180° about the y' axis. The $\mathbf{m}_i$ that are moving fastest and have moved farthest during τ are thereby moved farthest front into the positive quadrant of the $x'y'$ plane, as indicated. These nuclei naturally continue to move faster, as shown in (d), and at time 2τ all $\mathbf{m}_i$ come into phase along the positive y' axis, as shown in (e). The continuing precession of the $\mathbf{m}_i$ causes them again to lose phase coherence in (f).

The rephasing of the $\mathbf{m}_i$ causes a free induction signal to build to a maximum at 2τ, the so-called *spin echo*. If real transverse relaxation did not occur, the echo amplitude might be just as large as the initial value of the free induction following the 90° pulse. However, each $\mathbf{m}_i$ decreases in magnitude during the time 2τ because of the natural processes responsible for transverse relaxation characterized by T_2. Hence, T_2 may in principle be determined from a plot of peak echo amplitude as a function of τ. As in the measurement of T_1 by the inversion-recovery method, it would be necessary to carry out a separate pulse sequence for each value of τ and to wait between pulse sequences an adequate time (at least five times T_1) for restoration of equilibrium. In Chapter 9 we discuss other pulse sequences that eliminate the waiting time and circumvent several other shortcomings of this simple spin-echo method.

The spin echo was discovered in 1950 by Erwin Hahn[24] and is sometimes called a "Hahn echo." The real significance of the spin-echo method lies not in its use to measure T_2, but in the demonstration that an apparently irreversible dephasing of nuclear spins and decay of the FID (even to zero) can be reversed. As we see in Chapter 7, application of this concept is extremely important in obtaining narrow NMR lines in solids. We shall also encounter numerous examples in NMR of liquids where a spin echo is employed.

2.10 DYNAMIC PROCESSES

NMR spectra are sensitive to many dynamic ("exchange") processes. The rotating frame picture permits us to show qualitatively how NMR spectra respond to such processes, and the Bloch equations can be readily modified to account for them quantitatively. Many aspects of exchange will arise in later chapters.

By "exchange" we refer to some process that permits a given nucleus to move between two (or more) different environments in which it has different resonance frequencies. The different environments are usually, but not always, the result of different chemical milieus, as we shall see specifically in later chapters. For example, the conformational change of a cyclohexane ring (**I**) in which an equatorial hydrogen is converted to an axial hydrogen constitutes an exchange process:

I

Likewise, an intermolecular exchange of protons between an alcohol and a phenol is an exchange process;

$$ROH_a + PhOH_b \rightleftharpoons ROH_b + PhOH_a$$

For cyclohexane, no chemical bonds are broken in the process, whereas the phenol−alcohol example does involve bond disruption. The same theoretical treatment applies to both situations so far as chemical shifts are concerned. However, as we shall see in later chapters, breaking of chemical bonds disrupts spin−spin coupling and thus introduces other considerations.

Qualitative Evaluation

Suppose that a magnetic nucleus can exchange between two states in which it has, respectively, resonance frequencies v_A and v_B, with $v_A > v_B$. For simplicity, we shall assume that the nucleus has an equal probability of being in the two sites; hence the lifetime of the nucleus is state A, τ_A, must equal that in state B:

$$\tau_A = \tau_B \tag{2.56}$$

Consider a coordinate system that rotates about $\mathbf{B}_0$ in the same direction in which the nuclei process, at a frequency

$$v_0 = \frac{1}{2}(v_A + v_B) \tag{2.57}$$

In this rotating frame a nucleus at site A then precesses at $(v_A - v_0)$, while a nucleus at site B precesses at $(v_0 - v_B)$—that is, it appears in the rotating frame to be precessing in a direction opposite that of the nucleus in site A. We can now distinguish the following four cases regarding exchange rates:

1. *Very slow exchange.* The lifetime at each site, 2τ, is sufficiently long that a given nucleus enters site A and precesses many times at frequency $(v_A - v_0)$ before leaving site A and entering site B. The result is that interaction with the rf field occurs, and in the fixed laboratory frame of reference a resonance line appears at v_A. A similar situation occurs for the nucleus at site B. Thus the spectrum consists of two sharp lines at v_A and v_B, just as it would in the absence of exchange. (See Fig. 2.14a.)

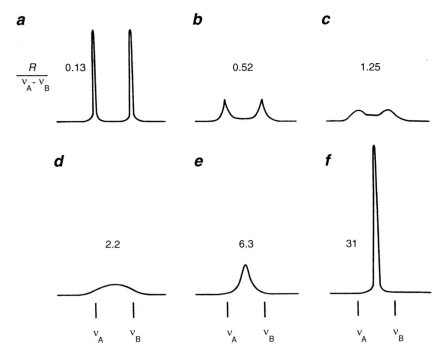

FIGURE 2.14 Calculated line shapes for various values of the exchange rate R relative to the difference in resonance frequency $(\nu_A - \nu_B)$ for two equally populated sites.

2. *Moderately slow exchange.* The lifetime 2τ is now somewhat smaller than the value in the preceding paragraph, so that the lifetime of the nucleus in each state is limited. The uncertainty principle (Eq. 2.30) imposes a minimum line width, as we found in our discussion of line widths from spin–lattice relaxation in Section 2.5. In the present case, $\nu_{1/2} \approx 1/\tau$, giving broadening of the sort shown in Fig. 2.14*b* and *c*.

3. *Very fast exchange.* A nucleus enters site A, where in the rotating frame it begins to precess at $(\nu_A - \nu_0)$. But before it can complete even a small portion of a single precession, its lifetime in site A expires, and it enters site B. It now begins to precess in the opposite direction in the rotating frame but again undergoes essentially no precession before it must again leave site B and reenter A. The result is that in the rotating frame the nuclear magnetization remains stationary, and thus in the laboratory frame it appears to be precessing at the frequency with which the frame rotates, ν_0. Hence, as shown in Fig. 2.14*f*, a sharp resonance line appears at ν_0, the average of the two Larmor frequencies, even though no nuclear magnetization actually precesses at that frequency.

4. *Intermediate exchange rate.* Between cases 2 and 3 there is a range of lifetimes that lead to an intermediate-type spectrum, a broad line spanning the frequency range $(\nu_A - \nu_B)$, as indicated in Fig. 2.14*d* and *e*.

Quantitative Treatments

The simplest theoretical approach to exchange is via the Bloch equations, to which terms are added to reflect the rate phenomena. The spectra shown in Fig. 2.14 are obtained from such a treatment. It is apparent that the line shapes depend on the ratio $R/(\nu_A - \nu_B)$, where the exchange rate $R = 1/\tau$. Thus "fast" and "slow" are measured with respect to differences in the nuclear precession frequencies in the two sites. Exchange rates can be measured by analysis of line shapes and by certain pulse experiments, as described in later chapters.

Exchange effects are important in understanding a number of aspects of NMR that we develop in the following chapters. It should be pointed out that exchange

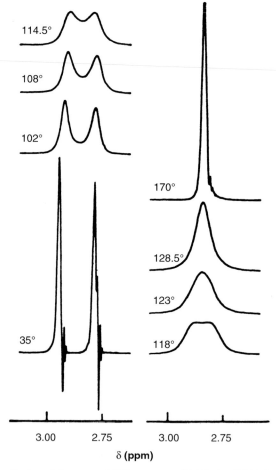

FIGURE 2.15   Portion of the proton NMR spectrum (60 MHz) of N,N-dimethylformamide at various temperatures. From Bovey.[25]

is not necessarily limited to two sites, and a number of examples will be encountered of multisite exchange. A particularly common situation, which is easily treated theoretically, is very fast exchange (case 3 above), where the single observed frequency is simply the weighted average of the individual precession frequencies:

$$\nu_{observed} = p_A\nu_A + p_B\nu_B + p_C\nu_C + \cdots \qquad (2.58)$$

where the p_i sum to unity.

The classical treatment using the Bloch equations is very useful but has limitations when quantum effects must be considered. Hence, when spin–spin coupling is present (which depends on the quantized spin states of the nuclear moments), the simple treatment is inadequate. In some instances, as we shall see in later chapters, consideration of spin states can be easily grafted onto the classical treatment, whereas in others a full density matrix treatment (Chapter 11) is needed.

A simple example of exchange occurs in *N,N*-dimethylformamide, as illustrated in Fig. 2.15. The C—N bond in an amide has partial double bond character; hence rotation about this bond is highly restricted but not entirely precluded. At room temperature the rotation rate is sufficiently slow that separate sharp lines are observed for the two methyl groups, which are in different environments because of proximity to the carbonyl group (see Chapter 4). With increasing temperature, however, the barrier to rotation is gradually surmounted, and the observed spectra follow the theoretically computed curves of Fig. 2.14.

2.11 TERMINOLOGY, SYMBOLS, UNITS, AND CONVENTIONS

We have already made a number of choices of terminology, symbols, and sign conventions, which are not universally followed in NMR literature. For a qualitative discussion of a particular aspect of NMR, choice of symbols and sign conventions may be unimportant, but in other areas it is essential that a consistent system be used. Unfortunately, there are different self-consistent systems that are employed in various books, so superficial comparison of mathematical expressions may lead to apparent contradictions. Here we state several of our choices explicitly and point out areas of divergence.

Symbols

In general, we use symbols for various quantities that are widely accepted and have been recommended by the International Union of Pure and Applied Chemistry (IUPAC).[26]

Magnetic Field and Magnetic Induction

Classical electromagnetism distinguishes among the magnetic field **H,** the magnetization **M,** and the magnetic induction inside a medium, **B:**

$$\mathbf{B} = \mu_0(\mathbf{H} + \mathbf{M}) \tag{2.59}$$

$$= \mu_0(\mathbf{H} + \kappa\mathbf{H}) \tag{2.60}$$

where μ_0 is the permeability of free space and κ is the volume magnetic susceptibility of the material. The magnitude of **H** is measured in oersteds in the older Gaussian units and in amperes/meter in SI units. $\mu_0 = 4\pi \times 10^{-7}$ mA^{-1} in SI units but is implicitly unity in Gaussian units. For diamgnetic materials κ is of the order of 10^{-6} and can be neglected here. However, we return to consideration of magnetic susceptibility in Chapter 4, under conditions where it is not negligible.

Some NMR books distinguish carefully between **H** and **B,** but most high resolution NMR literature refers to both **H** and **B** as a "magnetic field." Actually, it is the magnetic induction in the sample that is important, so **B,** measured in the SI unit of tesla (T) or in the older unit of gauss (G), is the appropriate quantity. (1 T = 10^4 G.) We use **B** in units of tesla. However, in common with almost all NMR literature, we use the jargon in which **B** is called the magnetic field, rather than the magnetic induction, in order to avoid having to say that we apply a magnetic field **H** in a particular direction, which then induces magnetization **B** in the same direction in the sample. Much of the older NMR literature (including previous editions of this book) used the symbol **H,** referred to as magnetic field, but made use of the fact that in the older units $\mu_0 = 1$, so that the induction in gauss has the same numerical value as field in oersteds. Thus the value of H was given in gauss.

In concert with all NMR literature, we choose the z axis for imposition of the static magnetic field $\mathbf{B}_0$. We apply $\mathbf{B}_0$ along the *positive* z axis, as do most (but not all) newer NMR books. Many earlier books (including previous editions of this book) apply the static magnetic field along the negative z axis in order to eliminate the minus sign in Eq. 2.3 and the many later equations that stem from it. A consequence of applying the magnetic field along the positive z axis, as shown in Eq. 2.5, is that the lowest energy levels have positive quantum numbers m, provided the magnetogyric ratio γ is positive. In particular, for $I = \frac{1}{2}$ and positive γ (the case for the vast majority of nuclei), spin state α has lower energy than β. One should be aware, however, that a number of authors have used $\mathbf{B}_0$ along the negative z axis, and nuclides with negative γ are occasionally encountered (^{15}N and ^{29}Si are common examples), either of which makes the β state lower in energy.

Signs of Rotations

The Larmor relation, Eq. 2.45, is the basis for determining the nuclear precession frequency. In many instances we are concerned only with the magnitude of the frequency, which is normally taken to be a positive quantity. However, the negative sign is important in dealing with quantitative aspects of NMR theory, not only in terms of nuclear precession in the field $\mathbf{B}_0$ but also in application of the Larmor equation to the rf field $\mathbf{B}_1$. The negative sign means that the rotation vector $\boldsymbol{\omega}$ is directed opposite to $\mathbf{B}$ (for the usual positive γ), which has consequences for the direction in which $\mathbf{M}$ rotates when an rf pulse is applied. Once more, there is no uniformity in the literature, either in the choice of the Cartesian axes to conform to a right-handed or left-handed coordinate system or in the definition of what is meant by a positive rotation. We use the following system endorsed by Ernst *et al.*[27] that now seems to be gaining favor, particularly in application of the product operator formalism.

First, we use a right-handed coordinate system, as shown in Fig. 2.3. A positive rotation, for both frequencies and angles, is defined as being clockwise if looking in the direction of the rotation vector (i.e., counterclockwise if looking along $\mathbf{B}$ for a nuclide with positive γ). We have already decided to orient $\mathbf{B}_0$ along $+z$, so $\boldsymbol{\omega}$ lies along $-z$, and the sense of precession in the laboratory frame (for positive γ) is that shown in Fig. 2.5. However, in the rotating frame, we have more flexibility for rotations about the three Cartesian axes. For the quantitative theory developed in Chapter 11, it turns out to be preferable to orient the rotation vectors along positive axes, because we actually treat rotations rather than magnetic fields. Thus, in the rotating frame we take $\boldsymbol{\Delta}\mathbf{B}_0$, the longitudinal component of the effective field (Eqs. 2.51 and 2.52), along the negative z axis and define a rotation vector Ω along $+z$:

$$\boldsymbol{\Omega} = \boldsymbol{\omega} - \boldsymbol{\omega}_{\mathrm{rf}} = -\gamma\boldsymbol{\Delta}\mathbf{B}_0 \tag{2.61}$$

A positive rotation about z then leads to magnetization precessing in the order $x' \rightarrow y' \rightarrow -x' \rightarrow -y'$.

We use the same rationale for applying $\mathbf{B}_1$ along the negative x' or y' axis, so that the rotation vector for a pulse lies along the corresponding positive axis. For example, a positive 90° pulse along $+x'$ (often called 90_x) results from a $\mathbf{B}_1$ oriented along $-x'$ and causes $\mathbf{M}$ to go from $+z$ to $-y'$. These conventions will be particularly useful, as we indicated, in later chapters, and we usually apply them in the vector diagrams in this and other chapters. However, in order to simplify illustrations, we depict $\mathbf{B}_1$ applied along positive axes in the rotating frame in several instances.

Frequency Units

We have already dealt with a number of expressions in which frequency appears, and we shall encounter many more such expressions. In some instances it is preferable to use the more fundamental unit of radians/second, for which we employ the symbol ω or Ω, while in other cases it is more convenient to use the measured unit of cycles/second, or hertz, designated ν (or occasionally F). Also, we often express energy terms such as $\mathcal{H}$ or E_n (an energy level) in units of frequency, rather than proper energy units, as we are concerned with the frequencies associated with differences in energy levels.

2.12 ADDITIONAL READING AND RESOURCES

Almost every NMR book begins with a discussion of basic theory. Short, elementary accounts of theory are given in such books as *Nuclear Magnetic Resonance* by P. J. Hore[28] and *Introduction to NMR Spectroscopy* by R. J. Abraham, J. Fisher, and P. Loftus,[29] as well as in books mentioned later that are directed toward biological applications or molecular structure elucidation.

Particularly lucid explanations of many fundamental aspects of NMR theory are given in the third edition of *Principles of Magnetic Resonance* by C. P. Slichter.[30] This book is directed toward physicists and assumes that the reader has a good mathematical background and a grounding in the density matrix formalism of the sort we provide in Chapter 11. However, it also includes among the equations excellent qualitative discussions of various topics.

Quantum Description of High-Resolution NMR in Liquids by Maurice Goldman[31] includes clear and concise reviews of the features of quantum mechanics and angular momentum that are essential for NMR.

A very good book at an intermediate level is *Nuclear Magnetic Resonance Spectroscopy* by Robin Harris,[32] which covers much of the material that we include in Chapters 2–8.

Several "classic" books in NMR, which are now out of date overall, are still authoritative in specific areas. For example, *Principles of Nuclear Magnetism* by Anatole Abragam,[33] *High Resolution Nuclear Magnetic Resonance* by J. A. Pople, W. G. Schneider, and H. J. Bernstein,[34] and *High Resolution Nuclear Magnetic Resonance Spectroscopy* by J. W. Emsley, J. Feeney, and L. H. Sutcliffe[35] are the sources for much material presented in this and other NMR books.

Dynamic NMR Spectroscopy by L. M. Jackman and F. A. Cotton[36] provides a good coverage of the theory of exchange processes, along with many applications.

2.13 PROBLEMS

2.1   Using the data in Appendix A, determine the following:
    (a) The resonance frequency for ^{1}H in a magnetic field of 7.0 T.
    (b) The magnetic field in which ^{13}C has a resonance frequency of about 100 MHz.
    (c) The relative values of the magnetic moments of ^{1}H, ^{31}P, ^{17}O, and ^{59}Co.

2.2 From Appendix A, find the relative receptivities (sensitivity × abundance) of ^1H, ^{13}C, ^{15}N, and ^{57}Fe.

2.3 Use the Pauli matrices in Eq. 2.10, together with the matrix representations of α and β as column or row matrices

$$\alpha = \begin{bmatrix} 1 \\ 0 \end{bmatrix} \text{ or } [1 \quad 0] \qquad \beta = \begin{bmatrix} 0 \\ -1 \end{bmatrix} \text{ or } [0 \quad -1]$$

to verify the relations given in Eqs. 2.8 and 2.9. Consult Appendix C if needed.

2.4 Use the relationships in Eq. 2.10 to find the values of the commutators $[I_x, I_z]$ and $[I_x, I_y]$, where $[A, B] \equiv AB - BA$.

2.5 Use the results of Eqs. 2.17, 2.20, and 2.22 to predict the dependence of the NMR sensitivity of a nucleus on its magnetogyric ratio γ. Verify your prediction with some values in Appendix A.

2.6 Find values for $(n_\alpha - n_\beta)/n_\alpha$ for ^{19}F, ^{31}P, and ^{15}N at (a) 1.4 T and 300 K; (b) 14 T and 5 K; and (c) 50,000 G and 300 K.

2.7 What value of B_1 is required to provide a 90° on-resonance pulse of 20 μs for ^{13}C?

2.8 Derive an expression in terms of T_2 for the width at half-height of a Gaussian-shaped line.

2.9 (a) Beginning with Eq. 2.29, derive an expression for the recovery of magnetization after an inversion-recovery experiment, as described in Fig. 2.12 (b) Because the rf field B_1 is not completely homogeneous, nuclear magnetizations in different parts of the sample may experience a pulse that is smaller than 180°. Modify the expression derived in (a) to account for the fact that **M** is incompletely inverted. (c) Show from the expression derived in (a) that the value of T_1 can be determined directly from the value of τ that gives zero signal. Show from the expression derived in (b) that this simple procedure gives erroneous results in the practical situation of an inhomogeneous B_1.

2.10 Find the value of B_{eff}/B_1 and the direction of B_{eff} relative to B_1 for $B_1 = 1$ G and protons 2 kHz off resonance.

2.11 Find the range of resonance frequencies over which a 90° pulse of 20 μs is effective (i.e., insensitive to resonance offset). If the pulse width is doubled, with B_1 remaining the same, what would you observe for a spectral line that is on resonance?

Instrumentation and Techniques

With the theoretical grounding in basic NMR phenomena that we developed in Chapters 1 and 2, we can now approach the practical problem of obtaining NMR spectra. We shall not consider any of the technical aspects of instrument design or electronic circuitry needed, because most NMR studies are carried out with commercially available apparatus. However, to use such equipment effectively, it is important to understand the fundamental instrumental requirements, to recognize the ways in which NMR spectra may be obtained, and to gain an appreciation of the many parameters that must be controlled to acquire satisfactory NMR data. In this chapter we cover the fundamentals of NMR spectrometers and explore the methodology of obtaining NMR spectra.

3.1 ADVANTAGES OF PULSE FOURIER TRANSFORM NMR

We are interested in systems of chemical interest, which usually consist of a number of resonance lines, so our discussion will be directed toward observation of such multiline spectra, principally in the liquid state. Most of the general features of instruments and techniques are also applicable to study of high resolution NMR spectra in solids, but we shall see in Chapter 7 that such studies demand additional specialized features. Although much of the instrumentation described here and many of the techniques can be used for continuous wave spectra (as described in Section 2.9), we concentrate here and in the remainder of this book on spectra excited by application of one or more rf pulses.

The use of a short, intense rf pulse to excite nuclei and to permit the observation of nuclear resonance was suggested in one of the first papers on NMR.[23] The technique was used extensively by physicists in studying simple systems consisting of only one (usually broad) NMR line. However, its use for complex molecules was virtually precluded by the experimenters' inability to disentangle signals from various chemically shifted nuclei. During the mid–1960s, however, two events brought pulse methods to the attention of chemists. First, Ernst and Anderson[17] showed, in their now classic paper, that slow passage spectra could be obtained with a great saving in instrument time by use of pulse NMR observations and Fourier transform (FT) mathematics. Second, during the same period practical minicomputers were developed for on-line laboratory service, and advances in computer software drastically reduced the time needed for FT computations.

The "Fourier transform revolution" changed the face of NMR. Slow passage cw methods are now largely relegated to lower field, lower cost instruments and to samples for which signal/noise ratio and speed of data acquisition do not present serious problems. The advantage of the pulse FT approach is that a sufficiently powerful rf pulse excites the entire spectrum simultaneously, as we showed in Section 2.9, and Fourier transformation permits us to disentangle the many frequencies that are present in the free induction decay that is the time response to the pulse.

Long before FT methods were applied to NMR, physicists and engineers had made extensive use of Fourier transformation to analyze signals that vary with time in order to extract the fundamental frequencies that are present. The basic relation between a time response $s(t)$ and its corresponding frequency spectrum is

$$S(\omega) = \int_{-\infty}^{\infty} s(t)e^{-i\omega t}\, dt \tag{3.1}$$

For NMR spectra, it is known that if $s(t)$ is the free induction decay following a pulse, $S(\omega)$ represents the slow passage spectrum. We shall also use the FT relationship in other ways—for example, for data processing (in Section 3.4) and for analyzing random molecular motions (in Chapter 8). We return to the use of pulse Fourier transform methods in Section 3.6.

3.2 BASIC NMR APPARATUS

The heart of an NMR instrument is a magnet of adequate field strength and adequate homogeneity. During the first 25 years of NMR studies, electromagnets were widely used, but since the early 1970s superconducting magnets have dominated except for some relatively low field, low cost instruments that are based on permanent magnets.

If narrow NMR lines are to be observed, the magnet must be sufficiently homogeneous that all molecules in the sample experience the same magnetic field and have the same resonance frequency. Here, "the same" means close enough that the variation in resonance frequencies is comparable with the line width to be observed. Usually it is highly desirable to employ the highest strength magnetic field compatible with the required homogeneity. Permanent magnets and electromagnets for NMR are generally restricted to about 2 tesla, whereas superconducting magnets of up to 18.7 tesla are now commercially available. The technology for producing homogeneous magnets of such high field strength is demanding, and high field magnets may cost on the order of $1 million. Miles of superconducting wire are used to fabricate the magnet, and the superconductors currently available require very low temperature; liquid helium (boiling at 4.2 K) is used as a refrigerant. Superconducting magnets, once energized, do not require a source of electrical power and can be fabricated to produce a remarkably stable magnetic field.

To excite NMR spectra we need radio frequency (rf) power, which is generated in the *transmitter*. The source of rf is normally an electronic device called a *frequency synthesizer*, which produces a very narrow band at one or more selected frequencies. As we shall find in later sections, most NMR studies require at least two or three separate frequencies, each of which is amplified in a separate transmitter. After amplification, the rf power is applied to the sample by electrical coils in a *probe*, illustrated schematically in Fig. 3.1. The probe holds the sample in the magnetic field. For most high resolution NMR studies, the sample is contained in a cylindrical glass tube, most commonly about 5 mm in diameter (but sometimes ranging down to 1 mm or up to about 10 mm, as discussed later). Figure 3.1 shows a single coil probe, but usually one or more additional coils are added to accommodate several radio frequencies simultaneously. The usual geometry of a superconducting magnet requires a long cylindrical probe, with the axis of the sample tube and probe coincident with that of the magnet, as illustrated in Fig. 3.1a, and thus parallel to the static magnetic field $\mathbf{B}_0$. On the other hand, the geometry of most electromagnets and permanent magnets, as shown in Fig. 3.1b, leads to the sample tube axis being perpendicular to $\mathbf{B}_0$. The axis of the rf coil must be perpendicular to $\mathbf{B}_0$ (defined as the z axis) in order to generate an rf field $\mathbf{B}_1$ along the x axis and tip $\mathbf{M}$, as described in Section 2.7.

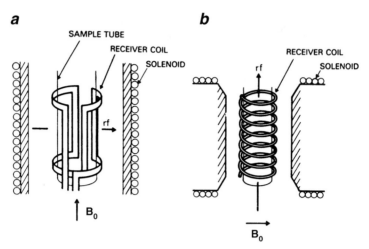

FIGURE 3.1   Typical probe geometry in (*a*) a superconducting magnet and (*b*) an iron core permanent or electromagnet. Note that the rf coil is fabricated to generate $\mathbf{B}_1$ that is perpendicular to $\mathbf{B}_0$.

The design and fabrication of a high quality probe are of crucial importance to the performance of an NMR spectrometer. Mechanical tolerances are severe, careful attention must be given to the magnetic susceptibility of all materials used in the probe, and electronics must often be provided for optimal probe performance at several frequencies concurrently. In addition to the "normal" probes for high resolution NMR in liquids, probes are available with the special requirements for high rf power and sample spinning needed for the study of solids. In addition, probes have been custom designed for the study of samples at very high pressures and very high temperatures. Commercial probes are also available for the study of flowing liquids and for coupling NMR to separation techniques such as high performance liquid chromatography.

As we saw in Section 2.7, the precessing magnetization induces an electrical signal in a coil placed in the probe. In most modern NMR spectrometers the coil used to transmit the exciting rf to the sample is also used to measure the nuclear induction signal, which is extremely weak. After substantial amplification in a preamplifier and further enhancement in the amplifier portion of the *rf receiver*, the rf signal is *detected*— a process that we discuss in more detail in Section 3.4. After detection, the resulting signal is in a much lower ("audio") frequency range, where it can be further treated by electronic circuitry, as we shall see.

3.3 REQUIREMENTS FOR HIGH RESOLUTION NMR

The components of an NMR system involve apparatus that is mechanically and electronically quite sophisticated. Particularly for the study of the narrow lines that make up the NMR spectrum of a liquid or solution (high resolution NMR), there are very stringent requirements on the magnet itself and the electronic systems. Moreover, several ancillary devices and techniques must be employed to ensure adequate homogeneity and stability.

Homogeneity

As we saw in Section 2.5, NMR lines from small molecules in the liquid phase often have a natural line width of less than 1 Hz (frequently a small fraction of a hertz), and as will be seen later, lines separated by 1 Hz or less must sometimes be resolved to obtain information of chemical interest. For proton NMR, rf frequencies of 200–800 MHz are usually used; hence a resolution of the order of 10^{-9} is often required. In addition to careful fabrication of the magnet, three approaches are used to obtain adequate homogeneity over the sample volume, as follows:

1. The residual inhomogeneity in the magnet can be thought of as arising from magnetic field gradients in various directions. Along the z axis, for example, the magnetic field may vary in a way that can be analyzed mathematically into several components: a first-order gradient (i.e., a linear variation of field along z), a second-order gradient (quadratic variation), etc. By placing into the bore of the magnet electrical coils that are precisely designed and fabricated, and by carefully adjusting small electrical currents through the coils, we can deliberately generate magnetic field gradients that are equal in magnitude but opposite in direction to the unwanted gradients, in principle completely eliminating their effects, but in practice reducing them to satisfactory levels. Linear gradients in the x and y directions, together with higher order gradients of many types (e.g., x^2, y^2, xy, xz, x^2y), can likewise be reduced. Typically, NMR spectrometers with superconducting magnets are equipped with several superconducting homogeneity coils and 20–40 room temperature coils. The currents are adjusted by the spectrometer operator, either manually or via a suitable iterative computer program, to achieve optimum homogeneity, as judged by the line width and overall shape of a sharp NMR line. Fortunately, most of the settings require only very infrequent adjustment.

In NMR jargon, these homogeneity adjusting coils are known as "shim coils," and the process of optimizing homogeneity is often called "shimming." This terminology is a carryover from early homogeneity adjustments of permanent magnets and electromagnets, in which the magnet pole pieces could be tipped very slightly to improve homogeneity by inserting thin pieces of metal (shims) behind them. As electrical coils were introduced to take over this function, they were widely called shim coils.

2. A considerable improvement in *effective* homogeneity across the sample is often achieved by spinning the sample tube about its axis. If the field gradient across the sample in a direction transverse to the spinning axis is given by ΔB,

then it is known that a spinning rate

$$R > \gamma \Delta B / 2\pi \text{ rotations/second} \qquad (3.2)$$

averages out much of the inhomogeneity by causing each portion of the sample to move periodically through the entire gradient. Qualitatively, this can be viewed as an example of the effect of "rapid exchange," which was discussed in Section 2.10. The sample thus behaves as though it experiences only the average field rather than the entire range of field strength. Sample spinning is usually accomplished by means of small air turbine mounted on the probe. Rotation of

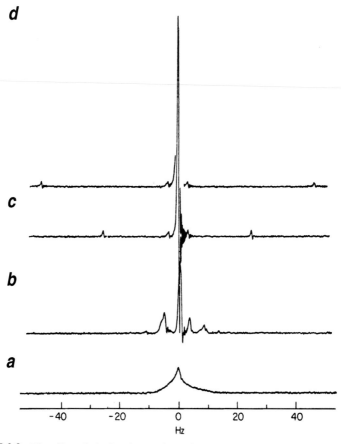

FIGURE 3.2   The effect of spinning the sample on the proton resonance spectrum of tetramethyl-silane (TMS) in a magnetic field with relatively poor homogeneity. (*a*) No spinning, showing a line with about 3 Hz half-width and a broad base. (*b*) Spinning at a very slow rate of 4 revolutions/second. (*c*) Spinning rate 25 rev/s. (*d*) 46.5 rev/s. The small "satellite" lines near the base of the strong TMS line, which do not depend on spinning rate, are due to TMS molecules containing ^{29}Si, as discussed in Section 6.16.

the order of 30 revolutions per second is typical. This procedure is so effective that it is very widely used in high resolution NMR studies of small molecules.

One drawback of this technique is that the periodic spinning modulates the magnetic field and leads to the appearance of *sidebands* i.e., "images" of the spectral peaks) symmetrically placed and separated by the spinning frequency and integral multiples of it, as illustrated in Fig. 3.2. Spinning sidebands can usually be reduced to less than 1% of the ordinary peak intensity by use of high-precision sample tubes and spinning apparatus and by proper adjustment of the electrical shim coils to reduce higher order gradients in the plane perpendicular to the spinning axis. Spinning sidebands can easily be recognized by their change in position when the spinning speed is altered. A higher speed not only causes them to move farther from the parent peak but also reduces their intensity. A second drawback occurs in two-dimensional NMR studies, where extremely small lateral motions in the spinning sample tube introduce additional noise, as we note in Chapter 10. For 2D NMR, sample tubes are often not spun if the loss in resolution can be tolerated. For studies of larger molecules, such as biopolymers with greater line widths, the absence of spinning does not hinder resolution.

3. Restriction of the sample volume can reduce substantially the overall inhomogeneity across the sample. In most instruments the "effective volume" of the sample i.e., that within the rf receiver coil) is restricted to about 0.1 ml, when the sample is placed in a 5-mm-diameter cylindical tube—a size that is regarded as the "standard" NMR tube. However, larger diameter tubes (8–15 mm) are sometimes preferred for studying dilute solutions or for observing nuclei of low sensitivity, even with some loss of homogeneity across the sample. (As we see in Section 3.11, much smaller diameter tubes are also valuable when the total amount of sample is limited.)

Stability

A highly homogeneous field is of little practical value in measuring sharp line NMR spectra if there are significant fluctuations or drift of field or frequency during the period of observation. The first real efforts to achieve adequate stability for high resolution NMR were aimed at stabilizing the radio frequency with a crystal-controlled oscillator and independently controlling the magnetic field strength with a flux stabilizer, which senses changes in the field and applies correction currents through the electromagnet power supply. However, it was soon recognized that for NMR it is the *ratio* of field to frequency that is important, not the separate values of each. Hence, the greatest stability is achieved by using the Larmor relation, through which an NMR signal itself can provide excellent field–frequency control by using the dispersion mode signal to provide a positive or negative signal in a feedback loop. Such a *field–frequency lock* is incorporated into all high resolution NMR instruments as essentially a very simple continuous wave spectrometer. Because deuterated solvents are widely used (see Section 3.12), most locks operate at the ^2H resonance frequency.

3.4 DETECTION OF NMR SIGNALS

An NMR experiment is designed to tip the macroscopic magnetization **M** away from its equilibrium position along the z axis and to generate M_{xy}, a component of **M** in the xy plane. As we have seen, the resultant precession of M_{xy} at frequency ω induces a current at this frequency in an electrical coil in the xy plane. Manipulation of the signal electronically and later conversion to the digital form needed for computations are made feasible only by detecting it in such a way that its frequency (in the range of usually hundreds of MHz) is first reduced to a more manageable frequency in the kHz range (sometimes called "audio" frequency because sound is also in this range). This is easily done by an electronic device termed a phase-sensitive detector, which multiplies the FID signal at ω with a reference signal from the radio frequency source at ω_{rf}. (In practice such frequency conversion is carried out in several steps.) The FID signal S can be represented by

$$S = Ce^{-t/T_2^*} \cos{(\omega t + \phi)} \tag{3.3}$$

or, in complex notation,

$$S = \tfrac{1}{2} Ce^{-t/T_2^*}[e^{i(\omega t + \phi)} + e^{-i(\omega t + \phi)}] \tag{3.4}$$

(In these equations and those following, we arbitrarily set the constants C to unity because we are concerned only with relative values of signal.)

Modern NMR spectrometers almost always use a *quadrature phase detection* arrangement to permit disentangling of signals and to provide an improvement in signal/noise by a factor of $\sqrt{2}$ relative to ordinary phase-sensitive detection. Quadrature detection can be carried out with a single phase-sensitive detector and *sequential* measurements of the signal with different reference phases, as we describe later, or with *simultaneous* measurements from two phase-sensitive detectors, with the reference signals phased at 90° relative to each other. For the latter case, the reference signals to the two phase detectors, R_A and R_B, are given by

$$\begin{aligned} R_A &= C' \cos{(\omega_{rf}t + \phi_{rf})} \\ R_B &= C' \sin{(\omega_{rf}t + \phi_{rf})} \end{aligned} \tag{3.5}$$

or in complex notation

$$R = R_A + iR_B = e^{i(\omega_{rf}t + \phi_{rf})} \tag{3.6}$$

The output of the phase-sensitive detectors is then the product of these signals

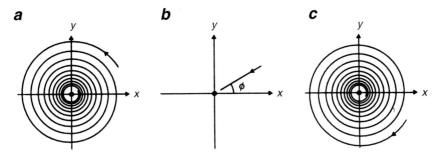

FIGURE 3.3   Depiction of the quadrature signal in Eq. 3.8 for (*a*) reference frequency $\omega_{rf} < \omega$; (*b*) $\omega_{rf} = \omega$; (*c*) $\omega_{rf} > \omega$. The inward spiral results from transverse relaxation at rate $1/T_2^*$. From Chen and Hoult,[37] *Biomedical Magnetic Resonance Technology* by C.-N. Chen and D. I. Hoult. Copyright 1989 IOP Publishing Limited. Reproduced with permission.

$$SR = e^{-t/T_2^*}[e^{i(\omega + \omega_{rf})t}e^{i(\phi + \phi_{rf})} + e^{-i(\omega - \omega_{rf})t}e^{-i(\phi - \phi_{rf})}] \tag{3.7}$$

It is a simple matter to apply a low-pass filter to remove the high frequency component at $(\omega + \omega_{rf})$ and leave only the component at $(\omega - \omega_{rf})$. The remaining signal is thus

$$SR = e^{-t/T_2^*}e^{-i(\phi - \phi_{rf})t}e^{-i(\omega - \omega_{rf})t} \tag{3.8}$$

A plot of Eq. 3.8 in the complex plane is shown in Fig. 3.3. Because the frequency of the output $(\omega - \omega_{rf})$ is just the frequency at which M_{xy} precesses in the rotating frame, Fig. 3.3 also provides a visual depiction of this precession as viewed along the z' axis. Note that the quadrature phase detection configuration permits us to differentiate between the frequencies $(\omega - \omega_{rf})$ and $(\omega_{rf} - \omega)$, whereas such frequencies are indistinguishable with a single phase detector. The phase angle $(\phi - \phi_{rf})$ can be chosen to provide the pure absorption mode on resonance, and in more complex experiments it can be adjusted as needed, as we shall see in detail later.

The alternative method for obtaining quadrature phase detection with the use of a single phase-sensitive detector, developed initially by Redfield[38] and used in some commercial spectrometers, has certain advantages in two-dimensional NMR. In this approach, the phase of the receiver is advanced by 90° after each measurement. The rationale can be better understood after we discuss digitization rates in Section 3.7.

3.5 PHASE CYCLING

Phase cycling refers to the repetition of a pulse sequence a number of times, with the phase of the rf pulses and/or the receiver reference phase incremented each time, usually by 90° or 180°. As an FID is acquired in each of these steps, it is processed by the two phase-sensitive detectors and added to (or subtracted from) the data already accumulated in two separate memory locations, as illustrated in the following. Phase cycling can be used either for the elimination of unwanted "real" resonances, for the suppression of artifacts, or for both purposes simultaneously. Phase cycling plays a critical role in 2D NMR, and we shall encounter a number of examples in later chapters. The fundamental assumption underlying the use of phase cycling is that spectrometer and magnet stability is sufficiently high to ensure that repetitions under the same conditions yield identical signals. For elimination of artifacts, we need the further assumption that the phase of the desired signal follows the phase of the excitation, whereas the phase of undesired signals does not.

We consider now an important example of phase cycling that is used in both 1D and 2D NMR, namely the suppression of artifacts resulting from imperfections in the hardware used for quadrature phase detection. We detail the principles and procedures involved in this example as a prototype for many more complex phase cycling procedures that we mention more briefly in later chapters.

As described in the Section 3.3, two supposedly identical detectors are arranged to sample the signal simultaneously along x' and y'. However, the two detectors are usually not quite identical, and the reference signals to the detectors may not differ by precisely 90°. Also, the sample-and-hold circuits that follow the detectors may have slightly different characteristics. Three types of artifacts result: (1) a DC offset between quadrature channels, (2) a gain difference between channels, and (3) a phase difference between channels. In terms of Eq. 3.5, artifact (1) means that there are constant terms of different magnitude added to the sine and cosine terms, while (2) and (3) imply different values of C' and ϕ_{rf} respectively, for the sine and cosine terms.

Artifact (1), a DC offset between the two channels, simply introduces a constant signal with no time dependence, which Fourier transforms into a peak at zero frequency. To see the effect of artifact (2), take an example in which the DC levels and reference phases are identical but the signal amplitudes differ, so that the normalized output of Eq. 3.8, expressed in terms of its two components, becomes

$$SR_A = \cos \omega t$$
$$SR_B = (1 + \epsilon) \sin \omega t \tag{3.9}$$

where ϵ may be positive or negative and $|\epsilon|$ is small but not negligible relative to unity. From simple trigonometric identities, we can rewrite Eq. 3.9 as

$$SR_A = (1 + \epsilon/2) \cos \omega t - (\epsilon/2) \cos(-\omega t)$$
$$SR_B = (1 + \epsilon/2) \sin \omega t - (\epsilon/2) \sin(-\omega t) \tag{3.10}$$

Thus, the effect of the amplitude discrepancies in the two channels is to alter slightly the amplitude of the correct signal at frequency ω but, more important, to introduce a small signal at frequency $-\omega$. The Fourier-transformed spectrum would then show a small "image" at $-\omega$. In a similar manner, it is easy to show that a discrepancy in phase also introduces an image at $-\omega$.

To eliminate these artifacts we make use of our ability to alter the phases of the pulse itself and of the reference signal and to select the portions of computer memory in which the digitized signals are stored. The standard method of eliminating these three types of artifacts is a four-step cycle [often called CYCLOPS (*cyclic observe phases*)]. To see the rationale of the method, consider first an example in which the discrepancy lies only in an amplitude imbalance. Suppose that in the first step we store the output of detector A in computer location I and the output from detector B in location II. From Eq. 3.9, we have, then, in storage

$$\text{I: } \cos \omega t$$
$$\text{II: } (1 + \epsilon) \sin \omega t \tag{3.11}$$

We now repeat the experiment with both receiver phases incremented by 90° and the pulse phase also incremented by 90° (i.e., $\mathbf{B}_1$ applied along the $-x'$ axis instead of the y' axis). The output signals then become

$$S'R_A = \cos (\omega t + \pi/2) = -\sin \omega t$$
$$S'R_B = (1 + \epsilon) \sin (\omega t + \pi/2) = (1 + \epsilon) \cos \omega t \tag{3.12}$$

For this second experiment we route data from detector A to memory location II, change sign, and coadd them to the data already there. Likewise, the data from detector B are routed to location I and coadded to data already there. Thus we obtain in memory after the two-pulse cycle

$$\text{I: } \cos \omega t + (1 + \epsilon) \cos \omega t = (2 + \epsilon) \cos \omega t$$
$$\text{II: } [(1 + \epsilon) \sin \omega t + \sin \omega t] = (2 + \epsilon) \sin \omega t \tag{3.13}$$

On Fourier transformation this provides the desired signal with no mirror image. It can be shown that a phase imbalance in the two channels is also eliminated by this cycle.

A DC offset between the two channels is not eliminated by this cycle but is eliminated by a two-pulse cycle in which the phase of the transmitter is altered by 180° on successive acquisitions and the resulting signals are alternately added and subtracted. Because the desired Fourier-transformed signal changes sign while the DC offset does not, subtraction cancels the offset but causes all real signals to add. To cancel both channel imbalance and DC offsets simultaneously, we must nest our original two-pulse cycle in a two-pulse phase-alternated cycle to produce the four-step CYCLOPS cycle:

Pulse phase :	x'	y'	$-x'$	$-y'$
Receiver phase :	x'	y'	$-x'$	$-y'$
Memory I :	$+SR_A$	$+SR_B$	$-SR_A$	$-SR_B$
Memory II :	$+SR_B$	$-SR_A$	$-SR_B$	$+SR_A$

As we see in later chapters, a number of types of phase cycling are critical to the execution of many 2D experiments. The procedures are similar to that used in CYCLOPS, but the details vary depending on the particular type of signal that must be suppressed. Meanwhile, in addition to any phase cycling unique to the 2D experiment, the complete four-step CYCLOPS cycle is often needed to suppress the quadrature detection artifacts, with the result that long cycles (16 to 64 steps) may be needed, with consequent lengthening of experimental time.

3.6 FOURIER TRANSFORMATION OF THE FID

Almost all molecular systems of interest give rise to many NMR frequencies, so that FIDs are typically complex interference patterns of a number of sine waves of differing amplitude and often of differing decay constants, T_2^*. The unraveling of these patterns and the display of amplitude as a function of frequency (the spectrum) are usually carried out by a Fourier transformation.

As indicated in Section 3.4, after detection the FID from a system giving a single NMR line consists of an exponentially decaying cosine wave of frequency $(\omega - \omega_{rf})$, as illustrated in Fig. 3.4a. If there are several nuclei that differ in Larmor frequency because of chemical shifts and/or spin–spin coupling, each line corresponds to a different frequency in the FID, and the interference among all these signals generates a response of the sort depicted in Fig. 3.4b. To extract those component frequencies we usually turn to Fourier transform procedures, which result in a clear display of the component frequencies present in the FID as an NMR spectrum.

In practice, there are a number of important points to be kept in mind. First, we wish to measure the FID over a short period of time, not between the infinite limits given in Eq. 3.1. Second, we must obtain a finite number of digital samples of the signal in order to carry out the calculation in a digital computer. And third,

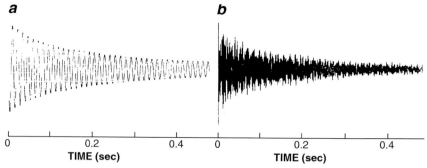

FIGURE 3.4 Oscillatory free induction decay following a 90° pulse applied to a sample with (a) a single resonance line and (b) a multiline spectrum.

there may be instrumental artifacts that must be accounted for and corrected in the calculation. Although useful for mathematical analysis, the formulation of Eq. 3.1 must be replaced by the discrete Fourier transform as the basic FT definition for computation in a digital computer, as follows:

$$S(\omega_n) = \frac{1}{N} \sum_{n=0}^{N-1} s(t_n)e^{-i\omega t_n} \tag{3.14}$$

where N is the total number of data points in the time domain (the FID). Computer algorithms to carry out this calculation rapidly are available and work most efficiently if N is an integral power of 2. The complex Fourier transform obtained in this computation consists of a real part (from the cosine part of the FT) and an imaginary part (from the sine part of the FT), each containing $N/2$ points in the frequency domain. Information in the real and imaginary transforms is largely but not completely redundant, as we shall see.

3.7   DATA ACQUISITION

The discrete Fourier transform provides a useful and efficient means of extracting information on the frequency components present in a time-varying signal and displaying the amplitudes of these components as a spectrum. However, a number of potential artifacts must be avoided if we are to obtain a faithful representation of the information actually present in the time domain signal.

Acquisition Rate

We saw in a previous section that the electrical signal that constitutes the FID is detected in one, or more commonly two, phase-sensitive detectors. Because the information is to be used in a digital computer, the signal from each detector is digitized by an *analog-to-digital converter* (ADC), a device that samples the signal at regular time intervals and provides an output that represents the signal at that time, as rounded off to the nearest bit. Most newer NMR spectrometers use 16- to 18-bit ADCs, which usually provide faithful representation of the signal intensity provided that the receiver gain is adjusted properly to utilize the full range of the ADC. However, in a sample with a very strong line (e.g., H_2O as a solvent), the signal may be more than 2^{16} times as large as noise, and the ADC cannot accommodate this large *dynamic range*. The problem is normally overcome by suppressing the solvent peak by methods we discuss in Chapter 9.

The minimum *sampling rate* of the ADC is critical to faithful representation of the signal frequencies. This rate, the number of measurements per second, is often also expressed as the *dwell time* of the ADC, that is, the time interval between measurements, $\Delta t = t_n - t_{n-1}$. It is well established in information theory that faithful representation of a sine wave requires the measurement of at least two points per cycle. Thus, a measuring rate of $2f_N$ data points per second properly represents frequencies up to and including frequency f_N, but frequencies greater than f_N are misinterpreted by the digitization process and *aliased* or *folded* to lower frequencies. For a measuring rate of $2f_N$ data points per second, f_N is called the *Nyquist frequency*.

The Nyquist principle is illustrated in Fig. 3.5a, which depicts the magnetization in the rotating frame, as sampled with a dwell time of $1/2W$ seconds, so that W becomes the Nyquist frequency for this example. With a single phase-sensitive detector, the measurements give only the projection along the x' axis. Frequencies $(W + \delta)$ and $(W - \delta)$ can be seen to give the same projections for both measurements within the cycle. Likewise, the clockwise magnetization moving at frequency $-(W + \delta)$ gives the same measurement as $+(W + \delta)$, a point we noted in Section 3.4 for the complete analog signals themselves. Sampling at $2W$ points per second with a single detector thus defines a *spectral width* that ranges from 0 (the pulse frequency in the rotating frame) to W Hz, and an undersampled frequency of $W + \delta$ appears to be folded back to a frequency $W - \delta$, as shown in Fig. 3.5b. Likewise, frequencies from 0 to $-W$ are folded into the 0 to W range.

Frequencies higher than W can be suppressed by a suitable low-pass analog filter with a sharp cutoff, so that such foldover can be largely eliminated. With a single phase-sensitive detector, the pulse frequency is normally set so that no signals are expected in the "negative" frequency range. Thus this potential source of confusion on signal frequency can readily be eliminated. However, in placing the pulse at one end of the desired range, rather than in the center, half of the pulse power is wasted. Moreover, *noise* in the "negative" frequency region folds into the 0 to W range and increases the noise level by $\sqrt{2}$. Almost all spectrometers now solve these problems by use of quadrature phase detection.

As we saw in Section 3.4, quadrature phase detection discriminates between frequencies higher and lower than the pulse frequency, but it does not prevent foldover from frequencies higher than the Nyquist frequency. For a desired spectral width W, there are two common methods for carrying out quadrature phase detection, as was indicated in Section 3.4. One method uses two detectors and samples each detector at W points per second, thus acquiring $2W$ data in the form of W complex numbers. The other (commonly called the "Redfield method") requires only a single detector and samples at $2W$ points per second while incrementing the phase of the receiver by 90° after each measurement. (In two-dimensional NMR studies, a variant of this method is usually called the *time-proportional phase incrementation*, or TPPI, method.) Because these methods result in quite different treatment of folded resonances, we now consider these approaches in more detail.

With two detectors we make simultaneous measurements in quadrature, which we may take to be along the x' and y' axes in the rotating frame. Because we can

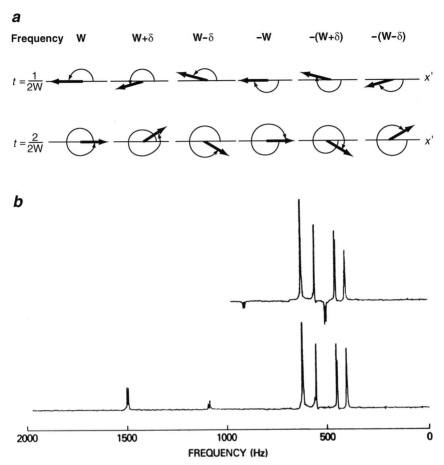

FIGURE 3.5 (a) Depiction of magnetization **M** precessing in the rotating frame at the frequencies indicated. W is the Nyquist frequency, and sampling is illustrated at $t = 1/2W$ and $1/W$. A single phase-sensitive detector measures only the projection of **M** along x'. (b)Example of foldover of spectral lines as a result of sampling the FID at too low a rate with a single phase-sensitive detector. The spectrum at the bottom, obtained with a spectral width of 2 kHz (4000 points/second), faithfully reproduces the true spectrum. The upper spectrum, obtained with a 1 kHz spectral width (2000 points/second) shows foldover of peaks at $1000 + \delta$ to $1000 - \delta$. Note that the folded peaks usually display phase distortion. Sample: 1,7-dimethylcytosine in DMSO-d_6 at 220 MHz.

discriminate between frequencies greater than and less than the pulse frequency with the quadrature phase detection scheme, it is advantageous to place the pulse (hence the reference frequency) at the center of the desired spectral window, which then runs from $+W/2$ to $-W/2$. Figure 3.6a again depicts **M** precessing in the rotating frame, with measurements now made with a dwell time of $1/W$ s.

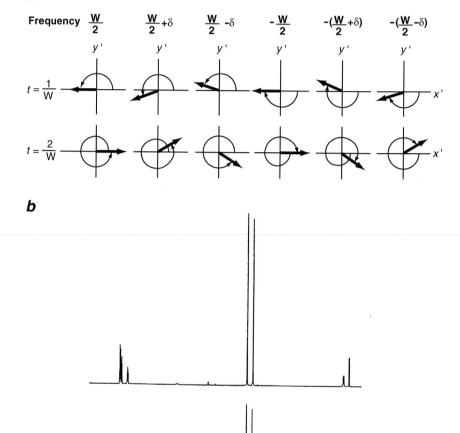

FIGURE 3.6 (*a*) Depiction of magnetization **M** precessing in the rotating frame at the frequencies indicated. *W*/2 is the Nyquist frequency, and sampling of both phase-sensitive detectors is illustrated at $t = 1/W$ and $2/W$ to obtain projections along both x' and y'.

(In this example $W/2$ becomes the Nyquist frequency.) Although the projections of **M** along the x' axis at the sampling times are identical for $(W/2) + \delta$ and $(W/2) + \delta$, the projections along y' are different, and these frequencies can thus be distinguished from each other. However, measurements for $(W/2) + \delta$ and $-[(W/2) - \delta]$ are identical along both x' and y'. Thus, there is again foldover, but as shown in Fig. 3.6b, the folded frequency appears near the other end of the spectral window. In general, as shown in the following, foldover occurs to a frequency that is W (or a multiple of W) lower; $(W/2) + \delta$ folds to $(W/2) + \delta - W$, or $-[(W/2) - \delta]$.

The general mathematical treatment of foldover, as illustrated in Fig. 3.6, is straightforward. Equation 3.8 provides the expression for the output signal from the quadrature phase-sensitive detector, which can be reformulated as

$$SR = \tfrac{1}{2} A e^{-\iota(\omega - \omega_{rf})t} \tag{3.15}$$

where A takes into account all the time-independent factors, including an arbitrary phase. At the Nyquist frequency $W/2$, $\omega - \omega_{rf} = \pi W$, and measurements are made at $t = 1/W$ and multiples thereof. A frequency above the Nyquist frequency, $(W/2) + \delta$, then leads to the following value of output signal at time $1/W$:

$$SR\left(\frac{W}{2} + \delta\right) = A e^{-2\pi i\left(\frac{W}{2} + \delta\right)\frac{1}{W}} \tag{3.16}$$

But

$$SR\left[-\left(\frac{W}{2} - \delta\right)\right] = A e^{-2\pi i\left[-\left(\frac{W}{2} - \delta\right)\right]\frac{1}{W}}$$

$$= A e^{-2\pi i\left[\left(\frac{W}{2} + \delta\right) - W\right]\frac{1}{W}}$$

$$= A e^{-2\pi i\left(\frac{W}{2} + \delta\right)\frac{1}{W}} e^{+2\pi i(W)\frac{1}{W}}$$

$$= A e^{-2\pi i\left(\frac{W}{2} + \delta\right)\frac{1}{W}}$$

$$= SR\left(\frac{W}{2} + \delta\right) \tag{3.17}$$

FIGURE 3.6 (*Continued*) (*b*) Example of foldover of frequencies above the Nyquist frequency with two phase-sensitive detectors. The lower spectrum (spectral width 5000 Hz) faithfully reproduces the true spectrum, but the upper spectrum (spectral width 3600 Hz) shows that peaks near + 2000 Hz now display an aliased frequency near − 1800 Hz and appear near the right-hand end of the spectrum. Sample: D-glucorono-6,3-lactone-1,2-acetonide in DMSO-d_6 at 500 MHz. Spectra courtesy of Joseph J. Barchi (National Institutes of Health).

a

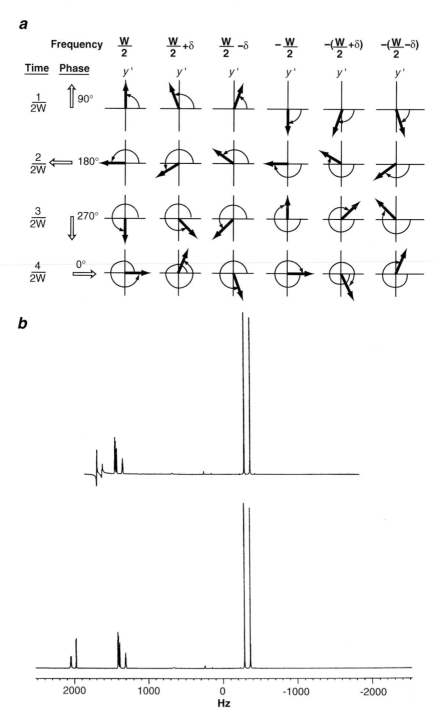

b

We make use of the identity $e^{2\pi i} = 1$ in the preceding treatment for $t = 1/W$ and in generalizing the result for measurements at multiples of $1/W$. It is sometmes useful to think of the spectrum as being displayed on a circle, rather than in a linear array, so that "foldover" is just a displacement around the circle.

The Redfield method of quadrature phase detection operates as shown in Fig. 3.7a. A spectral width from $W/2$ to $-W/2$ is defined by $2W$ measurements per second with a single detector, the phase of which is augmented by 90° after each measurement. Each signal magnitude, as illustrated in Fig. 3.7a, is the projection of **M** in the direction shown by the phase arrow. It is clear that measurements for frequencies $(W/2) + \delta$ and $(W/2) - \delta$ are identical, so that foldover occurs as indicated in Fig. 3.7b. Likewise, $-[(W/2) + \delta]$ folds into $-[(W/2) - \delta]$.

As indicated in Section 3.4, both methods of quadrature detection are used in commercial instruments. It is important to be aware of the method being used in a particular experiment, because foldover gives quite different results in the two approaches.

We have discussed the minimum sampling rate needed to prevent foldover. The question arises of whether there is any gain to *oversampling*, that is, sampling at higher than the minimum rate. Although oversampling does not provide additional precision in representing the frequencies, modest improvements in signal/noise, flatness of the spectral baseline, and dynamic range can be achieved by oversampling, together with an increase in the bandwidth of the low-pass analog filter and addition of digital filtering in the data processing. We will not discuss the rationale or operation of these techniques.

Acquisition Time

The FID must be sampled at a rate adequate to cover the spectral width, but how long should the data be acquired? That depends on both the desired resolution and considerations of signal/noise ratio. The FID for a given spectral component decays with a time constant T_2^*, which depends on the natural transverse relaxation time of the component and the inhomogeneity in the magnetic field. Thus, even acquisition of the FID for an unlimited time leads, on Fourier

FIGURE 3.7 (a) Depiction of magnetization **M** precessing in the rotating frame at the frequencies indicated. $W/2$ is the Nyquist frequency, as in Fig. 3.6, but the single phase-sensitive detectors is sampled as illustrated at multiples of $t = 1/2W$. A single projection is thus obtained for each sample, but the reference phase is incremented as shown to obtain projections sucessively along $+y'$, $-x'$, $-y'$ and $+x'$. (b) Example of foldover of frequencies above the Nyquist frequency with a single phase-sensitive detector whose phase is cycled in the Redfield–TPPI method. The lower spectrum is undistorted but the upper spectrum shows aliasing of the peaks near $+2000$ Hz so that they appear folded over to about 1600 Hz. Sample: D-glucorono-6,3-lactone-1,2-acetonide in DMSO-d_6 at 500 MHz. Spectra courtesy of Joseph J. Barchi (National Institutes of Health).

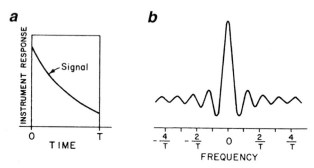

FIGURE 3.8   (*a*) Schematic representation of the exponential decay of the FID from a single NMR line truncated at time *T*. (*b*) Fourier transform of truncated signal.

transformation, to a line of width $1/\pi T_2^*$. Fourier transform theory makes it clear that by limiting the data acquisition to a period of length T s we can obtain a minimum line width of about $1/T$ Hz, as discussed in the following paragraph. Thus, truncation of the FID before πT_2^* can broaden the line, but continuation of data acquisition beyond about $3T_2^*$ leads to no improvement in resolution. Moreover, by $3T_2^*$ the FID has decayed to about 5% of its initial amplitude, and signal/noise is usually so low that continued data acquisition simply degrades the S/N in the Fourier-transformed line unless digital filtering procedures are used, as described in Section 3.8. As an initial guideline, then, we might take $3T_2^*$ as a reasonable data acquisition period, subject to later modification in order to take additional factors into account. In practice, of course, there may be spectral components with considerably different values of T_2 and these may not be known in advance, so that the acquisition time that is selected represents some compromise on the basis of reasonable estimates of T_2^*.

FT theory shows that the effect of abrupt truncation of the FID is to convolve the normal Lorentzian line shape with a function of the form

$$F(\nu) = \text{sinc } \nu = \frac{1}{\pi \nu} \sin \pi \nu \qquad (3.18)$$

illustrated in Fig. 3.8. As we pointed out, this results in a line that may be broader than the true width, and the shape is altered from that of a pure Lorentzian line. In particular, the sidelobes, or "wiggles," can present a problem if the FID is truncated before it has decayed into noise. We comment further on this potential problem in the following section.

3.8 DATA PROCESSING

Once the digitized data are acquired in the computer, a number of mathematical manipulations can be carried out to minimize or eliminate artifacts and to enhance resolution or signal/noise ratio.

Zero-Filling

With quadrature phase detection, we have seen that we must acquire data at a rate of W complex points/second to cover a spectral width of W Hz. In $3T_2^*$ seconds we thus acquire a total of $3WT_2^*$ complex data points, which appear over W Hz in the real part of the transformed spectrum (which we may take to be the absorption mode spectrum), or $3T_2^*$ points per Hz. Because a typical line width is $1/\pi T_2^*$, this provides only about one point to define a line. Although this is adequate to define the presence of the line and show its approximate frequency, it is usually necessary to have more points to locate the line frequency more precisely and to define its shape. Additional data points can be obtained without lengthening the acquisition time by simply adding zeros to the FID after it has been digitized. It can be shown that adding as many zeros as the number of experimental data points is particularly valuable, as this process retrieves information otherwise lost in the discarded imaginary spectrum. Additional sets of zeros do not add new information, but they further improve the appearance of the spectrum by simply interpolating between the other points on the basis of a sinc ωT interpolation function.

The total number of points to be transformed after zero–filling usually presents little problem in terms of data storage or computational capacity of modern computers for a 1D data set. However, as we shall see later, in 2D (and especially in 3D or 4D) NMR the approach outlined here that optimizes spectral width, resolution, and zero–filling independently often leads to data tables that are beyond the capacity of most computers to process in reasonable periods of time. In such cases, compromises must be made, as we discuss in later chapters.

Without zero–filling, data points in the transformed spectrum occur only at multiples of $1/T$ Hz, which correspond to zero-crossings of the sinc function, as shown in Fig. 3.8; hence the sidelobes are not observable. Doubling the number of points by zero–filling also places points only at zero-crossings for the absorption mode. Only when the number of points is quadrupled are the sidelobes observable.

Phase Correction and Spectral Presentation

If the phase-sensitive detectors are adjusted to give a phase angle (Eq. 3.8) $(\phi - \phi_{rf}) = 0$, the real part of the FT spectrum corresponds to pure absorption at the pulse frequency, but off-resonance lines display phase angles proportional to their off-resonance frequency as a consequence of limited rf power and nonzero pulse width (Eq. 2.55). However, acquisition of data as complex numbers from the two phase-sensitive detectors and subsequent processing with a complex Fourier transform permit us to obtain a spectrum that represents a pure absorption mode.

Because the real and imaginary parts of the spectrum always have exactly a 90° phase relative to each other at each resonance frequency, it is easy for the computer to generate a pure absorption spectrum $A(\omega)$ and a pure dispersion spectrum $D(\omega)$ by forming linear combinations of real ($\mathfrak{R}$) and imaginary ($\mathfrak{I}$) spectra that vary linearly with frequency:

$$A(\omega) = \cos(\phi - \phi_{rf})\, \mathfrak{R}[F(\omega)] - \sin(\phi - \phi_{rf})\, \mathfrak{I}[F(\omega)] \tag{3.19}$$

$$D(\omega) = \sin(\phi - \phi_{rf})\, \mathfrak{R}[F(\omega)] + \cos(\phi - \phi_{rf})\, \mathfrak{I}[F(\omega)] \tag{3.20}$$

The process of "phasing" the spectrum consists of choosing the coefficients of real and imaginary parts to optimize the appearance of the overall spectrum, either by manual adjustment or by application of one of several algorithms.

For some purposes in 2D NMR it is sufficient to obtain only the absolute value mode, that is, the square root of the sums of the squares of absorption and dispersion. The absolute value mode is independent of the phase of the signals, hence is simpler to compute and avoids the need for interactive phasing. However, as a composite of absorption and dispersion, it includes the long "wings" characteristic of the dispersion mode (see Fig. 2.10) and thus has substantially greater overlap of lines than pure absorption.

Interpolation by Linear Prediction

Often the initial points in the FID are not recorded faithfully, because the high power pulse causes distortion of the receiver over a finite amount of time, even though steps are taken electronically to minimize that time. It is well known from FT theory that errors in the first few time domain data points introduce artifacts such as "baseline roll" into the Fourier transformed spectrum. In some instances these are mere annoyances and can be eliminated adequately during phase correction and other spectral processing, but in other cases, especially if the spectrum contains one or more broad lines, a reliable flat baseline is essential. One method for interpolating missing or poorly defined data points is *linear prediction*. This technique may also be used to extrapolate the FID beyond the measured range in lieu of zero-filling in order to improve resolution and signal/noise, especially in two-dimensional NMR.

Linear prediction methods assume that the future time domain data can be represented as a linear combination of past time domain data, and vice versa—an assumption that is valid when the time domain data consist of exponentially damped sine functions and the signal/noise ratio is sufficiently large. FIDs from

FIGURE 3.9 Use of linear prediction to generate a flat baseline from corrupted FID data. (*a*) Fourier transform of observed FID, showing a badly "rolling" baseline that arises from an erroneous data point. (*b*) The first 50 data points in the FID, with the fourth point obviously damaged. (*c*) Result of deleting the first six data points and recalculating them by linear prediction (vertical scale expanded). (*d*) FT spectrum from the repaired FID. Adapted from Led and Gesmar,[39] *Encyclopedia of Nuclear Magnetic Resonance*, D. M. Grant and R. K. Harris, Eds. Copyright 1996 John Wiley & Sons Limited. Reproduced with permission.

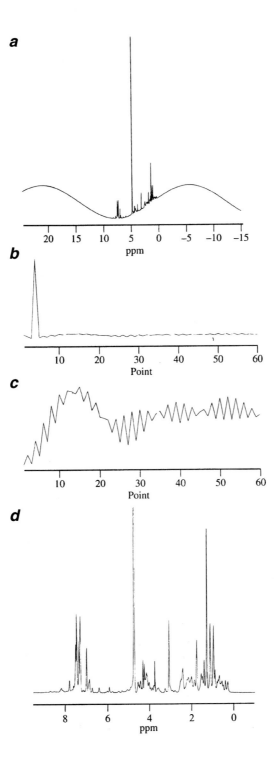

samples in liquids with Lorentzian lines often meet these criteria. The nth data point S_n can be expressed as a linear combination of the k preceding (or succeeding) points:

$$S_n = \sum_{j=1}^{k} a_j S_{n \pm k} \tag{3.21}$$

From N data points, $N - k$ such equations can be generated, and as long as $N - k$ is larger than the number of spectral lines, the problem is overdetermined and a set of coefficients a_j can be obtained by suitable mathematical procedures. With these coefficients the FID can be forward or back projected. For example, Fig. 3.9 illustrates the improvement in spectral quality as a result of replacement of a severely distorted data point.

3.9 DIGITAL FILTERING

In the previous sections we focused on steps that must be taken to transform data faithfully from the time domain to the final displayed spectrum. However, if we are willing to accept some modification in the true Lorentzian line shape, we have an opportunity to process data both before and after Fourier transformation in such a way as to make the final spectrum more useful. Clearly, such processing must be undertaken carefully in order not to lose important information or to introduce features that could lead us to erroneous conclusions.

Prior to Fourier transformation the time domain data may be modified by multiplication by a function $q(t)$, a process commonly called *digital filtering*. By suitable choice of $q(t)$, digital filtering may be used in NMR spectroscopy to enhance sensitivity, to improve spectral resolution, or to avoid truncation effects. Conceptually, there is very little difference between filtering of 1D and 2D NMR spectra, so the treatment here may later be extended readily to the two-dimensional case.

Sensitivity Enhancement

Because the signal decays as a function of time, while the noise is stationary, data points at the beginning of the FID contribute more to the resonance intensity than do points near the end of the FID. One common way to discriminate against later points with poor S/N without abruptly truncating the FID is to use an *exponential filter*, with

$$q(t) = \exp(-t/T_L) \tag{3.22}$$

The exponential factor in the time domain function then becomes

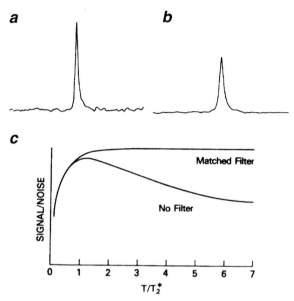

FIGURE 3.10 Use of an exponential filter to improve signal/noise ratio. (*a*) Acquisition time $T = 4T_2^\star$, with no filter. (*b*) $T = 4T_2^\star$, with a matched filter, showing better S/N but a broader line. (*c*) Relative S/N as a function of acquisition time and without a matched filter.

$$\exp(-t/T_2^\star)\exp(-t/T_L) = \exp[-t(1/T_2^\star + 1/T_L)] \tag{3.23}$$

so that the decay rate of the FID is increased to $(1/T_2^\star + 1/T_L)$. It has been shown that optimum S/N is obtained by selecting the filter constant $T_L = T_2^\star$, a so-called *matched filter*. By using an exponential matched filter we retain the shape of the transformed line as a Lorentzian, but the overall decay constant is now twice as large so that the line is broadened by a factor of 2. The use of an exponential filter is illustrated in Fig. 3.10. If the filter constant is chosen so that $T_L = \frac{1}{2}T_2^\star$, the broadening is only half as great, whereas sensitivity is still 94% $(2\sqrt{2}/3)$ as large as with a matched filter, as shown in Fig. 3.10. This "less-than-matched" filtering is often a good compromise between enhancing sensitivity and not losing too much resolution.

Gaussian multiplication is also an attractive alternative to increase sensitivity, with some concomitant alteration in line shape, as well as width. In this case

$$q(t) = \exp(-t^2/T_G^2) \tag{3.24}$$

In 2D spectroscopy the Gaussian function is often preferred over exponential multiplication because little broadening at the base of the resonance is induced by this filter, and significant sensitivity improvement can still be obtained.

Resolution Enhancement

The use of resolution enhancement filtering is occasionally valuable in ordinary 1D NMR but is more important in 2D ^{1}H spectra. A large number of different filtering functions are available for this purpose, all of which have the common aim of "slowing down" the decay of the signal. Again, an exponential filter may be used, but with a negative T_L, so that the decay rate $[1/T_2^* + 1/T_L]$ becomes smaller than $1/T_2^*$. However, this simple type of exponential filtering is often difficult to use, because it leads either to truncation effects or to poor sensitivity.

A more useful filter for resolution enhancement is the *convolution difference* filter, given by

$$q(t) = \exp(-t/T_A) - k \exp(-t/T_B) \tag{3.25}$$

which provides on Fourier transformation the difference between two processed spectra, A and B. Time constant T_A is selected to be relatively long (or even negative), leading to relatively narrow lines in spectrum A, while T_B is made very short to obtain broad lines. A fraction $k(k \approx 0.5-1)$ of spectrum B is then subtracted from spectrum A, leading to a resolution-enhanced difference spectrum. The convolution difference filter is particularly useful in removing broad signal components from a spectrum that contains low intensity sharp signals.

3.10 ALTERNATIVES TO FOURIER TRANSFORMATION

Fourier transformation is a rapid and efficient method for converting data from the time domain to the frequency domain. It is by far the most commonly used method for carrying out such a conversion, but it is not the only method, and in some instances it may not be the optimum method.

The linear prediction (LP), method which we discussed in Section 3.8 as a means of interpolating or extrapolating data points in the FID prior to Fourier transformation, can be extended to permit the extraction of frequencies and other spectral parameters directly without Fourier transformation. This calculation sometimes has the advantage of separating overlapping spectral lines and providing better quantitation than FT. However, although the linear algebra involved is not especially complex, far more computer multiplications are required than for FT. Thus, except for small data sets, this use of LP is usually not feasible.

The *maximum entropy method* (MEM) is based on the philosophy of using a number of trial spectra generated by the computer to fit the observed FID by a least squares criterion. Because noise is present, there may be a number of spectra that provide a reasonably good fit, and the distinction is made within the computer program by looking for the one with the "maximum entropy" as defined in information theory, which means the one with the *minimum* information content. This criterion ensures that no extraneous information (e.g., additional spectral lines) is used unless it can be strictly justified by the data. Like LP, MEM prevents the appearance of artifacts (such as sinc lobes and baseline roll) that are inherent in the FT approach, but it also is computationally slower and uses much more computer memory than FT. As a result, MEM is used only for special purposes and must be employed with care, as its nonlinear computations can produce spectra in which noise is suppressed to an extent that is misleading.

Bayesian analysis provides another alternative (also computationally intensive) to deal with very weak signals and avoid FT artifacts. This method, which uses probability theory to estimate the value of spectral parameters, permits prior information to be included in the analysis, such as the number of spectral lines when known or existence of regular spacings from line splittings due to spin coupling. Commercial software is available for Bayesian analysis, and the technique is useful in certain circumstances.

3.11 SENSITIVITY AND SIZE OF SAMPLE

Several factors determine the sensitivity of NMR and hence limit the minimum size of sample that may be studied. We have already seen in Section 2.8 that an NMR signal for a given nucleus should in principle increase quadratically with field strength, so that large values of B_0 are preferred. The inherent sensitivity varies substantially from one nucleus to another, as indicated in Appendix A. The electronic circuits employed and the care used in manufacture of the probe are critical. The *filling factor*, that is, the fraction of the volume of the receiver coil that is actually filled by sample, not glass of the sample tube or air space, is especially important. The use of thin pieces of glass on which the coil is wound, as well as the use of thin-walled sample tubes, is mandatory to obtain a high filling factor.

Instrumental Sensitivity

It is easy to specify the signal/noise ratio for a given type of NMR spectrometer for a specific sample but difficult to infer from that figure just how much sample is required to carry out an NMR study, as many factors must be considered. Even the S/N specifications given here are likely to become rapidly outdated, as illustrated in Table 3.1. The figures apply to a single continuous wave scan for the 1960s and a single pulse FT scan in later years. The remarkable increase in sensitivity over the years stems partly from increase in magnetic field but primarily from significant improvements in electronics and in the design of probes. The most recent advance is the use of a superconducting rf coil, where the large reduction in probe resistance results in a corresponding decrease in thermal (Johnson) noise, the principal source of instrumental noise. Unfortunately, the improvement thus obtained is partially offset by a much poorer filling factor from the dewar needed to cool the coil to a superconducting temperature while

TABLE 3.1   Signal/Noise for Typical NMR Spectrometers

Year	Frequency (MHz)	S/N[a]
1961	60	0.6
1965	100	3
1969	220	8
1978	200	30
1978	360	80
1985	500	360
1988	600	600
1994	750	900
1995	500[b]	2000
1998	800.[c]	1800

[a] Strongest peak in the methylene proton signal of ethylbenzene, 0.1% by volume. Single pulse (or single scan for continuous wave). Figures before 1978 are scaled down a factor of 10 from the specification, which was then for a 1% solution of ethylbenzene.

[b] With a superconducting receiver coil.

[c] Triple resonance probe.

maintaining the sample near room temperature. Overall, a factor of about 4 improvement has been attained thus far, as indicated in Table 3.1.

The specifications for S/N are only a general guide to the actual sensitivity that can be obtained in practice from various types of samples. For aqueous solutions, particularly those containing appreciable quantities of salt, dielectric losses and probe detuning can significantly degrade performance if the probe is not optimized for such samples. For nuclei other than hydrogen, the amount of material needed is generally much higher, but methods that we describe later for transferring proton polarization to other nuclei can sometimes enhance the signal manyfold. Moreover, as we see in Chapter 10, techniques exist for obtaining spectra of less sensitive nuclides by transferring polarization from protons to the other nuclide and then back to protons, where NMR is observed by so-called *inverse detection* with essentially proton NMR sensitivity. Most two-dimensional NMR methods compete favorably with ordinary 1D techniques in sensitivity, but some 3D and 4D methods are significantly less sensitive.

Radiation Damping

Although we usually make every effort to maximize the NMR signal in order to be able study small samples, there are occasions on which the signal is so strong that it causes difficulties. We know that the bulk magnetization induces a current in the rf coil that provides the NMR signal. However, this current generates an additional magnetic field at the sample, which opposes the applied rf field and drives the magnetization back to equilibrium more rapidly than normal relaxation. Although this induced field is quite small, its effect is to reduce T_2^* and broaden the NMR line. This phenomenon, *radiation damping*, is significant only when the signal is large—for example, from a protonated solvent such as H_2O.

The intensity of a strong NMR signal can easily be reduced by detuning the probe or reducing the filling factor, but usually we want to maintain sensitivity in order to study samples at low concentration. Hence we must often accept some radiation damping, which broadens the strong solvent peak. In addition, radiation damping may contribute, along with other effects discussed in Chapter 11, to generation of unexpected *inter*molecular resonance peaks in some two-dimensional NMR experiments.

Time Averaging

Because the FID can usually be acquired in 1 second or less, it is feasible in almost all NMR studies to use the principle of *time averaging*. Coherent addition of n FIDs from repetition of the experiment n times results in a signal that is n times as large as for a single FID. However, it is well known that *random* noise adds only as $\sqrt{n}$ (and most noise in an NMR experiment is random, arising from thermal fluctuations in the receiver coil). Thus, the signal/noise ratio increases as $\sqrt{n}$.

However, the gain in sensitivity from time averaging can be limited by the time required for spin–lattice relaxation to restore the magnetization to equilibrium along the z axis. If $T_1 \gg T_2$, little restoration has occurred at the end of the data acquisition (e.g., $3T_2^*$), and repetition of a 90° pulse results in only a very weak FID. Under these circumstances one must either wait longer between pulses or use a smaller value of the "flip angle" than 90°, so that a substantial z component of **M** remains. For a series of 90° pulses the optimum repetition time is $1.27T_1$. Actually, slightly better results are obtained by pulsing again immediately after each data acquisition but using a flip angle (the "Ernst angle") determined from the relation

$$\cos \theta = e^{-T/T_1} \tag{3.26}$$

where T is the time between pulses (the acquisition time in this case). Of course, a knowledge of the T_1's to be expected is needed in order to apply these relations, but often an adequate estimate can be made from previous studies of similar molecules.

Sample Tubes

Normally, for proton resonance a sample of about 0.4 ml is contained in a precision, thin-walled glass tube of about 5 mm outer diameter. As we saw in Section 3.3, the "effective volume" of the sample within the height of the rf coil is only about 0.1 ml, but the additional volume is required to ensure no sharp discontinuities in magnetic susceptibility near the coil and to approximate an infinitely long cylinder. The sample volume may be reduced by using plugs above and below the liquid that are fabricated of material with approximately the same susceptibility as the solvent.

Larger diameter sample tubes (usually 8–10 mm in diameter), usually with susceptibility-matching plugs, may be used to gain some sensitivity at the expense of a larger total amount of solution. At the other extreme, smaller diameter tubes (principally 1.7 or 3 mm) reduce inherent sensitivity but require smaller samples. For tubes of 3, 5, and 8 mm, S/N for a 0.1% ethylbenzene solution varies approximately in the ratio 1:2:3, but the total *amount* of sample needed is far less for the smallest tube. Microcells, in which the sample is contained in a spherically shaped cavity of 25–50 µl can also be used to reduce the total amount of sample required, and capillary sample tubes of outer diameter 355 µm, containing only about 10 µl, have come into use with specially wound receiver coils of very small diameter. Special "flow-through" sample tubes are also used in

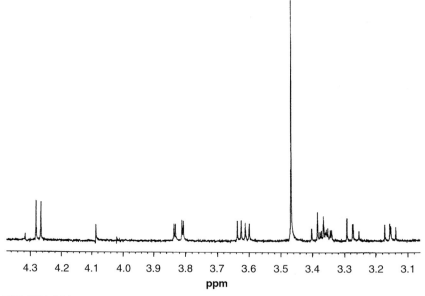

FIGURE 3.11   Proton NMR spectrum (500 MHz) of 1 µg of methyl-β-*d*-glucopyranoside in 80 µl of D_2O, obtained in 67 minutes. Courtesy of Nalorac Corporation

connection with separation methods, such as liquid chromatography and capillary electrophoresis.

Figure 3.11 shows an example of time averaging with a sample of only 1 microgram (about 5 nanomoles) of a sugar in 80 microliters of D_2O. With a 500 MHz spectrometer and a 3 mm diameter sample tube, the data were acquired in about 1 hour.

Control of Sample Temperature

Most NMR spectrometers are equipped to vary the temperature of the sample by passing preheated or precooled air or nitrogen past the spinning sample tube. The commercially available instruments are usually limited, by the materials used in the probe, to temperatures between $-100°$ and $+200°C$, but some high resolution probes have been constructed to operate at temperatures as low as $-190°$ and as high as $300°C$. At extreme temperatures some deterioration of magnetic field homogeneity is often found. It is important that adequate time be allowed for temperature equilibration of the system and that the homogeneity adjustments be optimized at each temperature. The uniformity of temperature through the sample and the stability of temperature control vary among instruments, and considerable care is required to ensure uniformity in large samples. Often it is convenient to measure sample temperature by substituting a sample tube containing a liquid with resonances that are known to be temperature dependent. For proton resonance the spectra of methanol and of ethylene glycol are usually used for low and high temperature regions, respectively, but much greater precision can be obtained by observing the ^{59}Co resonance of $K_3Co(CN)_6$, which varies by about 1.4 ppm/°C.

Changes in temperature of the sample alter the Boltzmann distribution of spins in the various energy levels in proportion to $1/T$, as indicated in Eq. 2.20. For example, a reduction in temperature from 27°C (300 K) to $-60°C$ causes an increase in signal intensity of 40%.

3.12   USEFUL SOLVENTS

In addition to the desired solvation property itself, the major considerations in selecting a solvent for NMR spectroscopy are the minimization of interaction or reaction between solvent and sample, the avoidance of strong interfering signals from the solvent, and usually the presence of deuterium in an adequate amount to serve as a field–frequency lock. A list of solvents commonly employed for ^1H and ^{13}C NMR is given in Table 3.2. The deuterated solvents are available commercially in 98–99.9% isotopic purity, but small signals from the residual protons must be expected in the region indicated. For ^{13}C NMR there are also interferences, but one effect of deuteration is to lengthen the T_1 of nearby ^{13}C

TABLE 3.2   Some Useful Solvents

Solvent	δ_H (residual)[a]	δ_C[a]
CDCl$_3$	7.24	77.0 (triplet)
CD$_3$OD	4.78	49.0 (septet)
	3.30 (quintet)	
Acetone-d_6	2.04 (quintet)	206.0
		29.8 (septet)
D$_2$O	4.65	
p-Dioxane-d_8	3.53	66.5 (quintet)
Dimethylsulfoxide-d_6	2.49 (quintet)	39.5 (septet)
Pyridine-d_5	8.71	149.9 (triplet)
	7.55	135.5 (triplet)
	7.19	123.5 (triplet)
Benzene-d_6	7.15	128.0 (triplet)
Acetic acid-d_4	11.53	178.4
	2.03 (quintet)	20.0 (septet)
SO$_2$	—	—

[a] Chemical shifts given relative to ^{1}H and ^{13}C resonances of tetra-methylsilane, as described in Section 4.3.

nuclei, so the interference is often not as severe as might be expected. For special applications, solvents depleted in ^{13}C are available. Chloroform-d is inexpensive and is probably the most commonly employed solvent for proton NMR. Samples with readily exchangeable hydrogen atoms (such as OH or NH groups) will, of course, lose these hydrogens to a solvent containing exchangeable deuterium atoms.

Many substances that are vapors at room temperature and atmospheric pressure may be used as NMR solvents in sealed tubes or at reduced temperature. For example, SO$_2$ has a vapor pressure of about 3 atm at room temperature and can be easily contained in sealed thin-walled, 5 mm diameter NMR sample tubes. Supercritical fluids are also used as NMR solvents in specialized sample tubes. For NMR studies of nuclei other than hydrogen and carbon, suitable solvents that do not contain the nucleus being studied are usually readily available. Frequently, the use of two or more solvents can provide valuable information on molecular structure, as indicated in Chapter 4.

3.13   ADDITIONAL READING AND RESOURCES

Most books concerned with application of high resolution NMR include discussion of many experimental aspects. *Recording One-Dimensional High Resolution Spectra*[40] in the *Encyclopedia of NMR* also provides a good overview. *150 and More*

Basic NMR Experiments—A Practical Course by S. Braun *et al.*[41] describes the operation of an NMR spectrometer and, as its title implies, gives guidance, with specific experimental parameters, for carrying out a variety of NMR procedures—from measuring the width of a 90° pulse to complex pulse sequences in two- and three-dimensional NMR.

Nuclear Magnetic Resonance: Concepts and Methods by Daniel Canet[42] contains particularly clear presentations on techniques and data processing for Fourier transform NMR and related methods. Articles in the *Encyclopedia of NMR* on *Fourier Transform Spectroscopy,*[43] *Fourier Transform and Linear Prediction Methods,* [39] and *Maximum Entropy Reconstruction*[44] are also very informative. *A Handbook of NMR* includes a very clear description of the maximum entropy method and its limitations.[19]

Biomedical Magnetic Resonance Technology by C.-N. Chen and D. I. Hoult[37] is aimed primarily at NMR imaging but contains excellent discussions of NMR instrumentation in general. Several articles in the *Encyclopedia of NMR* are also valuable sources of information, including *Spectrometers: A General Overview,*[45] *Probes for High Resolution,*[46] *Shimming of Superconducting Magnets,*[47] and *Concentrated Solution Effects.*[48]

NMR instrumentation is continually improving in sensitivity, sophistication, and ease of use. The sales literature from the major producers of NMR instruments, probes, and accessories often contains very useful technical information on recent advances and examples of current applications. The manufacturers of isotopically enriched materials also provide useful information on the availability of deuterated and other specialty solvents.

3.14 PROBLEMS

3.1 What minimum sampling rate is required to avoid aliasing over a spectral width of 5 kHz? If a digital resolution of 0.5 Hz is desired, how many data points are needed?

3.2 Assume a single ^{1}H NMR line on resonance with a line width of 1 Hz. Derive an expression for the relative signal/noise in the Fourier-transformed spectrum from an FID acquired for t seconds and evaluate for $t = 0.5$, 1, 2, and 5 seconds. Recall that noise accumulates with $\sqrt{t}$. Assume no digital filtering.

3.3 Suppose that an investigator inadvertently uses the parameters from an ^{1}H NMR study for an investigation of ^{17}O, where $T_2 \approx 0.05$ s. Use the results of Problem 3.2 to determine the loss in signal/noise at $t = 1$ s, relative to the optimum S/N for ^{17}O.

3.4 Repeat the calculation of Problem 3.3 for a matched filter.

3.5 Derive an equation analogous to that in Problem 2.9a to describe the recovery of the magnetization M_z after a 90° pulse.

3.6 It was pointed out in Section 3.11 that a series of 90° pulses should be repeated at $1.27T_1$ to obtain optimum S/N in a given total experimental time. Using the equation derived in the preceding problem to find the extent of recovery of M_z in the time $1.27T_1$, derive an expression for the relative S/N accumulated in 1000 T_1 seconds. Compare this result with values calculated for accumulations with 90° pulses repeated at $5T_1$, where virtually complete recovery from the pulse can be assumed. Recall that S/N increases with $\sqrt{n}$ for n repetitions.

3.7 The phases of the spectra in Figure 3.5 were adjusted using Eqs. 3.19 and 3.20. Why does this process not properly correct the phases of lines that are aliased?

Chemical Shifts

The chemical shift is the cornerstone of the chemical applications of NMR. As we noted in Chapter 1, this accidental discovery converted a technique designed initially for probing the structure of the atomic nucleus into one that can provide detailed information about the structure and dynamics of molecules. In this chapter we examine the theoretical underpinning of the chemical shift, explore empirical correlations between chemical shifts and functional groups in organic molecules, and describe simple physical models that can help us to understand and predict chemical shifts.

4.1 THE ORIGIN OF CHEMICAL SHIFTS

The chemical shift has its origin in the magnetic screening produced by electrons. A nucleus experiences not the magnetic field that is applied to the sample (B_0) but rather the field after it has been altered by the screening or shielding of the electrons surrounding the nucleus. Because electrons are also magnetic particles, their motion is influenced by the imposition of an external field; in general, the motion induced by an applied field is in a direction so as to oppose that field

(Lenz's law). Thus at the nucleus the magnetic field is

$$B_{nucleus} = B_0 - \sigma B_0 = B_0(1 - \sigma) \tag{4.1}$$

The screening factor, or *shielding factor,* σ is found to be small (roughly 10^{-5} for protons and $<10^{-3}$ for most other nuclei). Equation 4.1 emphasizes that the chemical shift is entirely the result of an effect induced by placing the molecule into a magnetic field. In the absence of that field the chemical shift does not exist.

The shielding factor is a property of the molecule, but as we see in later examples, the ability of the magnetic field to influence the motion of electrons depends on the orientation of the molecule relative to $\mathbf{B}_0$. Hence, $\boldsymbol{\sigma}$ is a second-rank tensor, not a simple scalar quantity. It is always possible to define three mutually orthogonal axes within a molecule such that σ may be expressed in terms of three principal components, σ_{11}, σ_{22}, and σ_{33}. In many instances molecular symmetry requires that two of the components of $\boldsymbol{\sigma}$ be equal (and in other instances it is possible to assume approximate equality), so that the components may be expressed relative to the symmetry axis as $\sigma_\parallel$ and $\sigma_\perp$ with a *chemical shielding anisotropy* defined $(\sigma_\parallel - \sigma_\perp)$.

In Chapter 7 we consider the observed effects of the shielding tensor in solids. However, as we shall see, rapid, random tumbling of molecules in liquids and gases leads to an averaging of the shielding interactions to an isotropic scalar quantity

$$\sigma_{iso} = \frac{1}{3}[\sigma_{11} + \sigma_{22} + \sigma_{33}] = \frac{1}{3}[\sigma_\parallel + 2\sigma_\perp] \tag{4.2}$$

Although the anisotropy is always present in the molecule and can manifest itself in other ways (e. g., relaxation—Chapter 8), NMR spectra of liquids are greatly simplified because only σ_{iso} is observed. When there is no ambiguity, we refer to this quantity simply as "the chemical shielding" and designate it σ.

4.2 THEORY OF CHEMICAL SHIFTS

For a single free *atom* in a spherically symmetric S electronic state, Lamb[49] showed that the effect of an imposed magnetic field is to induce an electron current that leads to a shielding factor,

$$\sigma_D = \frac{4\pi e^2}{3mc^2} \int_0^\infty r\rho(r) \, dr \tag{4.3}$$

Here $\rho(r)$ is the density of electrons as a function of radial distance from the nucleus, and e, m, and c are the usual fundamental constants.

For molecules the Lamb theory is inadequate because it assumes that the electrons are free to move in any direction, whereas in a molecule electronic motion is severely restricted. Ramsey[50] used second-order perturbation theory to develop a formula that in principle accounts for the shielding factor in molecules and has long provided a framework for qualitative understanding of the major factors contributing to chemical shielding. Ramsey's expression is the sum of two terms, $\sigma_D + \sigma_P$. σ_D is a first-order term that can be calculated from the ground electronic state. It is essentially the Lamb expression and is often called the diamagnetic term because it is positive and thus causes a reduction in the magnetic field at the nucleus and a shielding of the nucleus. σ_P is a second-order perturbation term that requires a knowledge of excited state wave functions and energies of excited states, including the continuum of electronic states above the ionization limit. σ_P is negative and is sometimes referred to as the temperature-independent paramagnetic term. (This distinguishes it from temperature-dependent paramagnetism that results from unpaired electrons.) σ_P corrects for the fact that the electrons in a molecule are not disposed with spherical symmetry about the nucleus in question. Thus the presence of p or d electrons near the nucleus is an important factor in determining the magnitude of σ_P. σ_P is small for hydrogen and can almost be ignored, but for all other nuclei σ_P turns out to be the dominant term. σ_D has a maximum value of about 20×10^{-6} (20 parts per million, ppm) for one electron, and this defines the range of chemical shifts for hydrogen, while σ_P can be as large as $-10,000$ ppm for some nuclei.

For hydrogen, the Ramsey theory is relatively easy to apply, but, as we see in Section 4.5, external perturbations from neighboring functional groups and solvent molecules are of comparable magnitude to the basic electronic terms, so that the theory is quantitatively of little practical value. For other nuclei, this theory could predict chemical shielding rather accurately if we had true wave functions and knowledge of the energies of all excited states. In practice, approximations must be made, but acceptable results can be obtained provided some fundamental aspects are kept in mind.

The magnetic effects are very small compared with electron interactions and are treated by perturbation theory, in which the external magnetic field is represented by a magnetic vector potential. The origin of the potential (*gauge origin*) can be placed on a specific atom or elsewhere in the molecule. σ_D and σ_P each contain terms that depend on the choice of gauge origin. However, these terms are of opposite sign and cancel in an exact calculation of the overall shielding, $\sigma = \sigma_D + \sigma_P$ because σ must be gauge invariant if it is to represent a measurable physical property. However, incomplete cancellation can occur when σ_P is calculated with approximate wave functions and estimates of excited state energies, thus leading to a unrealistic gauge-dependent and incorrect shielding.

Ab Initio Calculations of Chemical Shielding

Ab initio quantum mechanical methods are now able to provide reasonably good values for chemical shielding, but there are limitations to the accuracy that can be obtained. For the approximate wave functions that must be used, it is essential to use *gauge-invariant* methods to obtain meaningful shielding results. This require-

TABLE 4.1   Some Calculated and Experimental ^{13}C Shieldings[a]

	σ_{xx}	σ_{yy}	σ_{zz}	σ_{av}	Expt.
CH_4	195.8	195.8	195.8	195.8	195.1
C_2H_6	187.7	182.7	193.1	186.2	180.9
C_2H_4	177.9	−81.1	84.3	60.4	64.5
C_2H_2	39.0	39.0	279.4	119.1	117.2

[a]Shieldings in ppm, calculated from a common gauge origin.
From Webb.[51]

ment introduces practical problems in the quantum mechanical computations because it is not feasible, even with very fast computers, to use the large basis sets needed to assure accurate results. Several elegant methods have been devised to introduce gauge-including atomic orbitals (GIAO) or localized molecular orbitals (IGLO and LORG) with limited basis sets and to obtain rather good (and steadily improving) results—often accurate to about 3–4% of the shielding range for the lighter elements.

Probably the most useful aspect of the computations is not in trying to obtain quantitative agreement with experimental values of σ_{iso} but in demonstrating the contributions of the component values of σ and in determining the orientation in the molecule of the principal axes of σ. For example, Table 4.1 gives calculated and observed ^{13}C shieldings for four small hydrocarbons, where it is feasible to employ extended basis sets. By symmetry, σ in CH_4 is isotropic, and there is little shielding anisotropy in the sp^3 hybridized carbon atoms in C_2H_6 but the shielding components vary widely in the other molecules. The variation in observed isotropic shieldings with change in molecular structure often disguises much larger variations in individual components of σ. Likewise, Fig. 4.1 illustrates the significant variation of one component of σ with bond angle in three cyclic hydrocarbons while the other two components are relatively constant.

Absolute and Relative Shieldings

Theory provides calculated values of "absolute" shieldings, that is, the shielding relative to a bare nucleus with no electrons. As we see in Section 4.3, experimental measurements provide information on shieldings relative to some selected standard. For comparison between theory and experiment, additional data are needed. For example, it can be shown that σ_P calculated with the gauge origin at a particular nucleus in a small molecule is proportional to the molecular spin–rotation constant of that nucleus, which can be independently measured by microwave spectroscopy, because σ_D can be calculated precisely, this combination permits the establishment of an absolute experimental shielding scale for various nuclei. For hydrogen, simultaneous measurements of NMR and the electronic

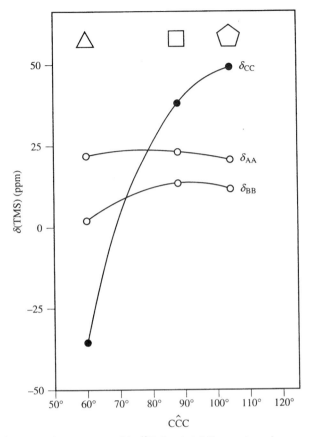

FIGURE 4.1   Principal components of the ^{13}C chemical shift tensor in cyclopropane, cyclobutane, and cyclopentane. Orientation of the principal axes, obtained from theoretical calculations of the shielding tensors, gives δ_{AA} in the HCH plane bisecting the HCH angle, δ_{BB} perpendicular to the HCH plane, and δ_{CC} perpendicular to the CCC plane. From Facelli,[52] *Encyclopedia of Nuclear Magnetic Resonance,* D. M. Grant and R. K. Harris, Eds. Copyright 1996 John Wiley & Sons Limited. Reproduced with permission.

spectrum in atomic hydrogen have permitted the determination of an absolute value of 25.790 ppm for liquid water at 34.7°C, and absolute values of shieldings of other nuclei have been determined by combinations of experiments involving atomic beams and standard NMR measurements.[53]

4.3   MEASUREMENT OF CHEMICAL SHIFTS

As we have seen, NMR lines in liquids are usually very narrow, and resonance frequencies in the range of tens or hundreds of MHz, can be determined with an accuracy of a small fraction of a Hz, whereas magnetic fields can be measured far less accurately. However, so long as the field is held constant with a field–frequency lock (Section 3.3), its value need not be known precisely if we wish to measure and interpret only the *difference* in resonance frequencies. Hence chemical shifts are measured and reported relative to some agreed *reference* substance. For the *ab initio* calculations just described, the "reference" is really a bare nucleus stripped of electrons—a condition not easily achieved experimentally. Rather early in the evolution of NMR techniques, the proton resonance in tetramethylsilane (TMS) was suggested as the reference for ^{1}H NMR, found widespread acceptance, and since 1972 has been the reference formally recommended for ^{1}H NMR by the International Union of Pure and Applied Chemistry (IUPAC).[54] The development of scales to express chemical shifts and the display of most NMR spectra are based on this fundamental role for TMS.

Chemical Shift Scales

NMR data are *measured* in frequency units (hertz) from the chosen reference, and in cases in which complex spectra occur or spin–spin coupling constants are to be given, the results should be *reported* in frequency units as well. For chemical shifts, however, the use of frequency units has the disadvantage that the reported chemical shift is dependent on the value of the magnetic field and through the Larmor relation (Eq. 2.14) on the rf frequency, because the chemical shift is induced by the field. Hence it is customary and highly desirable to *report* chemical shifts in the dimensionless unit of parts per million (ppm), which is independent of the rf frequency or magnetic field strength. We can see the relation between such a dimensionless scale and the shielding factor σ. With the inclusion of the shielding factor from Eq. 4.1, the Larmor equation becomes

$$\nu_0 = \frac{\gamma}{2\pi} B_0 (1 - \sigma) \tag{4.4}$$

In an applied magnetic field B_0 the sample and reference materials, with shieldings σ_S and σ_R, respectively, have resonance frequencies ν_S and ν_R. It is conventional (and in accord with IUPAC recommendations) for the chemical shift to be designated by the symbol δ and for the scale to increase to higher frequency from the reference. Thus δ is defined by

$$\delta \equiv \frac{\nu_S - \nu_R}{\nu_R} \times 10^6 \tag{4.5}$$

which becomes

$$\delta = \frac{(1 - \sigma_S) - (1 - \sigma_R)}{(1 - \sigma_R)} \times 10^6 = \frac{\sigma_R - \sigma_S}{1 - \sigma_R} \times 10^6$$
$$\approx (\sigma_R - \sigma_S) \times 10^6 \tag{4.6}$$

because $\sigma \ll 1$. Thus a larger chemical shift δ corresponds to a smaller shielding σ.

During the early days of NMR applications to chemistry, it was far easier instrumentally to hold the applied rf constant in frequency at ν_0 and to obtain the spectrum by sweeping the magnetic field through the various resonance lines sequentially. The resonances of sample and reference are then found at the values of the applied *magnetic field* given by

$$\nu_0 = \frac{\gamma}{2\pi} B_S (1 - \sigma_S) \qquad \nu_0 = \frac{\gamma}{2\pi} B_R (1 - \sigma_R) \tag{4.7}$$

To obtain consistency with the definition in Eq. 4.5, δ for field sweep is defined by

$$\delta = \frac{B_R - B_S}{B_R} \times 10^6 = \frac{\sigma_R - \sigma_S}{1 - \sigma_S} \times 10^6 \approx (\sigma_R - \sigma_S) \times 10^6 \tag{4.8}$$

Thus, a more highly shielded nucleus (larger σ) resonates at a higher applied field when the field is scanned. Although magnetic field sweeps are now mostly of historical interest in high resolution NMR, the jargon of "high field" and "low field" resonances persists in much of the NMR literature.

Some older literature (especially for nuclides other than hydrogen) uses a definition of δ in which ν_S and ν_R are reversed, so that the δ scale is reversed in sign. Also, some early ^{1}H NMR chemical shifts are reported on the "τ scale" in which $\tau \equiv 10 - \delta$. These arcane points are mentioned here only to clarify statements in older literature; we shall adhere to the definition and conditions of Eq. 4.5.

The δ scale is clearly dimensionless, but with the factor of 10^6, δ has the units of parts per million (ppm) and a value should be stated, for example, as $\delta = 4.13$ ppm. Column headings in tables may be expressed as "δ/ppm." In some instances when small differences in chemical shifts are being discussed, it may be more convenient to use a multiplicative factor of 10^9, with the units of parts per billion (ppb).

Internal and External References

Two types of reference are used in NMR — *Internal* and *external*. An internal reference is a compound giving a sharp NMR line that is dissolved directly in the sample solution under study. The reference substance is then dispersed uniformly at a molecular level through the sample. The magnetic field acts equally on the sample and reference molecules, so that Eq. 4.6 and the other relations derived before are completely valid. Provided the reference compound does not react chemically with the sample, the only serious drawback of an internal reference is the possibility that intermolecular interactions might influence the resonance frequency of the reference. Usually, by careful choice of relatively inert compounds this effect can be made small enough to be disregarded, as we discuss later. Internal references are used whenever possible, and the vast majority of NMR data are reported relative to an internal reference. Occasionally, however, it is not feasible to commingle the reference and sample materials, and an external reference is used.

An external reference is a compound placed in a separate container from the sample. For liquid samples, an external reference compound is often placed as a neat (undiluted) liquid either in a small sealed capillary tube inside the sample tube or in the thin annulus formed by two precision coaxial tubes. In either case, the usual rapid sample rotation (Section 3.2) makes the reference signal appear as a sharp line superimposed on the spectrum of the sample. An external reference is advantageous in eliminating the possibility of intermolecular interactions or chemical reaction with the sample. Also, there are no problems with solubility of the reference in the sample solution. There is, however, a serious difficulty raised by the difference in bulk magnetic susceptibility between sample and reference.

Effect of Bulk Magnetic Susceptibility

It is important to understand the concept of bulk magnetic susceptibility, not only to assess its impact when an external reference is used but also to appreciate its role in other applications of NMR, such as certain imaging studies (Chapter 14). In all substances with completely paired electrons, the motion of the electrons in a magnetic field is such as to make the substance diamagnetic, that is, repellent to a magnetic field, while materials with unpaired electrons are paramagnetic and attractive to a magnetic field. As a result, the magnetic induction B_0 in a substance is altered by the magnetization per unit volume induced in the sample, which is given by

$$M = \kappa B_0 \tag{4.9}$$

where κ, the *volume magnetic susceptibility*, is negative for all diamagnetic materials and of the order of 10^{-6}. Electromagnetic theory shows that the magnetic induction in a bulk material (solid or liquid) is given in cgs units by

$$(B_0)_{\text{bulk}} = B_0 \left[1 + \left(\frac{4\pi}{3} - \alpha \right) \kappa \right] \tag{4.10}$$

There are two variables to consider, κ and α (the *shape factor*). If the material being studied and the reference material are molecularly dispersed in a single sample tube (i.e., an internal reference), then there is only one value of κ, and both sample and reference molecules experience the same magnetic field. Hence, the effect of magnetic susceptibility can be ignored.

For an external reference, however, the sample and reference are physically separated, and Eq. 4.10 applies to each bulk component:

$$(B_0)_S = B_0[1 + (\tfrac{4}{3}\pi - \alpha)\kappa_S]$$

$$(B_0)_R = B_0[1 + (\tfrac{4}{3}\pi - \alpha)\kappa_R] \tag{4.11}$$

Now the shape factor must be considered. If the interface between sample and reference is a sphere, $\alpha = 4\pi/3$, so that the susceptibility correction reduces to zero. Spherical sample cells have been used, but they are generally inconvenient, and imperfections in the glass wall can introduce spurious effects. For a cylindrical sample tube that is long relative to its diameter, $\alpha = 0$ for the usual geometry used in superconducting magnets (axis of the sample tube parallel to that of the magnet), $\alpha = 2\pi$ for the two axes perpendicular (the geometry common in permanent and electromagnets), and $\alpha = 4\pi/3$ if the sample tube axis is inclined at the "magic" angle of 57.4° to the magnetic field.

For $\alpha \neq 4\pi/3$, the sample and reference molecules experience slightly different effective magnetic fields, and the apparent chemical shift determined according to Eq. 4.6 is in error and must be corrected. κ_S is approximately the weighted average of the values for solute and solvent; for sufficiently dilute solutions this is nearly the value for the solvent. In general, the external reference has a value $\kappa_R \neq \kappa_S$. By combining Eqs. 4.4, 4.5, and 4.11 and simplifying, we obtain

$$\delta_{measured} = [(\sigma_R - \sigma_S) - (\tfrac{4}{3}\pi - \alpha)(\kappa_R - \kappa_S)] \times 10^6 \tag{4.12}$$

where $\delta_{true} = (\sigma_R - \sigma_S) \times 10^6$ is the desired quantity. Here we have used the facts that $\sigma \ll 1$ and that $\kappa_R - \kappa_S \ll 1$, because κ's are of the order of 10^{-6} Note that the shape factor is applicable to a cylinder that is very much longer than its diameter. Although this condition is not fully met in a typical NMR sample tube, Eq. 4.12 is usually adequate. The small range of proton chemical shifts makes it mandatory to correct for susceptibility if an external reference is used, but for other nuclei, with large chemical shift ranges, an error of 1 ppm or so in measured chemical shifts can often be tolerated.

Equations 4.10 to 4.12 are written for cgs units. In SI units the constant $4\pi/3$ and all values of α must be divided by 4π (see Section 2.11).

Substitution of Sample and Reference

Rather than record a single spectrum that contains both sample and reference, the frequency of the reference compound can be measured in a separate experiment. This technique, the *substitution method*, depends on the high stability of modern NMR spectrometers when a deuterated solvent is used to provide a field/frequency lock, as described in Section 3.3. This method requires two measurements but is quite versatile. If the sample and the reference compound are each dissolved in the same solvent at low concentration, the substitution method is equivalent to use of an internal reference. If the reference compound is a nearly neat liquid with only a small amount of the deuterated solvent to serve as a lock, the measured chemical shifts may be very slightly different from those obtained with an internal reference because of differing molecular interactions (see Section 4.5).

It might appear that a magnetic susceptibility correction would be needed if the susceptibilities of sample and reference differ, but this is not the case. With the field/frequency lock established via the deuterated solvent, the applied magnetic field (H_0) simply shifts slightly to maintain the magnetic induction (B_0) inside the sample tube constant so as to keep the ^2H on resonance. If different deuterated solvents are used for sample and reference, a simple correction must be made for the difference in their ^2H chemical shifts.

Reference Compounds

Tetramethylsilane became the established internal reference compound for ^1H NMR because it has a strong, sharp resonance line from its 12 protons, with a chemical shift at low resonance frequency relative to almost all other ^1H resonances. Thus, addition of TMS usually does not interfere with other resonances. Moreover, TMS is quite volatile, hence may easily be removed if recovery of the sample is required. TMS is soluble in most organic solvents but has very low solubility in water and is not generally used as an internal reference in aqueous solutions. Other substances with references close to that of TMS have been employed, and the methyl proton resonance of 2,2-dimethylsilapentane-5-sulfonic acid (DSS) at low concentration has emerged as the reference recommended by IUPAC for aqueous solutions.[55] Careful measurements of the DSS–TMS chemical shift difference when both materials are dissolved at low concentration in the same solvent have shown that for DSS $\delta = +0.0173$ ppm in water, and $\delta = -0.0246$ ppm in dimethyl sulfoxide. Thus, for most purposes, values of δ measured with respect to TMS or DSS can be used interchangeably.

Figure 2.6 illustrates the approximate resonance frequencies for a number of nuclides at a particular value of B_0. On this scale, the range of chemical shifts for each nuclide is smaller than the width of the vertical line denoting the nuclide. It is clear that the frequency range is a continuum, so the ^1H resonance of TMS can in principle serve as the reference for all nuclei. In a modern NMR spectrometer all frequencies are derived from a single source, so all chemical shifts can be determined relative to TMS (or relative to DSS in aqueous solutions). IUPAC has formally recommended that chemical shifts of ^2H, ^{13}C, ^{15}N, and ^{31}P in biopolymers measured in aqueous solution be reported relative to the ^1H resonance of DSS.[55] It seems likely that this recommendation will be broadened to use TMS as the reference for all nuclides in nonaqueous solutions.

Chemical shifts can be given as "Ξ values," where Ξ (Xi) is defined as the resonance frequency in a magnetic field in which TMS has a resonance frequency of exactly 100 MHz. However, the data typically must be reported to eight or nine significant figures, which is satisfactory for tabulations but can be cumbersome for discussion. Hence, chemical shifts are commonly expressed relative to secondary references—each containing the nuclide being studied. In fact, virtually all NMR chemical shifts obtained prior to the last few years were *measured* with respect to such secondary references (internal or external), and some are still measured in that way. Table 4.2 gives the Ξ values for a number of frequently used secondary references.[56] These data can be used to interconvert chemical shifts for nuclides

TABLE 4.2   External References for Chemical Shifts[a]

Nuclide	Spin	Reference	Ξ (MHz)[b]
^{1}H	1/2	TMS [(CH$_3$)$_4$Si] in CDCl$_3$	100.000000
^{2}H	1	(CD$_3$)$_4$Si, neat liquid	15.350608
^{6}Li	1	LiCl in D$_2$O	14.716105
^{7}Li	3/2	LiCl in D$_2$O	38.863786
^{10}B	3	BF$_3$·O(C$_2$H$_5$)$_2$	10.743657
^{11}B	3/2	BF$_3$·O(C$_2$H$_5$)$_2$	32.083972
^{13}C	1/2	TMS in CDCl$_3$	25.145004
		DSS in water (internal reference)	25.144953[c]
^{14}N	1	CH$_3$NO$_2$ in CDCl$_3$	7.226329
^{15}N	1/2	CH$_3$NO$_2$ in CDCl$_3$	10.136783
		NH$_3$ (liquid)	10.132912[c]
^{17}O	5/2	D$_2$O	13.556429
^{19}F	1/2	CCl$_3$F	94.094000
^{23}Na	3/2	NaBr in D$_2$O	26.451919
^{29}Si	1/2	TMS in CDCl$_3$	19.867187
^{31}P	1/2	H$_3$PO$_4$ in D$_2$O (85%)	40.480737
		(CH$_3$O)$_3$PO in water (internal reference)	40.480864[c]
^{57}Fe	1/2	Fe(CO)$_5$ in C$_6$D$_6$	3.237778
^{59}Co	7/2	K$_3$[Co(CN)$_6$] in D$_2$O	23.727072
^{113}Cd	1/2	(CH$_3$)$_2$Cd	22.193173
^{119}Sn	1/2	(CH$_3$)$_4$Sn in C$_6$D$_6$	37.290629
^{129}Xe	1/2	XeOF$_4$	27.811005
^{199}Hg	1/2	(CH$_3$)$_2$Hg in C$_6$D$_6$	17.910323
^{205}Tl	1/2	Tl(NO$_3$)$_3$ in D$_2$O	57.683833
^{207}Pb	1/2	(CH$_3$)$_4$ Pb in C$_6$D$_6$	20.920597

[a] Examples for some frequently observed isotopes, taken from an extensive tabulation in the *Encyclopedia of NMR*,[56] with minor amendments given by the author of that tabulation in a personal communication.

[b] NMR frequency in a magnetic field in which the ^{1}H resonance of TMS is exactly 100 MHz.

[c] From Markley *et al.*[55]

other than hydrogen expressed in Ξ values and expressed relative to secondary references.

4.4 EMPIRICAL CORRELATIONS OF CHEMICAL SHIFTS

As we saw in Section 4.2, the bases for chemical shifts can be accounted for theoretically in principle, but *ab initio* calculations cannot at present provide exact values for these quantities in complex molecules. For predictions of chemical shifts for nuclei in particular chemical environments, it is therefore necessary to rely very largely on empirical correlations between observed shifts and chemical structure. Figures 4.2–4.6 provide a general orientation on the orders of magnitude involved and the effects of various functional groups for the chemical shifts of five widely studied nuclides. These figures are meant to be illustrative of the usual values observed. There may be individual compounds that fall outside the ranges given; likewise, it is often possible to restrict the range more narrowly if specific classes of compounds are considered. For example, in proteins ^1H, ^{13}C, and ^{15}N chemical shifts in planar peptide groups correlate with dihedral angles that define the secondary structure (e.g., α helix, β sheet) and can be used for predicting such structure (see Chapter 13).

The most striking feature of Figs. 4.2–4.6 is the very small range of chemical shifts for hydrogen nuclei (≈ 12 ppm) relative to the much larger ranges for the heavier nuclides, which is in line with our discussion in Section 4.1 of the relative importance of σ_P and σ_D. We discuss other features of these figures in the following sections.

4.5 SOME ASPECTS OF PROTON CHEMICAL SHIFTS

Because of the importance and wide utility of ^1H NMR, we look particularly at factors that determine ^1H chemical shifts. These features can be used semiquantitatively to predict and/or rationalize specific ranges of ^1H chemical shifts.

Effect of Electron Density

Although the Lamb expression in Eq. 4.2 applies strictly to an isolated atom, it can be evaluated approximately for the *local* contribution of a proton in a molecule to give

FIGURE 4.2   Approximate chemical shift ranges for protons in various functional groups. Reference: TMS (internal). *Chemical shift highly dependent on hydrogen bonding.

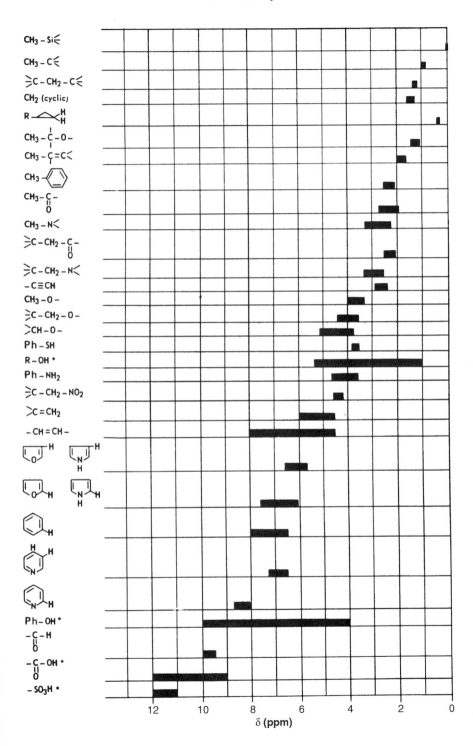

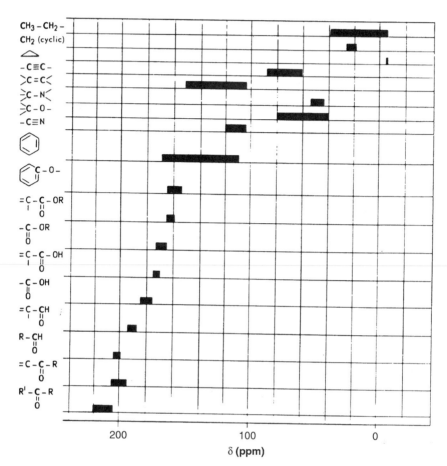

FIGURE 4.3   Approximate chemical shift ranges for ^{13}C (indicated in boldface) in various functional groups. Reference: ^{13}C in TMS (internal).

$$\sigma_D(\text{local}) = 20 \times 10^{-6}\lambda \tag{4.13}$$

where λ is the effective number of electrons in the 1s orbital of the hydrogen. For a completely screened hydrogen atom, λ would approach 1, and for a hydrogen ion, it would be 0. Thus, the local diamagnetic effect is in the range of a few parts per million, which is just the range of chemical shifts observed for protons.

Equation 4.13 suggests that there ought to be some sort of correlation between the shielding factor and the electron density around the hydrogen. For example, a more acidic proton, such as the OH proton in phenol, should be less shielded than the corresponding less acidic proton in an alcohol. This is indeed found to be the case, the chemical shift for the OH proton of phenol occurring

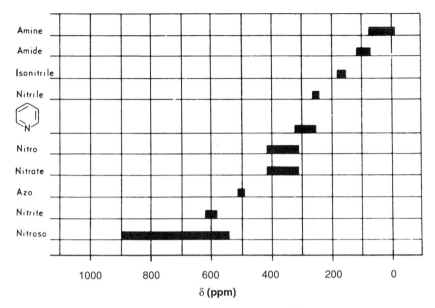

FIGURE 4.4   Approximate chemical shift ranges for ^{14}N and ^{15}N in various functional groups. Reference: Liquid NH_3 (external).

about 4 ppm to higher frequency than that of ethanol. (Hydrogen bonding can alter these chemical shifts substantially, as we see later.)

The presence of a formal charge near a magnetic nucleus leads to a substantial shielding or deshielding. For example, comparison of the chemical shifts in the

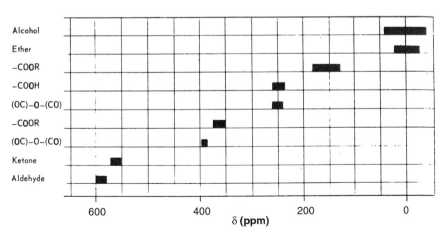

FIGURE 4.5   Approximate chemical shift ranges for ^{17}O in various functional groups. Reference: ^{17}O in water (external).

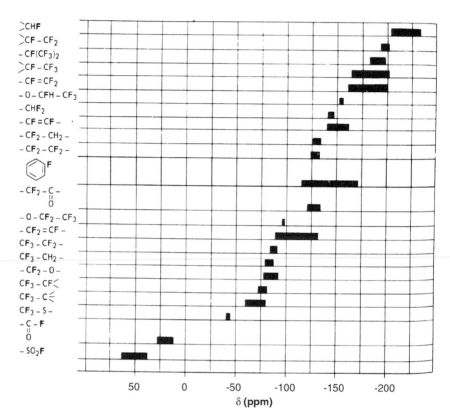

FIGURE 4.6   Approximate chemical shift ranges for ^{19}F in various functional groups. Reference: CCl_3F (external).

symmetric molecules $C_5H_5^-$, C_6H_6, and $C_7H_7^+$ shows that a charge localized on one of the carbon atoms of benzene would result in a shielding or deshielding of the attached proton by about 9 ppm/electron. This value may be used to calculate from NMR data the approximate change in charge density resulting from the introduction of an electron-donating or electron-withdrawing substituent. In general, electron densities obtained this way have been in rather good agreement with those calculated from molecular orbital treatments.

A strongly electronegative atom or group attached to or near a magnetic nucleus has the expected effect of deshielding the nucleus. Thus, rough correlations are found between chemical shift and electronegativity of substituents. Figure 4.7 shows typical results for CH_3CH_2X. Note the parallelism between the correlations for ^1H and ^{13}C chemical shifts but the vastly different ranges that are encompassed.

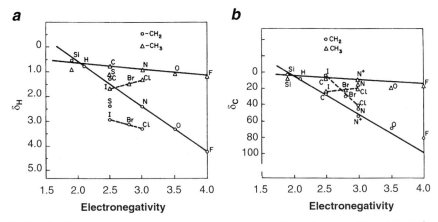

FIGURE 4.7   Chemical shifts in CH_3CH_2X as a function of the electronegativity of X. (*a*) ^1H chemical shifts. (*b*) ^{13}C chemical shifts. Adapted from Spiesecke and Schneider.[57]

Substitution on an aromatic ring causes changes in shielding of protons resulting from addition or withdrawal of charge. The generalizations used by chemists in predicting electron density at *ortho*, *meta*, and *para* positions apply in large measure to NMR spectra, as indicated for some typical substituents in Fig. 4.8. To a large extent, substituent effects are approximately additive for aromatic systems. Such empirical methods of predicting chemical shifts are quite useful, but their limitations should be recognized. In aromatic systems, for example, additivity provides surprisingly good results for many *meta* and *para* disubstituted benzenes but gives only fair agreement with experiment for substituted benzenes containing appreciable dipole moments.

Magnetic Anisotropy

Although variation in electron density around a proton is probably the most important factor influencing its chemical shift, many exceptions are found to a correlation between δ and electron density. One important factor is the "neighbor anisotropy" effect arising from anisotropy in the magnetic susceptibility of nearby atoms or portions of a molecule. For example, consider a proton that is bonded to or spatially close to Y, where Y is an atom or a more complex part of a molecule. If a magnetic field $\mathbf{B}_0$ is imposed on the molecule, the electrons around Y move in such a way as to generate a magnetic dipole moment at Y:

$$\boldsymbol{\mu}_Y = \boldsymbol{\chi}_Y \mathbf{B}_0 \qquad (4.14)$$

where $\boldsymbol{\chi}_Y$ is a tensor describing the magnetic susceptibility of Y. We can obtain better insight into the effect of $\boldsymbol{\mu}_Y$ on the local field at the proton if we look

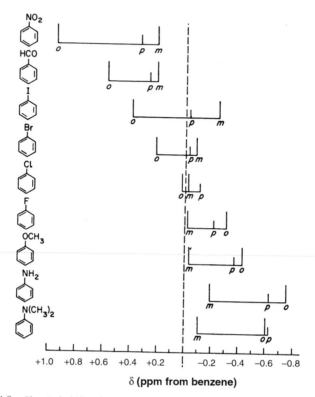

δ (ppm from benzene)

FIGURE 4.8   Chemical shifts of protons *ortho*, *meta*, and *para* to the substituent in a series of monosubstituted benzenes. Chemical shifts are expressed relative to benzene as zero. Adapted from Spiesecke and Schneider.[58]

separately at the three principal components of χ_Y, depicted in Fig. 4.9 as three orthogonal orientations of the H–Y vector. For a diamagnetic Y, the induced magnetic dipole components at Y are pointed opposite $\mathbf{B}_0$, and at the proton, situated a distance R from Y, the local field augments or detracts from $\mathbf{B}_0$, as illustrated. Because the field strength arising from a point magnetic dipole varies inversely as the cube of the distance from the dipole, the increment of magnetic field experienced by the proton H due to the induced moment at Y is the average of the three contributions shown in Fig. 4.9, and the change in shielding is

$$\Delta\sigma = -\frac{1}{3R^2}(2\chi^{\|} - \chi^{\perp} - \chi'^{\perp}) \qquad (4.15)$$

The notations $\|$ and $\perp$ refer to the direction of the H–Y vector relative to $\mathbf{B}_0$, with the two perpendicular components generally being different unless there is axial symmetry around the H–Y vector. Equation 4.15 shows that the magnitude

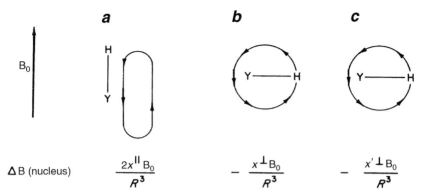

FIGURE 4.9   Illustration of secondary magnetic field generated at the hydrogen nucleus as a result of diamagnetism of a neighboring group Y. Arrows indicate closed lines of magnetic flux from a point magnetic dipole at the center of electric charge in Y. (*a*) H−Y axis parallel to $\mathbf{B_0}$; (*b, c*) H−Y axis perpendicular to $\mathbf{B_0}$. (*b*) and (*c*) are equivalent if Y has axial symmetry.

of the field increment at the proton depends on a *magnetic anisotropy* in Y, that is, a lack of equality of the three susceptibility components.

In general, where the proton in question does not lie on one of the principal axes of magnetic susceptibility, the shielding varies according to both distance and angle:

$$\Delta\sigma = \frac{1}{3R^3}(1 - 3\cos^2\theta)(\chi^{\|} - \chi^{\perp}) \tag{4.16}$$

where the proton is at a distance R from the center of the anisotropic group or bond in a direction inclined at an angle θ to the direction of $\chi^{\|}$, and $\chi^{\perp} = \chi'^{\perp}$.

One particularly simple example of the neighbor anisotropy effect occurs in HC≡CH, where the anisotropy arises from the freedom of the electrons in the triple bond to circulate at will around the axis of the molecule. If Y in Fig. 4.9 represents the C≡CH fragment, then $\chi^{\|}$ is large in magnitude because the flow of electrons around the bond generates a moment along the H−Y axis. On the other hand, the electrons are less likely to circulate perpendicular to the H−Y axis because they then would cut through chemical bonds. Consequently $\chi^{\perp}$ and $\chi'^{\perp}$ are small (and equal because of the axial symmetry). Keeping in mind that both $\chi^{\|}$ and $\chi^{\perp}$ are negative, we expect from Eq. 4.15 a large *positive* $\Delta\sigma$. Thus, the resonance is predicted to be more shielded than it would be in the absence of this large neighbor anisotropy effect. A comparison of the proton chemical shifts of the series C_2H_6, $CH_2{=}CH_2$, and HC≡CH shows that the HC≡CH line is only 0.6 ppm less shielded than the line of C_2H_6, while C_2H_4 lies 4.4 ppm less shielded than C_2H_6. On electronegativity grounds alone C_2H_2 should be less shielded than C_2H_4.

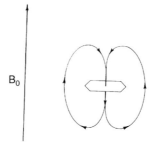

FIGURE 4.10   Secondary magnetic field generated by ring current in benzene.

The C=O group is a common source of magnetic anisotropy, because the three components of χ_Y are quite different from each other.

Ring Currents

A special, and quite important, type of magnetic anisotropy effect occurs when it is possible to have interatomic circulation of electrons within a molecule. A circulation of π electrons around the periphery of a benzene ring, for example, gives rise to a "ring current" and resultant induced shielding effects. In this case $\chi_\parallel$ is parallel to the sixfold symmetry axis and is much larger than the in-plane components $\chi_\perp = \chi'_\perp$. The resultant magnetic moment μ_Y may be pictured as a point magnetic dipole at the center of the ring, with the dipole field falling off as the cube of the distance. Closed lines of magnetic flux, as in Fig. 4.10, show that the sign of the ring current effect is highly dependent on geometry. An aromatic proton in the plane of the ring experiences a ring current field that enhances the applied field; hence its resonance occurs at a higher frequency (i.e., lower shielding) than might otherwise be anticipated. In fact, the chemical shift of benzene is about 1.7 ppm less shielded than that of ethylene, which is very similar from the standpoint of hybridization and electronegativity. A proton constrained to lie over the aromatic ring, on the other hand, is expected to experience increased shielding, and many examples of such effects are known.

The magnitude of the shielding due to a ring current may be estimated from a point dipole calculation or more accurately from a model that treats the π electrons as rings above and below the atomic plane. The results of such a calculation have been tabulated and are illustrated in the contour diagram of Fig. 4.11.

The ring current model may also be used to interpret data on such molecules as porphyrins, whose large ring currents lead to substantial shifts to higher frequency for protons outside the ring and a very large shift to lower frequency for the NH protons inside the electron ring.

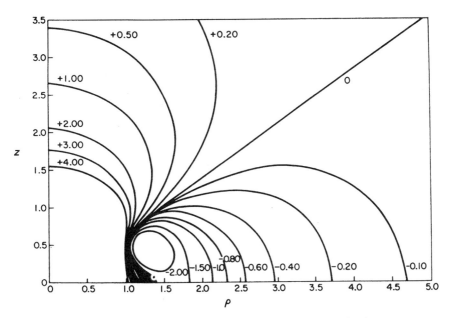

FIGURE 4.11   Shielding ($\Delta\sigma$) arising from the ring current in benzene. The plot represents one quadrant of a plane passing through the center of the ring, which lies horizontally. ρ and z are expressed in multiples of the C—C bond distance in benzene, 1.39 Å. From Johnson and Bovey.[59]

Solvent Effects

Because most high resolution NMR spectra are obtained in dilute solutions, solvent molecules may influence the chemical shifts of a solute. For example, collisions and repulsive interactions between solute and solvent molecules can distort the electronic environment of a given nucleus, hence alter its chemical shielding σ by an amount $\Delta\sigma_W$. Theory predicts that $\Delta\sigma_W$ should be negative, and experimental results indicate that a change in proton shielding of the order of 0.1–0.2 ppm might be expected from this effect. Generally, large polarizable halogen atoms in the solvent lead to increased negative values of $\Delta\sigma_W$. For nuclei other than hydrogen this term can be much larger, but it is still generally small relative to the ranges of chemical shifts found for such nuclides.

A second effect of solvent on chemical shielding, $\Delta\sigma_A$, arises from nonrandom orientational averaging of magnetically anisotropic solvent molecules relative to solute molecules. The interaction may be thought of as a very transitory complex in which the solvent molecule provides the neighboring magnetic anisotropy discussed before. Again, aromatic rings and groups such as C=C, C=O, C≡C, and C≡N cause especially large effects, with aromatic

TABLE 4.3   Effect of Hydrogen Bonding on Proton Chemical Shifts[a]

Compound	δ(bonded)[b]	δ(bonded) $- \delta$(free)
C$_2$H$_5$OH$\cdots$O$-$C$_2$H$_5$ ($\mid$ H)	5.3	4.6
C$_6$H$_5$OH$\cdots$O$=$P(OC$_2$H$_5$)$_3$	8.7	4.3
Cl$_3$CH$\cdots$O$=$C(CH$_3$)$_2$	8.0	0.7
Cl$_3$CH$\cdots$⟨benzene ring⟩	6.4	−0.9
C$_6$H$_5$C(O$-$H$\cdots$O / O$\cdots$H$-$O)CC$_6$H$_5$	14.0	7.8
C$_6$H$_5$NH$_2$$\cdots$N⟨pyridine ring⟩	5.3	2.0
C$_6$H$_5$NH$_2$$\cdots$O$=$S(CH$_3$)$_2$	5.0	1.7
C$_6$H$_5$SH$\cdots$O$=$C$-$N(CH$_3$)$_2$ ($\mid$ H)	3.8	0.5
CH$_3$C$=$CH$-$CCH$_3$ (O$-$H$\cdots$O)	15.5	—

[a]Data compiled from work in the author's laboratory and from various literature sources. These data are illustrative but should not be regarded as accurate because of variations caused by differences in solvent, concentration, and temperature.
[b]In parts per million with respect to TMS (internal).

solvents usually leading to positive values of $\Delta\sigma_A$ of about 0.5 ppm and solvents containing triple bonds contributing negative values of about 0.2–0.4 ppm. However, the values of $\Delta\sigma_A$ vary from one part of a large molecule to another because of steric and other effects that influence the averaging. This selectivity, especially in aromatic solvents, has been widely exploited in separating chemical shifts that accidentally coincide in a particular solvent. In spectra with a number of lines lying close together, it is often helpful to study the spectrum as a function of changing proportions of a mixed solvent, so that each line can readily be tracked.

A polar molecule dissolved in a dielectric medium induces a "reaction field," the effect of which is usually to reduce the shielding around a proton in the solute. This effect, $\Delta\sigma_E$ might be as large as 1 ppm for polar molecules in solvents of high dielectric constant.

Hydrogen Bonding

In addition to the general solvent effects discussed earlier, *specific* solute–solvent interactions may occur, the most important of which is hydrogen bonding. Hydrogen bonding is recognized as a relatively strong, specific interaction of the form A—H$\cdots$B between molecules or between suitable portions of the same molecule. Proton chemical shifts are extremely sensitive to hydrogen bonding. In almost all cases, formation of a hydrogen bond causes the resonance of the bonded proton to become deshielded—often by 10 ppm or more. In general, there is a rough correlation between the NMR shift and the strength of the bond as measured by the enthalpy of its formation. Table 4.3 shows typical changes in proton resonance frequency on hydrogen bonding. Intramolecularly bonded enols and phenols display resonances at especially large values of δ.

The shift to high frequency on hydrogen bonding results from a decrease in diamagnetic shielding around the hydrogen nucleus, as demonstrated by molecular orbital calculations that show reduced electron density near the hydrogen. This redistribution of electrons in the A—H bond is probably brought about by repulsion from the nonbonding electrons in B.

A hydrogen bond to π electrons in aromatic rings leads to exactly the opposite observed shift, namely to low frequency. The reason is clear: the hydrogen atom bonds to the π electron–rich face of the aromatic ring, where it is subject to a ring current effect that far outweighs any deshielding from the formation of the hydrogen bond itself. In this geometrical arrangement, then, the proton is subject to a net additional diamagnetic shielding, hence shifts to lower frequency.

Proton NMR has proved to be one of the most sensitive methods of studying hydrogen bonding, both qualitatively and quantitatively. The large proton deshielding on hydrogen bonding is an important factor in interpreting ^{1}H NMR spectra, as OH and NH chemical shifts are often found to be solvent dependent. In addition, the chemical shift of an OH or NH resonance can be of diagnostic value in structural elucidations when intramolecular hydrogen bonds are present.

Because intermolecular hydrogen bonds between small molecules are usually made and broken very rapidly relative to the chemical shift difference between bonded and nonbonded forms (expressed in hertz), separate lines are generally not observed for different species, and the frequency ν of the single observed line is given by Eq. 2.58. In suitable systems the variation of ν with concentration and temperature can be analyzed to give equilibrium constants and other thermodynamic properties for hydrogen bond formation.

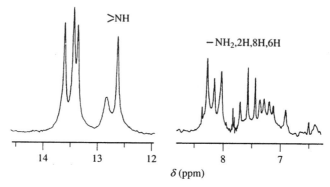

FIGURE 4.12 ^{1}H NMR spectrum (500 MHz) of a self-complementary decadeoxynucleotide, d(GCATTAATGC)$_2$, in H$_2$O at a concentration of 20 millimolar. From Wüthrich.[60]

On the other hand, many intramolecular hydrogen bonds in small molecules and in biopolymers and oligomers exist in the slow exchange regime, where separate lines are observed for each species. For example, in transfer ribonucleic acids (tRNAs) hydrogen bonds between the purine and pyrimidine bases are quite strong, and chemical exchange with protons in the aqueous solvent is slow on the NMR time scale. Figure 4.12 provides an illustration in which the equilibrium so favors the hydrogen-bonded form that no lines due to "free" NHs are observed, whereas the bonded hydrogens show resonances at very high frequencies (~12–14 ppm). Likewise, in native proteins amide hydrogens are found in the 8–10 ppm range because of strong hydrogen bonding.

NMR has become an indispensable analytical technique for studying rates of H $\rightleftharpoons$ D exchange at specific sites when a protein is deuterated. Such exchange rates provide information on secondary protein structure and are useful in investigating the mechanism of protein folding.

In the A—H···B system the chemical shifts of both A and B are also found to change on hydrogen bond formation. The magnitude of shifts on hydrogen bonding for an ^{17}O in a carbonyl group that serves as a hydrogen bond acceptor (B) can be as large as 50–60 ppm to lower frequency, whereas an oxygen in a single bond in B (such as an ether) displays a low frequency shift of only a few ppm. ^{17}O in the donor A—H, such as an alcohol or phenol, shifts ~5–20 ppm to higher frequency. ^{15}N has been studied less extensively but shows somewhat similar behavior. In proteins, *ab initio* computations of ^{15}N chemical shifts indicate that typical hydrogen bonding interactions (N—H···O) cause deshielding of the ^{15}N of about 13 ppm. Although these hydrogen bond shifts are quite significant, they nevertheless represent only a relatively small fraction of the total chemical shift range for these nuclei, in contrast to the situation for ^{1}H chemical shifts, where the hydrogen bond effect may be the dominant factor in determining a chemical shift.

4.6 NUCLEI OTHER THAN HYDROGEN

The factors affecting proton chemical shifts are also applicable to other nuclei. However, the paramagnetic effect σ_P is much larger than for hydrogen, as pointed out in Section 4.2, and accounts qualitatively for the much larger ranges of chemical shifts shown in Figs. 4.3–4.6. For example, the difference in chemical shift between F_2 and F^- of approximately 500 ppm is attributed to σ_P. For the spherically symmetric ion, σ_P can be ignored, but in F_2, with considerable p bond character, σ_P is very large.

The large range of chemical shifts for nuclei other than hydrogen suggests that their chemical shifts should reflect often subtle differences in chemical structure, and this is indeed found to be true. For example, ^{19}F provides a sensitive probe of changes in the electron density in many molecules, as illustrated in Fig. 4.6. Because ^{19}F has a range of chemical shifts greater than 250 ppm and a large magnetogyric ratio, the chemical shift range in Hz is larger than that for any other commonly studied nucleus. The chemical shift range of ^{31}P exceeds 300 ppm, which divides into largely separate portions for trivalent and pentavalent phosphorus.

Some less frequently encountered nuclides, such as ^{57}Fe and ^{59}Co, have chemical shift ranges of more than 10,000 ppm, hence encompass a very large frequency range in spite of smaller magnetogyric ratios. The chemical shift of monatomic ^{129}Xe gas is very sensitive to its environment, and hyperpolarized xenon (Section 2.5) can be used to probe adsorption sites in solids.

Nitrogen

As illustrated in Fig. 4.4, nitrogen chemical shifts cover a range of about 1000 ppm and make ^{14}N and ^{15}N very useful nuclides for distinguishing structural features. Both nuclides have very low inherent sensitivity, about 10^{-3} as great as that for ^1H. ^{14}N is over 99% naturally abundant, but it has large quadrupole moment, which often leads to rapid relaxation and very broad lines, as we shall see in Chapter 8. Nevertheless, in many compounds line widths are narrow enough to allow discrimination between chemically shifted environments. ^{15}N has a spin of $\frac{1}{2}$, hence no quadrupole moment, but its natural abundance of less than 0.4% makes direct observation difficult at natural abundance. However, isotopic enrichment and/or the use of indirect detection methods (discussed in Chapter 10) permits relatively facile study of ^{15}N, particularly in two- and three-dimensional NMR.

Carbon-13

^{13}C NMR is second only to proton NMR in popularity. ^{13}C has a natural abundance of 1.1% and sufficient inherent sensitivity to make study relatively easy with modern high field instruments. Figure 4.3 shows that the chemical shift range is about 200 ppm, about 20 times as great as that for ^{1}H, so that ^{13}C NMR is extremely useful in the investigation of organic molecules. Although the paramagnetic shielding term is dominant, the chemical shifts in Fig. 4.3 show a remarkable qualitative similarity in sequence to those of ^{1}H in Fig. 4.2. For example, alkanes are highly shielded, alkenes and aromatics much less so, and carbonyl carbons are quite deshielded, with chemical shifts about 160–220 ppm from TMS. Polar substituents generally have a large deshielding effect on adjacent carbons. Magnetic anisotropy and ring current effects, as we have seen, seldom exceed 1 ppm, hence are of much less relative importance in determining ^{13}C chemical shifts than for proton shifts. However, other factors, such as steric and stereochemical effects, have been identified in many instances.

As high resolution NMR methods have been applied to solids (see Chapter 7), information has become available on the individual components of the shielding tensor, not only on its trace, as observed in liquids. These data, along with *ab initio* calculations (Section 4.2), provide a deeper understanding of the factors responsible for ^{13}C chemical shifts. For example, 1s electrons make a substantial diamagnetic contribution to the shielding, but the effect of these innermost electrons is relatively constant, regardless of chemical environment of the carbon. As expected, the paramagnetic shift σ_P, related to the 2p electrons, dominates the diamagnetic contributions from 2s and 2p electrons. As noted in Table 4.1, the anisotropy of the chemical shielding from the three tensor components is small for tetrahedral carbons in saturated hydrocarbons but much larger for carbons with double or triple bonds.

4.7 COMPILATIONS OF SPECTRAL DATA AND EMPIRICAL ESTIMATES OF CHEMICAL SHIFTS

It is worthwhile reiterating the point made in Section 4.4 that theory alone is insufficient for predicting accurate chemical shifts for most molecules and that correlations of the sort shown in Figs. 4.2–4.6 are derived from experimental observations of thousands of known compounds. It is outside the scope of this book to include extensive tabulations of data beyond the few examples in the preceding sections. However, large compilations of ^{1}H and ^{13}C spectra are available commercially in computer databases, and extensive tabulations of chemical shifts were given in many early books on NMR. Some sources of useful data are given in Section 4.11.

Several elaborate algorithms have been developed to relate ^{1}H and ^{13}C chemical shifts to structural fragments in organic molecules and are incorporated in several of the large spectral databases mentioned in Section 4.11. These treatments are very effective for ^{13}C. As we saw in Section 4.5, ^{1}H chemical shifts are influenced drastically by neighboring molecular moieties, hence are more difficult to predict. However, the currently available algorithms, together with large compilations of chemical shifts and coupling constants, can often simulate very good ^1H spectra.

4.8 ISOTOPE EFFECTS

Small but significant changes in chemical shifts are often found on isotopic substitution. Such a change between two *isotopomers* (molecules that differ only in substitution of an isotope) is called an *isotope effect* and is given the symbol Δ. The *primary* isotope effect refers to a change of chemical shift in the nuclide being observed, for example, $^1H/^2H/^3H$ or $^{14}N/^{15}N$. The primary isotope effect is usually negligible except for strong hydrogen bonds (hydrogen bond energy of about 50–100 kJ/mole), where $\Delta = \sigma(^2H) - \sigma(^1H)$ is 0.5–0.9 ppm, and very strong hydrogen bonds (>100 kJ/mole), where Δ is -0.3 to -0.7 ppm.

The *secondary* isotope effect refers to a chemical shift of a nuclide other than the one isotopically substituted. The theory, which has been developed most extensively for a diatomic system, such as ^{13}C—$H(D)$, shows that the principal factor in altering the chemical shift in the isotopomer is the change in the character of anharmonic vibrations. Isotope effects are commonly characterized by the number of chemical bonds intervening between the observed and substituted nuclei and can be quite large. For example, $\delta(^{13}C)$ decreases by about 0.3 ppm on going from $^{13}CHCl_3$ to $^{13}CDCl_3$ (a one-bond effect), and $\delta(^{19}F)$ decreases by 0.47 ppm in going from CF_2=CH_2 to CF_2=CD_2 (a three-bond effect). Secondary isotope effects are smaller in ^1H resonance, in keeping with the smaller total range of chemical shifts, but even here significant effects can be seen. For example, $\delta(^1H)$ decreases by 0.019 ppm between CH_4 and CH_3D (a two-bond effect), and the three-bond effect in going from CHF=CHF to CHF=CDF is only 0.005 ppm (or 5 ppb). Long-range effects through as many as 12 bonds are known. The mechanism of transmission of such effects appears to be related to electron delocalization through conjugated systems.

Secondary isotope effects through two or three bonds are sometimes helpful in structural elucidation, particularly when an exchangeable hydrogen can easily be replaced by deuterium and its effect on the chemical shift of a nearby carbon nucleus observed.

4.9 EFFECTS OF MOLECULAR ASYMMETRY

As we have seen, the chemical shift of a nucleus depends on its immediate environment. When two nuclei are in environments that are not equivalent by symmetry, we can anticipate that their chemical shifts will be different, with the magnitude of the difference depending on details of the environment. In many instances it is clear from the molecular structure whether two or more nuclei are in equivalent or nonequivalent environments, but in many other instances certain asymmetric features are not as easily discernible. For example, in a substituted ethane CH_2X—$CPQR$, where X, P, Q, and R are any substituents, we find experimentally that the two hydrogen nuclei may have different chemical shifts. To see how such chemical nonequivalence arises, consider the three stable (staggered) conformers **I**–**III**.

which are pictured along the C—C bond. At very low temperature or under conditions of severe steric hindrance, rotation about the C—C single bond might be so restricted that conversion of one conformer to another is very slow. Under such circumstances, as we have seen in Chapter 2, the observed spectrum is just the *superposition* of the spectra of the three individual conformers. However, when rotation is rapid, as is normally the case with molecules of this sort near room temperature, the observed spectrum represents the *average* of the chemical shifts and coupling constants found the individual conformers, each weighted according to the fraction of time the molecule spends in that conformation.

Suppose the three conformations pictured are equally populated, as they would be if completely free rotation occurs about the C—C bond. The chemical shift of H_A in conformation **I** is probably influenced particularly by the groups P and R adjacent to it. It might at first appear that there is an equal contribution to the chemical shift of H_B in conformation **III**, where groups P and R are adjacent to H_B. However, closer examination of the entire molecule shows that differences exist between the conformations: groups P, X, and Q are neighbors in **I,** and R, X, and Q are adjacent in **III**. Thus, although the "immediate" environments of H_A in **I** and of H_B in **III** are the same, steric or electronic effects from the remainder of the molecule can in principle lead to different chemical shifts for H_A in **I** and H_B in **III**. A similar analysis applies to all other potentially equivalent pairs. Hence we conclude that the average chemical shift of H_A is not necessarily equal to that of H_B. Whether such a difference is observed depends, of course, on the net result of the magnetic effects involved and on the experimental conditions. The point is that we should always expect such differences and regard equivalence in the observed chemical shifts of H_A and H_B as fortuitous.

The asymmetry responsible for the nonequivalence of the chemical shifts of H_A and H_B need not be due to an immediately adjacent asymmetric carbon atom. For example, the CH_2 protons in

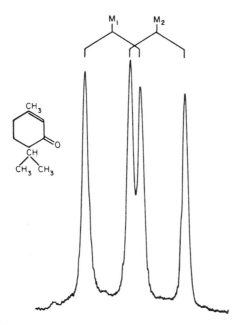

FIGURE 4.13 ^{1}H NMR spectrum (60 MHz) of the isopropyl group of piperitone in CDCl$_3$. Each of the nonequivalent methyl groups, M$_1$ and M$_2$, is spin coupled to the isopropyl proton and thus gives two lines as indicated and discussed in Chapter 5.

$$C_6H_5-\underset{\underset{O}{\|}}{S}-O-CH_2-CH_3$$

have different chemical shifts. Many examples are known of chemical shift nonequivalence among protons and other nuclei many bonds away from a site of asymmetry. Figure 4.13 gives a simple example of nonequivalence, in this case of two methyl groups in a small molecule, where the chemical shift difference of the methyl protons is readily observable even at very low magnetic field (60 MHz ^{1}H resonance frequency).

It should be emphasized that the presence of some sort of asymmetry is a necessary condition for chemical nonequivalence of two protons (or two methyl groups, etc.). This does not mean, however, that the molecule must be completely devoid of a plane of symmetry. For example, in the situation we have been considering, suppose that R is the group CH$_2$X, giving the molecule **IV**:

$$X-\underset{\underset{H_B}{|}}{\overset{\overset{H_A}{|}}{C}}-\underset{\underset{Q}{|}}{\overset{\overset{P}{|}}{C}}-\underset{\underset{H_B}{|}}{\overset{\overset{H_A}{|}}{C}}-X \qquad Q-\underset{\underset{R}{|}}{\overset{\overset{P}{|}}{C}}-\underset{\underset{H_B}{|}}{\overset{\overset{H_A}{|}}{C}}-\underset{\underset{R}{|}}{\overset{\overset{P}{|}}{C}}-Q$$

$$\textbf{IV} \qquad\qquad\qquad \textbf{V}$$

Although this molecule has a plane of symmetry, H_A and H_B are, as we have seen, nonequivalent. Alternatively, suppose that X is CPQR, giving the molecule **V**. This can exist as a d, l pair or as a *meso* compound. In the d and l forms, H_A and H_B are equivalent but in the *meso* compound they are nonequivalent.

The relationships existing within molecules possessing some asymmetric characteristics have been treated in detail by several authors. Protons H_A and H_B in a molecule such as **I–III** (with P, Q, and R different) are said to be *diastereotopic*, whereas if P = Q, they would be *enantiotopic*. In the latter case, as in the case of two enantiomeric molecules, the NMR spectrum normally does not distinguish between the two, and the chemical shifts of H_A and H_B are the same. However, in a *chiral* solvent the interactions between the solvent and the two protons are not necessarily equivalent, and chemical shift differences (usually small) may be found.

Finally, we should point out that our entire discussion of nonequivalence has been predicated on the assumption of equal populations for all conformers. In most cases in which asymmetry is present, there are differences in energy of the conformers, hence in their populations. Such differences can significantly enhance the magnitude of the nonequivalence.

4.10 PARAMAGNETIC SPECIES

Metallo-organic compounds in which the metal is diamagnetic display chemical shifts for proton resonance that cover a range only slightly larger than that found for other organic molecules. If the metal is paramagnetic, however, chemical shifts for protons often cover a range of 200 ppm, and for other nuclei the range can be much greater. These large chemical shifts arise from a *contact interaction* and/or a *pseudocontact interaction*. The former involves the transfer of some unpaired electron density from the metal to the ligand. This unpaired spin density can cause positive or negative chemical shifts, depending on the electron distribution and electron spin correlation effects.

The pseudocontact interaction (perhaps more appropriately called a dipolar interaction) arises from the magnetic dipolar fields experienced by a nucleus near a paramagnetic ion. The effect is entirely analogous to the magnetic anisotropy discussed in Section 4.5. It arises only when the g tensor of the electron is anisotropic; that is, for an axially symmetric case, $g_\parallel \neq g_\perp$. The g value for an electron is defined as

$$g = \nu_0 / \beta_0 B_0 \tag{4.17}$$

where β_0 is the Bohr magneton and ν_0 and B_0 are the electron resonance frequency and magnetic field, respectively. This anisotropy in g leads to anisotropic magnetic susceptibility, $\chi_\parallel \neq \chi_\perp$, and by Eq. 4.16 a nucleus experiences a shift inversely proportional to R^3. Both contact and dipolar shifts from unpaired electrons are temperature dependent, normally varying approximately as $1/T$. The presence of unpaired electrons usually causes rapid nuclear relaxation and leads to line broadening (see Sections 2.5 and 8.7). High resolution NMR in paramagnetic complexes can be observed only in cases in which the relaxation time is favorable.

Lanthanide Shift Reagents

The large chemical shifts caused by paramagnetic species have been exploited in *shift reagents*, which contain a paramagnetic ion attached to a ligand that can in turn complex with the molecule being studied. The object is to induce large alterations in the chemical shifts of the latter molecule, while minimizing paramagnetic line broadening. The most successful ions in this regard are certain lanthanides, which have such a short relaxation time for the unpaired electron that little line broadening occurs (see Chapter 8). The mechanism of action of the lanthanides is principally the pseudocontact mechanism, which falls off in a predictable manner with distance $(1/R^3)$.

The most commonly used shift reagents for organic compounds employ Eu^{3+}, Pr^{3+}, or Yb^{3+} as the paramagnetic ion in a chelate of the form **VI**

VI

with $R = C(CH_3)_3$ (usually abbreviated *dpm*) or $R = CF_3CF_2CF_2$ (abbreviated *fod*). The important properties of the ligand are adequate solubility in organic solvents and significant complexing ability with nucleophilic functional groups. In some case other factors may be overriding in the choice of a reagent, for example, ligand chirality when it is desired to form a complex with only one of a pair of optical isomers. In addition to use with organic compounds, other lanthanide shift reagents have been employed to complex extracellular sodium ions in *in vivo* NMR studies.

As indicated in Eq. 4.16, the direction of shift depends on the anisotropy in the susceptibility, but it also depends on the angle between the principal axis of susceptibility and the vector **R** to the nucleus. For Eu^{3+} the induced shifts are normally to higher frequency and for Pr^{3+} they are to lower frequency. On occasion some nuclei may lie at an angle $\theta > 54.7°$, so that the factor $(3 \cos^2 \theta - 1)$ changes sign, and shifts occur in the directions opposite those given before.

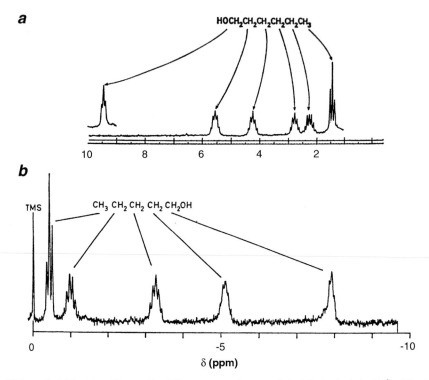

FIGURE 4.14   Use of lanthanide shift reagents to alter chemical shifts in alcohols. (*a*) ^{1}H NMR spectrum (100 MHz) of *n*-hexanol in CCl_4 with the reagent Eu(*dpm*)$_3$. (*b*) ^{1}H NMR spectrum (100 MHz) of *n*-pentanol in CCl_4 with the reagent Pr(*dpm*)$_3$. From La Mar *et al.*[61]

Figure 4.14 shows typical changes in chemical shifts that are found for organic compounds complexed with shift reagents. As anticipated, the magnitude of the shift is largest for nuclei (in this case protons) that are closest to the site of binding of the reagent (here, the oxygen). Because coupling constants are generally unaffected by shift reagents, spectra with complex, overlapping groups of lines are often disentangled. Before the advent of high field NMR, such spectral simplification as an aid to analysis was probably the most common purpose for using shift reagents. Currently, shift reagents are employed primarily to obtain information on molecular conformation from quantitative consideration of the relative shifts of different nuclei.

4.11 ADDITIONAL READING AND RESOURCES

Probably the most extensive source of information on various aspects of chemical shift theory is the *Encyclopedia of NMR*, which contains a large number of articles under titles beginning with *Chemical Shift, Shielding Calculations,* and *Isotope Effects.* Relatively little on this subject appears in most NMR books, but the serial publications listed in Section 1.3 contain some helpful articles.

Paramagnetic molecules and lanthanide shift reagents are discussed in detail in *NMR of Paramagnetic Molecules* by G. N. La Mar *et al.,*[61] *NMR of Paramagnetic Molecules in Biological Systems* by Ivano Bertini and Claudio Luchinat,[62] and *NMR Shift Reagents* by R. E. Sievers.[63]

Data on chemical shifts and discussions of empirical and semiempirical correlations with molecular structure are given in many books, including *NMR Spectroscopy* by Harald Günther,[64] *NMR Spectroscopy* by F. A. Bovey,[65] *Interpretation of Carbon-13 NMR Spectra* by F. W. Wehrli and T. Wirthlin,[66] *^{15}N NMR Spectroscopy* by G. C. Levy and R. L. Lichter,[67] and *NMR of Newly Accessible Nuclei* by Pierre Laszlo.[68]

Experimentally observed ^{1}H spectra and tabulations of chemical shifts for several common nuclides are given in many older NMR books that we shall not list here. The best current sources of such information are the commercially available databases. Of particular interest are the following:

- ACD/Labs (www.acdlabs.com) markets several extensive NMR databases. The ^{1}H database now exceeds 600,000 experimental chemical shifts and 110,000 coupling constants from 81,000 molecules, and that for ^{13}C is based on more than 900,000 chemical shifts. Databases for ^{19}F and ^{31}P each have more than 20,000 chemical shifts. The databases are linked to programs that predict NMR spectra for given molecular structures and substructures.
- Chemical Concepts (www.wiley-vch.de/cc/) markets similar NMR databases in conjunction with vibrational spectral and mass spectral collections, with a total of more than 600,000 entries.
- Aldrich Chemical Company markets a spectral library of 11,000 spectra (^{1}H spectra at 300 MHz and ^{13}C spectra at 75 MHz) in book form and, in conjunction with ACD/Labs, on CD-ROM in computer-searchable form.
- Bio-Rad Laboratories includes programs for predicting ^{13}C spectra from data based on more than 40,000 compounds in its Sadtler Suite (on CD-ROM), which also covers infrared spectra and chemical structure software.
- A large tabulation of chemical shifts in proteins is included in the BioMagResBank database (www.bmrb.wisc.edu). The form of the database is described by Seavey *et al.*[69]

4.12 PROBLEMS

4.1 The methylene protons of ethanol dissolved in CCl_4 have a chemical shift measured as 215 Hz from TMS (internal reference) at 60 MHz. Express the chemical shift as δ in parts per million. At 500 MHz, what is the

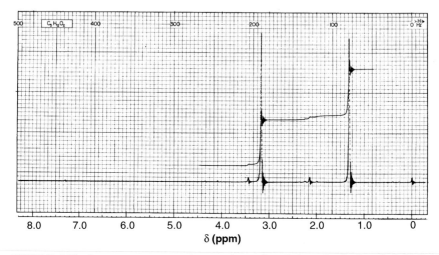

FIGURE 4.15 ^{1}H NMR spectrum at 60 MHz of a sample with the molecular formula $C_5H_{12}O_2$. Small peaks near 2.2 and 3.4 ppm are from impurities.

chemical shift in ppm and in Hz? What is the chemical shift in ppm and in Hz in a magnetic field of 7.00 tesla?

4.2 The difference in chemical shift between the α and β protons of naphthalene in dioxane solution has been reported as 0.36 ppm. Do you expect the α or β protons to resonate at lower frequency (i.e., greater shielding)? Why?

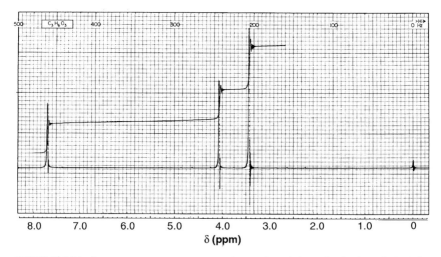

FIGURE 4.16 ^{1}H NMR spectrum at 60 MHz of a sample with the molecular formula $C_3H_6O_3$.

4.3 The ^1H chemical shift of $CHCl_3$, measured with respect to external neat liquid benzene, is 49.5 Hz at 60 MHz ($CHCl_3$ at higher frequency). The ^{13}C chemical shift of $CHCl_3$ is 52 ppm more shielded than that of neat liquid benzene. What percentage of each of these reported chemical shifts is due to magnetic susceptibility effects? (Values of volume magnetic susceptibility are given in the *Handbook of Chemistry and Physics* and in other sources.)

4.4 From Fig. 4.11, what is the effect of the ring current on a proton at a distance 3.1 Å above the plane of a benzene ring and with a projected distance in the plane of 1.8 Å from the center of the ring?

4.5 With the aid of the ^1H chemical shift correlation chart, Fig. 4.2, deduce the molecular structures of the molecules with the spectra shown in Figs. 4.15 and 4.16.

4.6 The chemical shift of an ^{15}N nucleus in a sample in aqueous solution is reported as $\Xi = 10.134027$ MHz. From Table 4.2, express this chemical shift in ppm relative to an external reference of liquid ammonia. From Fig. 4.4, what type of functional group is probably responsible for this resonance?

Coupling between Pairs of Spins

Our discussion of NMR thus far has focused on the interaction between a single nuclear spin and an imposed magnetic field. In Chapters 2 and 3 we examined the behavior of such a spin as radio frequency energy is applied, and in Chapter 4 we discussed the consequences of the shielding induced by the magnetic field in the electrons surrounding the nucleus. We have considered an ensemble of nuclei in terms of the populations among energy levels and we have alluded to interactions between nuclei that lead to relaxation. However, we have not yet dealt explicitly with the possible interactions (usually called *coupling*) between a pair of nuclear spins. In this chapter we investigate the origin and consequences of two different types of coupling and describe methods by which the interactions can be modified for certain purposes.

5.1 ORIGIN OF SPIN COUPLING INTERACTIONS

The two distinctly different types of interactions between pairs of nuclear spins that we consider here both depend on the orientation of the coupled spins and on the product of the magnitudes of the magnetic moments (or the magnetogyric

ratios) of the coupled spins. However, there are very important differences in the mechanisms by which the spins exchange information with each other.

Magnetic Dipolar Interactions

A nuclear spin behaves as a magnetic dipole (i.e., a small classical magnet with north and south poles), with a field that decreases with the cube of the distance from the nuclear magnet. Hence, a second nearby nuclear magnet experiences a "local" field from its neighbor that may add to or subtract from the externally imposed magnetic field B_0, according to the orientations of the spins relative to B_0. It is easy to show that a proton (^1H nucleus) gives rise to a field of about 1 gauss (10^{-4} T) at a distance of 1 A (10^{-10} m). Thus, a nucleus at approximately the distance of a chemical bond experiences a field of about ± 1 G, which is very small compared with the usual magnitude of B_0 ($\sim 10^5$ G). However, from the Larmor equation (Eq. 2.14), we see that such a field alters the resonance frequency of a proton by about ± 4 kHz or that of a ^{13}C nucleus by about ± 1 kHz, a very large shift relative to a typical line width of about 1 Hz and a very large effect relative to most chemical shifts. In general, there are a number of nearby magnetic nuclei at various distances and orientations. There is thus a wide range of resonance frequencies, whose envelope usually creates an apparently broad NMR line.

We shall examine magnetic dipolar coupling in more detail in Chapter 7, where we show that when the interacting nuclei are in molecules that are in rapid, random motion, as are most small molecules in solution, this interaction averages almost completely to zero. *In this chapter and in Chapter 6, we treat only situations in which magnetic dipolar interactions can be ignored.*

Electron-Coupled Spin–Spin Interactions

There is a second type of coupling between nuclear spins that persists in spite of rapid molecular tumbling. This type of interaction is commonly called *spin–spin coupling*. Because it is normally manifested in solution, it must arise from a mechanism that is independent of the rotation of the molecule.

Ramsey and Purcell[70] initially suggested the basic mechanism for the coupling interaction, which involves the electrons that form chemical bonds. Consider, for example, two nuclei, A and X, each with $I = \frac{1}{2}$, which are connected by a chemical bond (e.g., ^1H and ^{19}F in the molecule HF). Suppose nucleus A has its spin oriented parallel to B_0. In very slightly over half the molecules spin X is also oriented parallel to B_0, and in very slightly less than half, spin X is antiparallel to B_0 at a higher energy, as we saw in Section 2.4.

Now consider the pair of electrons in the bond between A and X. By the Pauli exclusion principle, these electrons are oriented antiparallel to each other, as

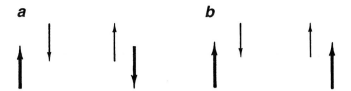

FIGURE 5.1 Schematic representation of the origin of electron-coupled spin–spin interaction. Nuclear spins are denoted by boldface arrows, electron spins by lightface arrows. (*a*) Antiparallel orientation of nuclear spins; (*b*) parallel orientation of nuclear spins.

illustrated in Fig. 5.1. Although electrons are delocalized, there is a high probability that one electron is near nucleus A and the other is near nucleus X. For a molecule in which the spins of A and X are antiparallel, there are energetically favorable interactions between A and its nearby electron *and* between X and its nearby electron. Thus, the overall energy of the state in which spins A and X are antiparallel to each other is reduced by this interaction mediated by the paired electrons. As we shall see, the reduction in energy is very small (of the order of Hz to kHz), much less than the energy change from the flip of a nuclear spin, which we know causes an energy change in the range of many MHz.

For a molecule in which spins A and X are parallel to each other, one nuclear spin and one electron have an energetically favorable relation, but the other is energetically unfavorable. Again, the magnitude of the energy change is small, but the result of this electron-mediated interaction is that approximately half the molecules have an energy slightly greater than they would have had in the absence of the interaction and approximately half have a slightly lower energy. Thus, when nucleus A undergoes resonance and "flips" its spin orientation with respect to B_0, the energy of its transition depends on the initial orientation of X relative to A, and two spectral lines result, the difference in their frequencies being proportional to the energy of interaction (coupling) between A and X.

Because the coupling interaction energy is usually quite small, the two lines lie close to each other, thus causing NMR spectra of molecules in liquids or solutions to appear as chemically shifted lines that seem to be split into doublets or, more generally, into multiplets when several nuclei are coupled. Although we use the terminology of "splitting" of lines, we should not lose sight of the fact that each of the observed spectral lines actually originates from molecules in which there are different relative orientations of nuclear spins. We saw examples of spin–spin coupling in Fig. 1.4 (several multiplets with equally spaced lines but of different amplitudes) and Fig. 1.5 (a complex multiplet that defies simple description in terms of spacings and amplitudes). To understand these coupling patterns and predict when each will occur, we develop a simple quantum mechanical treatment in Chapter 6. We defer further discussion until then and in this chapter concentrate on the nature and magnitude of the coupling interaction itself.

Our explanation of the origin of spin–spin coupling does not depend on the molecule being in an external field. Unlike the chemical shift, which is induced by and hence proportional to the applied field, spin–spin coupling is characteristic of the molecule itself. The magnitude of the interaction between nuclei A and X is given by a spin–spin coupling constant J_{AX}, which is always expressed in frequency units (hertz or occasionally radians/second). This is a unit of convenient magnitude, which is directly proportional to energy. Mathematically, as we see in more detail in Chapter 6, the fact that the coupling depends on the relative orientation of the spins of nuclei A and X can be expressed by using the scalar product, $J_{AX} I_A \cdot I_X$. As a result, this indirect coupling is often called *scalar coupling*. This term is precise in describing the spin interaction but is sometimes misconstrued to suggest that J_{AX} is a scalar quantity. It is not. J is actually a tensor, just like the chemical shielding tensor, σ. As with σ, when molecules tumble rapidly, only the average is observed; hence, in liquids it is correct to treat J as a simple scalar quantity or "constant." We shall look into the tensor aspects of J in Chapter 7.

5.2 GENERAL ASPECTS OF SPIN–SPIN COUPLING

Spin coupling can occur where two nuclei are bonded together, such as ^{13}C—H or ^{31}P—H, or where several bonds intervene, such as H_A—^{12}C—^{12}C—H_B. In general, the spin coupling information is carried by electrons through chemical bonds, not through space. (A few exceptions, especially with nuclei other than hydrogen, are known.) From the electron spin polarization mechanism the magnitude of the coupling is *generally* expected to decrease as the number of intervening bonds increases, but there are other important factors, as described in later sections.

For nuclei with $I > 1$, more lines result from spin interactions, as there are $2I + 1$ possible orientations of a nuclear spin I relative to the applied field. Thus for the molecule HD (deuterium has $I = 1$), the proton resonance consists of three lines, while the deuterium resonance consists of two lines. We recall from Section 2.1, however, that nuclei with $I > \frac{1}{2}$ possess a nuclear electric quadrupole moment. In an asymmetric electrical environment nuclei with large quadrupole moments usually relax rapidly and, as shown in Chapter 8, such relaxation can be considered approximately as rapid transitions between two spin states. By a process similar to fast exchange (see Section 2.10), the rapidly relaxing nucleus behaves as though it has the average energy of the two states and can thus "decouple" from its spin partner and lead to loss of multiplet structure. The halogens Cl, Br, and I almost always relax rapidly, as do most of the heavier nuclei with $I > \frac{1}{2}$. ^{14}N and ^{2}H sometimes relax fast enough to be partially or completely decoupled.

Signs of Coupling Constants

In our discussion of the simple mechanism of electron-coupled spin–spin inter-actions, we argued that the state in which two coupled nuclei have antiparallel spin orientations is lowered slightly in energy by the coupling and the state in which the spins are parallel is raised slightly in energy. Chemical bonding and the interactions of nuclear spins through one or more bonds are not always so simple, however, and in some cases the state in which the spins of the coupled nuclei are parallel is lowered in energy and the antiparallel state raised in energy. We distin-guish between these two situations by referring to the first system as possessing a *positive* coupling constant ($J > 0$) and to the second as having a *negative* coupling constant ($J < 0$). As we see in Chapter 6, the *relative* signs of the various coupling constants within the molecule sometimes influence the appearance of the spectrum and hence can be determined from the observed spectrum or from the use of double resonance experiments. The *absolute* signs of J's cannot be found from ordinary high resolution NMR spectra, but there are overwhelming reasons (based on both theory and more sophisticated NMR experiments) for believing that all one-bond $^{13}C-H$ coupling constants are positive. Absolute values of other coupling constants can then be determined from their signs relative to that of the $^{13}C-H$ coupling constant.

Some Observed Coupling Constants

As we shall see, *ab initio* and semiempirical theories are able to account semiquan-titatively for most aspects of spin coupling, but theory alone is unable to predict accurate values for coupling constants in most molecules. Hence, our knowledge of the range of coupling constants found for different molecular systems rests largely on observations and empirical correlation. Most values of J can be ob-tained from observed NMR spectra, frequently just by direct measurement of line splittings but in some instances by analysis based on considerations in Chapter 6.

Table 5.1 lists typical values of proton–proton coupling constants for various molecular species. This tabulation is meant to be illustrative, not exhaustive, with respect to both the types of molecules included and the overall ranges listed. We have adopted the commonly used notation for coupling constants, in which the number of chemical bonds intervening between the coupled nuclei is given as a superscript and the identity of the coupled nuclei as a subscript (e.g., $^3J_{HH}$ for a cou-pling between protons on adjacent carbon atoms, sometimes called *vicinal* coupling). The superscript or the subscript or both are deleted when there is no ambiguity.

Proton–proton couplings through single bonds are usually attenuated rapidly, so that generally $^4J < 0.5$ and is usually unobservable. However, couplings are

TABLE 5.1 Typical Proton–Proton Spin Coupling Constants

Type	J_{HH}(Hz)	Type	J_{HH}(Hz)
H—H	280	$\begin{array}{c}\diagdown\\C=C\diagup\end{array}\begin{array}{c}C-H\\ \diagup\ \ \diagdown H\end{array}$	7
$\begin{array}{c}C\diagdown\ \ \diagup H\\ C\\ C\diagup\ \ \diagdown H\end{array}$	−12 to −15	$\begin{array}{c}\diagdown\\C=C\diagup\end{array}\begin{array}{c}C-H\\ \diagup\\ H\end{array}$	−1.5
H—C—C—H (free rotation)	7	$\begin{array}{c}H\diagdown\\C=C\diagup\end{array}\begin{array}{c}C-H\end{array}$	−2
$H-C-\overset{\mid}{\underset{\mid}{C}}-C-H$	~0	C=C—C=C ‖ ‖ H H	10
⌇ —H / —H (cyclohexane)	ax-ax 8–10 ax-eq 2–3 eq-eq 2–3	$\begin{array}{c}H\diagdown\ \ C\\ \ \ \diagup\ \ \diagup\\ C\end{array}\begin{array}{c}C\ \ \diagdown\\ \diagdown\ \ \diagdown\\ C\ \ H\end{array}$	±1
—H / —H (cyclopentane) (cis or trans)	4–5	$-N=C\diagup\begin{array}{c}H\\ \diagdown H\end{array}$	7–17
—H / —H (cyclobutane) (cis or trans)	8	$O=C\diagup\begin{array}{c}H\\ \diagdown H\end{array}$	42
—H / —H (cyclopropane) (cis) (trans)	8–10 4–6	H—C—C≡C—H	−2
		H—C—C≡C—C—H	2
H—C—O—H	5	$\begin{array}{c}H\diagdown\\C=C\\ \diagup\end{array}\begin{array}{c}H\ 5\ mem.\\ \ \ 6\ mem.\\ (ring)\ 7\ mem.\end{array}$	6 10 12
$H-\overset{\overset{\displaystyle O}{\|}}{C}-\overset{}{C}-H$	+3	$X=C\diagup\begin{array}{c}C-H\\ \diagdown C-H\end{array}$ (X = C, O)	0 to ±2
$H-\overset{\overset{\displaystyle C}{\|}}{C}-\overset{\overset{\displaystyle O}{\|}}{C}-H$	8	benzene (ortho) (meta) (para)	8 2 ~0.5
$\begin{array}{c}H\diagdown\\C=C\diagup\end{array}\begin{array}{c}\\ \diagdown H\end{array}$	12 to 19	pyridine (2–3) (3–4) (2–4) (3–5) (2–5) (2–6)	5 8 0–3 1.5 1 ~0
$\begin{array}{c}\diagdown\\C=C\diagup\end{array}\begin{array}{c}H\\ \diagdown H\end{array}$	−3 to +2		
$\begin{array}{c}H\diagdown\\C=C\diagup\end{array}\begin{array}{c}H\\ \diagdown\end{array}$	7 to 11		
$\begin{array}{c}H-C\diagdown\\C=C\diagup\end{array}\begin{array}{c}C-H\\ \diagdown\end{array}$	1–2		

(continues)

TABLE 5.1 (*continued*)

Type	J_{HH}(Hz)	Type	J_{HH}(Hz)
(2–3)	2	(2–3)	5
(3–4)	4	(3–4)	4
(2–4)	1	(2–4)	1
(2–5)	±1.5	(2–5)	3

stereospecific, as we see in Section 5.4, so that certain geometric arrangements of nuclei result in values of ^4J and even ^5J that are observable. Coupling through more than three single bonds is called *long-range coupling* and is of considerable interest in stereochemistry.

Generally, couplings are transmitted more effectively through double and triple bonds than through single bonds. For example, vicinal couplings for ethane derivatives are usually <10 Hz, while in ethylene derivatives $^3J(cis) \approx 10$ and $^3J(trans) \approx 17$ Hz. In aromatic systems also the coupling is transmitted more effectively than through single-bonded systems. *Ortho* coupling constants (^3J) are generally of the order of 5–8 Hz, *meta* couplings (^4J) are about 1–3 Hz, and *para* couplings (^5J) are quite small, often <0.5 Hz. In general, there is no coupling observed between different rings in fused polycyclic systems, so such long-range couplings must be less than about 0.5 Hz. Substituents have only small effects on the magnitudes of aromatic couplings, but introduction of heteroatoms can cause significant alterations, as indicated in Table 5.1.

Geminal proton—proton coupling constants (^2J) depend markedly on substituents, as indicated in Table 5.1. The trend of ^2J with substitution has been treated successfully by theory and will be discussed in Section 5.3.

Table 5.2 lists some representative coupling constants between protons and other atoms. ^1J is usually quite large, but other coupling constants cover a wide range. The large values of ^1J for ^{13}C—H often lead to unwanted complexity when ^{13}C spectra are observed. As we see in Section 5.6 this problem can be eliminated by double resonance *decoupling*. Geminal and vicinal couplings between protons and other nuclei, such as ^{13}C and ^{15}N, are generally small (often 1–5 Hz), and those between H and F may be quite large.

Table 5.3 gives a small, illustrative selection of coupling constant not involving protons. Again, couplings involving ^{19}F are noticeably large.

Reduced Coupling Constants

The strength of the coupling interaction, J, depends on the magnitudes of the two coupled magnetic moments, hence is proportional to the product of the

TABLE 5.2 Typical H–X Coupling Constants

Type	$|J|$ (Hz)	Type	$|J|$ (Hz)
^{13}C–H (sp^3)	125	(ortho) (meta) (para)	8 4–7 2
(sp^2)	160		
(sp)	240		
^{13}C–C–H	−5 to +5		
^{13}C–C–C–H	0 to 5		20–45
$^{15}NH_3$	61		1–20
^{15}N–CH$_3$	1–3		
C=^{14}N–CH$_3$	3		80
$\underset{Ph}{\overset{Ph}{>}}$C=$^{15}NH$	51		
	91		500–700
^{17}O–H	80		200
$\underset{F}{\overset{H\quad H}{>}}$C	45		10
H–C–C–F	5–20		<5
H–C–C–C–F	1–5		

magnetogyric ratios of the coupled nuclei. For comparison of the magnitudes of coupling constants between different nuclei and to compensate for the negative sign introduced in some cases by negative magnetogyric ratios, a *reduced coupling constant* K_{AB} can be defined:

$$K_{AB} = \frac{2\pi}{\gamma_A \gamma_B} J_{AB} \qquad (5.1)$$

TABLE 5.3 Typical Coupling Constants Not Involving Hydrogen

Type	J(Hz)
^{13}C—F	−280 to −350
—^{13}C—^{13}C—	35
—^{13}C—^{13}C≡N	50–55
^{13}C=^{13}C	70
—^{13}C≡^{13}C—	170
^{15}N=^{15}N	5 to 15
—^{13}C—^{15}N	−4 to −10
—^{13}C≡^{15}N	−17
C with two F	160
F—C—C—F	−3 to −20
(fluorobenzene)	(ortho) −17 to −22
	(meta) 11 to −10
	(para) 14 to −14
C=C with F cis/trans	−120
C=C with two F	30–40
P—P	100
O=P—P=O	500
F—P=O	1000

The dependence of J on γ means that the coupling constant itself (but not the reduced coupling constant) changes with isotopic substitution even though the electron distribution in the molecule is unchanged. For example, on deuterium substitution,

$$J_{HX}/J_{DX} = \gamma_H/\gamma_D = 6.51 \qquad (5.2)$$

5.3 THEORY OF SPIN–SPIN COUPLING

For proton–proton coupling it has been shown that the spin interaction arises principally from the electron spin–electron spin interaction, not from orbital interaction of electrons. This simplifies the theory somewhat. For some other nuclei, orbital interaction may also come into play. We shall summarize a few of the conclusions applicable to spin–spin coupling without going through the details of the theory.

From our qualitative description of the origin of spin coupling, it is clear that the interaction depends on the proximity between electron and nuclear spins, and a quantitative treatment verifies that the interaction depends on density of electrons at the nucleus. It is well known that only s electrons have density at the nucleus, so we expect a correlation between the magnitude of the coupling and the s character of the bond. Such a relation is found, as we see in the following section.

The theoretical development of the electron-coupled spin–spin interaction has been carried out principally by second-order perturbation theory, just as is done with shielding by electrons. In Section 4.2 we mentioned some of the problems in applying *ab initio* methods to calculation of shielding. For calculation of spin coupling, many of these problems remain, and it is also necessary to use basis sets that correctly represent the electron wave function at the nucleus. Unfortunately, Gaussian functions, which are widely used because of the relative ease of computing large basis sets, do not meet this requirement, so special basis sets must be employed. *Ab initio* calculations of spin coupling in selected spin systems have been carried out, but generally it has so far proved impractical to deal with the very large basis sets required to encompass the essential singlet and triplet states.

As a result, the Ramsey–Purcell formulation has been extended principally by semiempirical methods, which provide some insight into the relevant interactions and permit useful correlations with other physical properties. Exact calculation of coupling constants is virtually impossible with the present limited knowledge of electronic excitation energies and wave functions, but both valence bond and molecular orbital treatments have been applied successfully to small molecules or molecular fragments in predicting the general magnitude of coupling constants and their dependence on various parameters, as we see in the next section.

5.4 CORRELATION OF COUPLING CONSTANTS WITH OTHER PHYSICAL PROPERTIES

Theory suggests and experiment confirms that coupling constants can be related to a number of physical parameters. Among the most important are (1) hybridization, (2) dihedral bond angles, and (3) electronegativity of substituents.

The dependence of J on electron density at the nucleus suggests a relation between J and amount of s character in the bond. Such a relation is indeed found for $^{13}C-H$ couplings in sp, sp^2, and sp^3 hybridized systems, as indicated in Table 5.2. Similar rough correlations are found for other $X-H$ couplings ($X = {}^{31}P, {}^{15}N,$ ^{119}Sn), for $^{13}C-{}^{13}C$ couplings, and for some other $X-X$ couplings. In all cases, including $^{13}C-H$, however, addition of substituents may well cause large changes in effective nuclear charge or uneven hybridization in different bonds, so that exact correlations should not be expected. For example, the nominally sp^3 hybridized molecules CH_4, CH_3Cl, CH_2Cl_2, and $CHCl_3$ have values of $J(^{13}C-H)$ of 125, 150, 178, and 209 Hz, respectively.

One of the most fruitful theoretical contributions to the interpretation of coupling constants has been the valence bond treatment by Karplus[71] of $^3J_{HH}$ in ethanelike fragments, $H_a-C_a-C_b-H_b$. The most interesting conclusion is that this coupling depends drastically on the dihedral angle between the H_a-C_a and the C_b-H_b bonds. The calculated results were found to fit approximately the relation

$$^3J = A + B\cos\theta + C\cos 2\theta \qquad (5.3)$$

The solid line in Fig. 5.2 is a plot of Eq. 5.3 with the parameters $A = 7$, $B = -1$, and $C = 5$ Hz. It is apparent that large values of J are predicted for cis ($\theta = 0°$) and $trans$ ($\theta = 180°$) conformations and small values for gauche ($60°$ and $120°$) conformations. These predictions have been amply verified, and the Karplus relation is of great practical utility in structure determinations.

The Karplus equation was derived strictly for ethane. Substitution, especially with strongly electronegative atoms such as oxygen, can cause substantial changes in coupling. Quantitative application of the Karplus equation depends critically on the values of the parameters, which are obtained empirically from some set of experimental data. Hence, caution should be used not to overinterpret the predictions for molecules that are somewhat different from those used in the parameterization. On the other hand, within one set of molecules—proteins where the dihedral angles have been determined by x-ray crystallography—the equation has been found to be very reliable. In fact, there is evidence that from very precisely measured values of J involving peptide and side chain residues, the predictions of the Karplus relation may be more accurate than many dihedral angles determined by x-ray crystallography.[72]

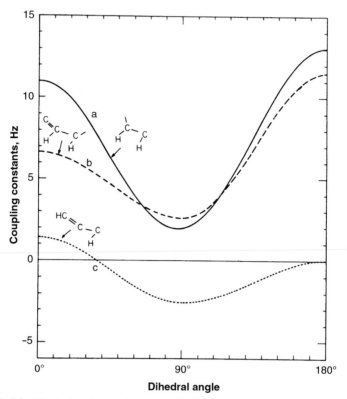

FIGURE 5.2 Calculated variation of vicinal and allylic proton–proton coupling constants with dihedral angle between the C—H bonds shown.

The success of the theoretical treatment of the angular dependence of ^3J in an ethane fragment has served as the rationale for empirical calculations of the angular dependence of $^3J_{ab}$ and $^4J_{ac}$ in the allylic system **I**.

$$\underset{H_c}{\overset{}{\diagup}}C=C\overset{\diagup C-H_a}{\underset{\diagdown H_b}{\diagdown}}$$

I

Plots of $^3J_{ab}$ and $^4J_{ac}$ as functions of dihedral angles between C—H$_a$ and C—H$_b$ or C—H$_c$ are given in Fig. 5.2. Although these relationships are only approximate, there is generally rather good agreement with experimental data. Although all three curves in Fig. 5.2 have a similar angular dependence, the change in sign for ^4J leads to a maximum in $|J|$ near 90°, rather than a minimum as in the two curves for ^3J.

H–H couplings of 1 Hz or more through four single bonds have been observed almost exclusively in systems of the sort in which the four bonds are in a planar "W" conformation (**II**).

II   **III**

Usually this fragment is part of a cyclic or polycyclic system, which determines the conformation. Coupling constants of 1–2 Hz are commonly found in such cases, but larger values (3–4 Hz) have been observed in bicyclo [2.2.1]heptanes and related molecules. Conjugated systems in a planar zigzag configuration (**III**) display five-bond H–H coupling constants of 0.4–2 Hz. Again, the fragment is often part of a cyclic molecule, and one of the carbon atoms may be substituted by nitrogen. The values of ^4J and ^5J in these two types of systems fall off rapidly with departures from planarity.

Vicinal *heteronuclear* coupling constants also depend on dihedral angles. Curves similar to those in Fig. 5.2 have been developed for $^3J(HCOP)$ and $^3J(NCCH)$. In molecules in which unshared (nonbonding) electrons are localized in one direction, as in some nitrogen- or phosphorus-containing compounds, geminal couplings are also dependent on angular orientation. For example, in oxaziridines $^2J(H-^{15}N) = -5$ Hz in **IV** but is nearly zero in **V**.

IV   **V**

The molecular orbital approach has also been used to calculate proton–proton coupling constants, for example, in explaining the variation of $^2J_{HH}$ in both sp^2 and sp^3 systems with change of substituents. The most interesting feature of the theory is that electronegative substituents that remove electrons from the *symmetric* bonding orbital of the CH_2 fragment cause increases in $^2J_{HH}$, whereas substituents that remove electrons from the *antisymmetric* bonding orbital cause decreases in ^2J. The former corresponds to inductive withdrawal of electrons, and the latter arises from hyperconjugation. In some cases the effects add, rather than oppose each other. In formaldehyde, for example, the electronegative oxygen causes an inductive withdrawal of electrons from the C—H bonds, thus increasing $^2J_{HH}$. The two pairs of nonbonding electrons of the oxygen are donated to the

C—H orbitals (hyperconjugation), thus further increasing $^2J_{HH}$. Formaldehyde is predicted, then, to have a large, positive $^2J_{HH}$. The measured value, $+42$ Hz (see Table 5.1), is the largest H—H coupling constant known, except for the directly bonded protons in the hydrogen molecule.

5.5 EFFECT OF EXCHANGE

In Section 2.10 we described the effect of exchange of a nucleus between sites where it has different Larmor frequencies, and we illustrated exchange among two or more sites differing in chemical shift. However, the relations derived there are quite general and can be applied to exchange involving spin coupling. For example, we have seen in Section 5.4 that $^3J_{HH}$ depends on dihedral angle, so that each of the three conformers about a C—C bond might be expected to display different values of $^3J_{HH}$. In most noncyclic molecules, however, there is rapid rotation about the C—C bond at room temperature, and the observed value of J is often about 7 Hz, the average of the values for the individual conformers.

On the other hand, if a chemical bond is broken in the exchange process, a different result is obtained. For example, in CH_3OH the methyl and hydroxyl protons are spin coupled with $J \approx 5$ Hz, and under suitable conditions the CH_3 resonance consists of a doublet, as shown in Fig. 5.3. As we noted in Section 5.1, the lines arises from molecules in which the OH protons have opposite spin orientations. If the OH proton exchanges between molecules, the methyl resonance may be affected. Suppose a given molecule of methanol contains an OH proton whose spin is oriented parallel to $\mathbf{B}_0$. If this proton is lost and is replaced by another proton, either from another CH_3OH molecule or from somewhere else (e.g., an H^+ or H_2O impurity), the CH_3 Larmor frequency is unchanged if the replacing proton has the same spin orientation but changes by J Hz if the new proton is oriented antiparallel to $\mathbf{B}_0$. In the usual *random* process, then, half of all exchanges result in a change in the CH_3 Larmor frequency, while the other half do not. With this statistical factor taken into account, the results of Section 2.10 and Fig. 2.14 apply and account for the collapse of the spin multiplets of both the CH_3 and OH rsonances in CH_3OH with increase in temperature, as shown in Fig. 5.3. In the fast exchange limit, the observed resonance is at the frequency of the chemical shift, which is just the average of the frequencies of the spin components. (In situations in which spin wave functions mix, as described in Chapter 6, the result is qualitatively similar but must be treated quantitatively with a more sophisticated theory.)

Proton exchanges in OH or NH groups are often catalyzed by H^+ or OH^- and so are highly pH dependent. Sometimes exchange at an intermediate rate broadens the resonance line so much that it may pass unobserved.

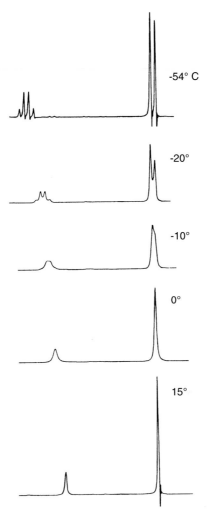

FIGURE 5.3   Collapse of spin splitting in methanol with increase in temperature. Both the CH$_3$ and OH proton resonances are shown.

5.6 SPIN DECOUPLING
AND DOUBLE RESONANCE

The presence of spin coupling provides much of the richness of NMR spectra, not only because of the information on bonding and conformation that we can infer from the magnitudes of coupling constants but also because spin information can be transferred from one nucleus to another by means that we shall explore in later chapters. Likewise, in solids, magnetic dipolar coupling provides valuable information on internuclear distances and molecular motions, as we discuss in Chapter 7. Nevertheless, these couplings can be a nuisance and can complicate the interpretation of NMR spectra in many instances. Fortunately, by applying radio frequency pulses or continuous wave rf, we can "fool" the spin system into displaying a spectrum that would be obtained in the absence of certain couplings.

On a qualitative level, these methods can be thought of as causing one nuclear spin to change its orientation so rapidly that, by rapid exchange, its coupled neighbor senses only an average orientation, hence appears to be "decoupled" from the neighbor. For example, if nucleus A is coupled to X, we can apply a continuous rf field B_2 at the resonance frequency of X while studying the spectrum of A by pulse or continuous wave methods with an rf field B_1. If B_2 is sufficiently large, it can be thought of as causing so many transitions among X nuclei that a given nucleus rapidly changes its spin state.

An alternative semiclassical explanation can also be used to explain decoupling. In a frame rotating at ν_x, μ_x can be viewed as being oriented along the large rf field $\mathbf{B}_2$ that is applied in the x' direction, as we discuss in Chapter 9 under *spin locking*. With μ_x thus forced to be orthogonal to μ_A, the scalar product becomes zero. As we shall see in later chapters, the situation is somewhat more complicated that these simple pictures imply, but the spin Hamiltonian can be altered to accomplish the desired decoupling.

Spin decoupling is one aspect of what is more generally termed *double resonance*, whereby two nuclides of the same species (*homonuclear*) or different species (*heteronuclear*) are concurrently irradiated. When B_2 (expressed in Hz) is much greater than J, decoupling occurs. However, if B_2 is of the same order of magnitude as J, the Hamiltonian is only slightly perturbed (termed *spin tickling*) and certain resonance lines are split. If B_2 is still smaller (of the order of a line width), the energy levels are not affected but populations change (the nuclear Overhauser effect, NOE). These *selective* double resonance methods with continuous wave rf have long played an important role in elucidating molecular structure and in understanding the basis of spin interactions, but they have largely given way to two-dimensional NMR methods based on sequences of rf pulses, as discussed in Chapters 10−13.

5.7 ADDITIONAL READING AND RESOURCES

Many collections of coupling constants have been compiled in tabular form, and many other approximate values can be obtained from inspection of first-order splittings in published spectra. The most useful sources of such data are the databases listed in Section 4.12 from ACD/Labs, Chemical Concepts, Aldrich

Chemical Company, and Sadtler Laboratories and the books by Bovey, Günther, and Levy and Lichter also given there.

Articles on theoretical aspects of spin coupling appear from time to time in the serial publications listed in Section 1.3. Also, several articles in the *Encyclopedia of NMR* provide further information on theory and correlations with molecular structure, including *Indirect Coupling: Semiempirical Calculations*,[73] *Indirect Coupling: Theory and Applications in Organic Chemistry*,[74] *Indirect Coupling: Intermolecular and Solvent Effects*,[75] and *Stereochemistry and Long Range Coupling Constants*.[76] An article in *Concepts in Magnetic Resonance* gives a good review of parameterization of the Karplus equation.[77]

5.8 PROBLEMS

5.1 The ^{1}H spectrum at 200 MHz from two spin-coupled protons consists of four lines of equal intensity at 72, 80, 350, and 358 Hz, measured with respect to TMS. Predict the spectrum at 300 MHz and state the values of δ (ppm) and J(Hz).

5.2 The magnitude of the geminal H–H coupling in CH_4 has been determined as 12.4 Hz. As the spectrum of CH_4 consists of only a single line, how could this figure have been obtained?

5.3 If $J(^{14}N-H)$ for NH_3 is $+40$ Hz, what is the sign and magnitude of $J(^{15}N-H)$ in $^{15}NH_3$? Of $J(^{15}N-D)$ in $^{15}ND_3$?

5.4 Using reduced coupling constants, compare the magnitudes of the X–H coupling in CH_4 (125 Hz) and $^{14}NH_4{}^+$ (55 Hz).

5.5 In the pair of tautomers **VI** and **VII**, which are in rapid exchange, the resonance of H* is split into a doublet of 35 Hz at 25°C and 52 Hz at -50°C as a result of coupling to ^{15}N. Account for the splitting and its change with temperature. In Fig. 5.3 we saw that rapid exchange caused the OH coupling to disappear. Why does the ^{15}N–H coupling persist with rapid exchange between **VI** and **VII**?

VI   **VII**

5.6 Consider the ^{13}C NMR spectrum in Fig. 5.4, which was obtained with broadband ^{1}H decoupling. (a) The resonance centered near 79 ppm arises from the solvent $CDCl_3$. Why is this a triplet? (b) With the help of the

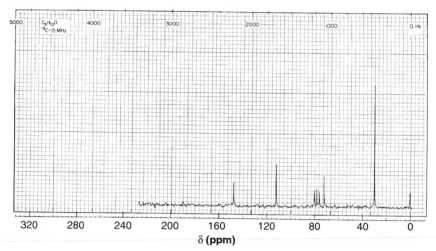

FIGURE 5.4 ^{13}C NMR spectrum at 15 MHz with broadband ^1H decoupling of a sample with the molecular formula $C_5H_{10}O$. ^{13}C chemical shifts: 146.14, 110.77, 70.99, and 29.38 ppm. The triplet centered near 78 ppm is from the solvent $CDCl_3$.

chemical shift correlation chart, Figure 4.3, deduce the structure of the molecule giving rise to the spectrum.

5.7 Deduce the structures of the molecules giving rise to the ^1H NMR spectra in Figs. 5.5 to 5.9.

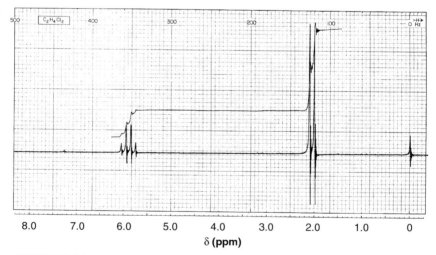

FIGURE 5.5 ^1H NMR spectrum at 60 MHz of a sample with the molecular formula $C_2H_4Cl_2$.

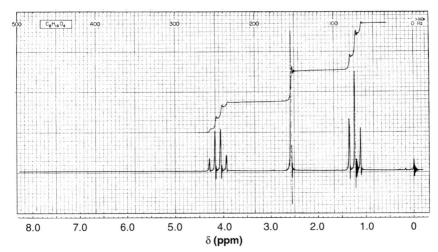

FIGURE 5.6 ^{1}H NMR spectrum at 60 MHz of a sample with the molecular formula $C_8H_{14}O_4$.

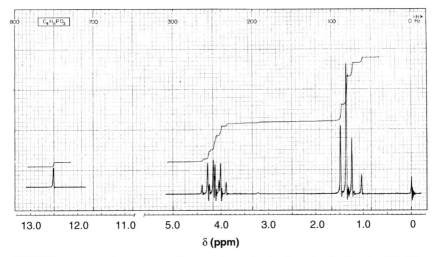

FIGURE 5.7 ^{1}H NMR spectrum at 60 MHz of a sample with the molecular formula $C_4H_{11}PO_3$.

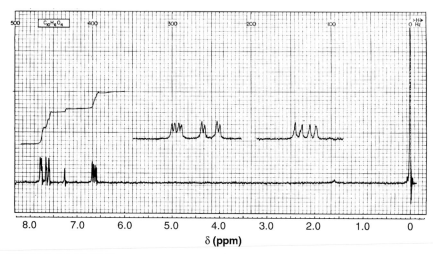

FIGURE 5.8 ¹H NMR spectrum at 60 MHz of a sample with the molecular formula $C_{10}H_6O_4$.
Inset: Abscissa scale, 1 Hz per division.

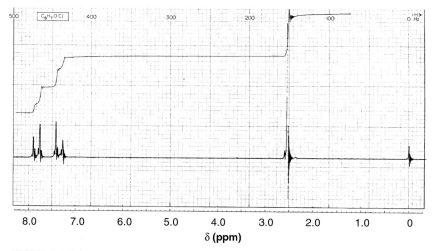

FIGURE 5.9 ¹H NMR spectrum at 60 MHz of a sample with the molecular formula C_8H_7OCl.

Structure and Analysis of Complex Spectra

In previous chapters we have described the origin of the chemical shift and of indirect spin coupling, and we have seen a number of illustrations of high resolution NMR spectra. We now need to look more carefully at the way in which the effects of chemical shifts and spin coupling can be added to the basic treatment of spin physics that we studied in Chapter 2. In this chapter we explore the ways in which nuclei interact not only with the applied magnetic field but also with each other. The steady-state quantum mechanical approach of Chapter 2 can easily be expanded by using a Hamiltonian that includes chemical shifts and couplings.

The results can often be expressed in simple algebraic form and can be depicted graphically.

Before embarking on the theoretical treatment, it is important to clarify some aspects of symmetry and equivalence, which are critical in any quantum mechanical calculation. In the nuclear spin system we must consider both equivalence in chemical shifts and equivalence in coupling constants.

6.1 SYMMETRY AND EQUIVALENCE

Nuclei are said to be *chemically equivalent* when they have the same chemical shift, usually as a result of molecular symmetry (e.g., the 2 and 6 protons or the 3 and 5 protons in phenol) but occasionally as a result of an accidental coincidence of shielding effects. Nuclei in a set are *magnetically equivalent* when they all possess the same chemical shift *and* all nuclei in the set are coupled equally with *any* other single nucleus in the molecule. Thus, in the tetrahedral molecule difluoromethane (**I**) H_a and H_b *are* magnetically equivalent because by symmetry they must have

the same chemical shift, and they are by symmetry equally coupled to F_a and equally coupled to F_b. On the other hand, in 1,1-difluoroethylene (**II**) H_a and H_b are chemically equivalent but *are not* magnetically equivalent, because H_a and F_a are coupled by $J(cis)$, while H_b and F_a are coupled by $J(trans)$, and in general $J(cis) \neq J(trans)$. Note carefully that the test for equal coupling is to be made for *each* nucleus in the set being tested to a *specific* nucleus outside the set. Thus, the fact that the coupling between H_a and F_a is the same as that between H_b and F_b is irrelevant. Figure 6.1 shows markedly different spectra for **I** and **II**. With the simple quantum mechanical treatment that we develop in the next sections, we shall find that it is easy to account for these and other NMR spectra.

The term "equivalent nuclei" has been widely used, in some cases to denote chemical equivalence and in others magnetic equivalence. Other terms have been suggested to attempt to avoid confusion, for example, "isochronous" for nuclei with the same chemical shift. We shall always specify "chemically equivalent" or "magnetically equivalent" when there is any ambiguity.

When there is *rapid* internal motion in a molecule, such as internal rotation or inversion, the equivalence of nuclei should be determined on an overall average basis, rather than in one of the individual conformations. For example, in CH_3CH_2Br, the three CH_3 protons are magnetically equivalent because they couple equally on the average with each of the methylene protons, even though

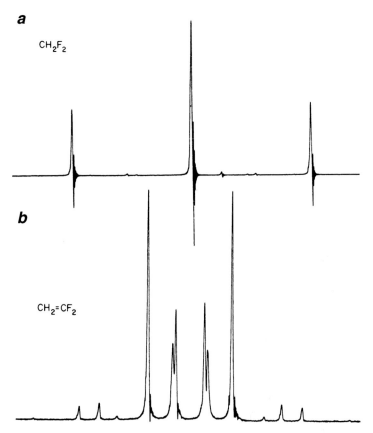

FIGURE 6.1 ^{1}H NMR spectra of (*a*) CH_2F_2 and (*b*) CH_2═CF_2. These spectra are obtained at 60 MHz but would be identical at 600 MHz. [The very weak lines in (*a*) and the two very weak lines in (*b*) are due to spinning sidebands and an impurity.]

in any one of the three stable conformations they would be magnetically non-equivalent. "Rapid" rotation refers to the NMR time scale for a dynamic process, as we discussed in Section 2.10.

Note, however, that in a molecule with an asymmetric center, rapid internal rotation does not guarantee magnetic equivalence even when nuclei are rendered chemically equivalent by the rotation. For example, in a 1,2-disubstituted ethane, the two protons attached to a given carbon atom are enantiotopic, hence chemically equivalent in an ordinary achiral solvent, as we saw in Section 4.9. Of the three conformations **III–V**, **III** and **IV** are mirror images, the average of which must have $\nu_A = \nu_B$ and $\nu_X = \nu_Y$, while **V** contains a plane of symmetry, so that the same equalities hold here. On the other hand, H_A and H_B are not magnetically equivalent, because the vicinal coupling constants are not necessarily averaged to the same value. For example, in conformation **III**, where H_A and H_X are coupled by a *trans* coupling constant, substituents P and R are adjacent, whereas in **V**, where H_A and H_Y are *trans* coupled, P and R are far apart. Thus J_{AX}(*trans*) is not necessarily equal to J_{AY}(*trans*), and similar inequalities hold for the other conformations. In many individual cases the differences in the average J's are so small that deviations from the simpler spectrum characteristic of magnetically equivalent nuclei are not observed.

6.2 NOTATION

Our treatment is considerably simplified by restricting ourselves to nuclei with $I = \frac{1}{2}$, as we can cover most of the general principles without tedious algebraic manipulations. In addition, nuclei with $I = \frac{1}{2}$ are studied far more extensively than others. It will be helpful to use the widely employed system of notation in which each nucleus of spin $\frac{1}{2}$ is denoted by some letter of the alphabet: A, B, X, etc. We choose letters of the alphabet representative of relative chemical shifts; that is, for two nuclei that have a small chemical shift relative to each other, we choose two letters of the alphabet that are close to each other, and for nuclei that have large relative chemical shifts, we use letters from opposite ends of the alphabet. (We define "small" and "large" chemical shifts later.) Different nuclear species (e.g., H and F) are also represented by letters from opposite ends of the alphabet, because their "chemical shifts" differ by many megahertz.

If two or more nuclei have identical chemical shifts, they must be denoted in one of two different ways, depending on whether they are magnetically equivalent nuclei or chemically equivalent. Magnetically equivalent nuclei are denoted by a subscript (e.g., A_2X). Nuclei that are chemically equivalent but not magnetically equivalent are denoted by repeating a letter with a prime or a double prime (e.g., $AA'X$). Nuclei with $I = 0$, as well as those that relax so rapidly that they behave as though they are not magnetic, such as Cl, Br, and I (see Section 5.2), are not given any designations.

A few examples should clarify this notation:

(a) $CH_2\!\!=\!\!CCl_2$ A_2
(b) CH_2F_2 A_2X_2
(c) $CH_2\!\!=\!\!CF_2$ $AA'XX'$
(d) CH_3OH A_3B or A_3X (depending on magnetic field and hydrogen bonding effect on δ_{OH}—see Section 4.5)
(e) $CH_2\!\!=\!\!CHCl$ ABX (or ABC at low magnetic field)

(f) $AA'BB'$ or $AA'XX'$ (depending on magnetic field and assuming no coupling between rings)

(g) $CH_3CH\!\!=\!\!CH_2$ A_3MXY
(h) $^{13}CH_2F_2$ A_2M_2X
(i) $CH_3CH_2NO_2$ A_3X_2 (assuming rapid internal rotation)
(j) $CH_2ClCH_2NO_2$ $AA'XX'$

Generally, no significance is attached to the order in which the nuclei are given. For example in (b) either the hydrogen or fluorine nuclei could be designated A; likewise, (d) could be called an A_3B or AB_3 system. Nuclei with different magnetogyric ratios (e.g., ^1H and ^{13}C) clearly should be denoted by letters far apart in the alphabet. For nuclei of the same species the notation depends on their chemical shieldings, the magnetic field, and the magnitude of the spin coupling between the nuclei. Specifically, nuclei 1 and 2 are said to be *weakly coupled* and constitute an AX system when $\nu_1 - \nu_2 \gg J_{12}$, where ν_1 and ν_2 are the chemical shifts expressed in Hz, not ppm. The nuclei are said to be *strongly coupled* and constitute an AB system when $\nu_1 - \nu_2 \leq J_{12}$. Thus, in an ABX system nuclei A and B are strongly coupled, while A and X are weakly coupled. Several additional examples will appear later in this chapter and in the problems at the end of the chapter. For molecules containing several sets of chemically equivalent nuclei, the notation becomes unwieldy, and other notations have been used.[78] We do not need such notation for the spin systems that we treat here.

6.3 ENERGY LEVELS AND TRANSITIONS IN AN AX SYSTEM

Before deriving quantitative expressions for general spin systems, we shall examine qualitatively the energy levels and transitions arising from two spin $\frac{1}{2}$ nuclei that are not coupled or are only weakly coupled (AX system). As usual, we take the static imposed magnetic field to lie along the z axis, and we express the orientation of the z component of nuclear spin I_z as α or β for $I_z = \frac{1}{2}$ or $-\frac{1}{2}$, respectively. A system of N nuclei of spin $\frac{1}{2}$ is described by the 2^N possible product functions; for the two-spin system the four functions are

$$\phi_1 = \alpha_1\alpha_2 = \alpha\alpha \qquad f_z = 1$$

$$\phi_2 = \alpha_1\beta_2 = \alpha\beta \qquad f_z = 0$$

$$\phi_3 = \beta_1\alpha_2 = \beta\alpha \qquad f_z = 0 \tag{6.1}$$

$$\phi_4 = \beta_1\beta_2 = \beta\beta \qquad f_z = -1$$

The shorter notation $\alpha\alpha$ is often used in place of the notation $\alpha_1\alpha_2$ when there is no chance of ambiguity. In such cases it is understood that the nuclei are always given in the same order.

The values f_z given in Eq. 6.1 are eigenvalues of an operator F_z, which is defined as the sum of the I_z operators for all spins:

$$F_z = (I_z)_1 + (I_z)_2 + \cdots = \sum_{i=1}^{N} (I_z)_i = \sum_{i=1}^{N} I_{zi} \tag{6.2}$$

The parentheses in $(I_z)_i$ make it clear that the notation refers to the z spin component of nucleus i, but most literature omits the parentheses, as indicated, for simplicity. We shall usually use the shortened notation in this and similar expressions. The effect of F_z on a product function is to generate a $+\frac{1}{2}$ every time an α appears and $-\frac{1}{2}$ every time a β appears, so that the eigenvalue f_z represents the total z component of spin of the function. The classification of basis functions according to total spin greatly simplifies the calculation of energy levels, as we shall see.

Case I. No Coupling between A and X

If we consider the imposed magnetic field as lying in the positive z direction, so that an α state has lower energy than a β state, the energies of the four states in Eq. 6.1 lie in the order given in the *center* portion of Fig. 6.2, which is drawn for no spin–spin coupling between the nuclei. States ϕ_1 and ϕ_3 differ in the "flipping" of spin A, as do states ϕ_2 and ϕ_4. Likewise, ϕ_1 and ϕ_2 and ϕ_3 and ϕ_4 differ in the spin orientation of X. Thus, transitions between the states may be labeled as A or X transitions, and these correspond to the NMR transitions discussed in Chapter 2. (We shall see later that the "double flip" transitions between ϕ_2 and ϕ_3 and between ϕ_1 and ϕ_4 are forbidden by selection rules.) If we arbitrarily take the resonance frequency of A to be greater than that of X, then the energy levels are spaced as indicated in Fig. 6.2, and the spectrum consists of two lines, as shown, with frequency increasing to the left. Each line arises from two separate transitions, but the transitions are degenerate, that is, coincident in energy.

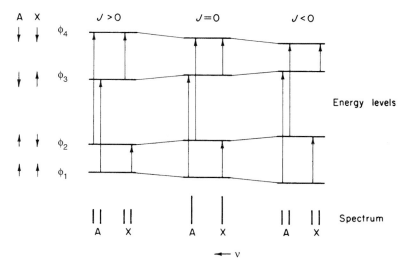

FIGURE 6.2 Schematic representation of the energy levels and spectrum of an AX system with and without spin coupling.

Case II. Weak Coupling

Suppose there is a coupling $J > 0$ between A and X. Because a positive coupling constant implies that antiparallel spin orientations possess less energy than the corresponding parallel orientations (see Section 5.2), ϕ_2 and ϕ_3 are lower in energy than they were in the absence of coupling, and ϕ_1 and ϕ_4 are higher in energy. This situation is depicted in the left portion of Fig. 6.2, but the magnitude of the changes in energy relative to the separations between levels is greatly exaggerated. The transitions no longer have the same energy, and a doublet results for the A portion of the spectrum; a similar doublet appears in the X portion. If $J < 0$, the energy levels change as shown in the right portion of Fig. 6.2. Again, two doublets appear in the spectrum. In this case the lower frequency A line results from a transition between ϕ_3 and ϕ_1, rather than between ϕ_4 and ϕ_2 as it did in the case $J > 0$, but there is no *observable* difference in the spectrum with change in the sign of the coupling constant.

With this qualitative picture in mind, we are now ready to address the problem of coupled spins quantitatively—without any restrictions to only weak coupling.

6.4 QUANTUM MECHANICAL TREATMENT

The standard quantum mechanical approach to any problem is to begin with a set of basis functions that could describe the spin system under certain specialized circumstances. The basis functions represent only a first approximation to the true wave functions. For example, simple molecular orbital treatments may start with hydrogen atom wave functions as the basis, even though they are likely to be a rather poor approximation. Fortunately, for nuclear spins, the problem is much simpler, and we can begin with basis functions that are a very good first approximation by using products of the exact wave functions for individual spins, as described subsequently.

The next step is to define the Hamiltonian for the system. For NMR it is easy to write down the Hamiltonian for a set of nuclei in a molecule that is tumbling rapidly. We know that each nucleus interacts with the magnetic field as we discussed in Chapter 2, that direct dipolar interactions average to zero and need not be considered, and that indirect spin coupling between any pair of nuclei can be handled as a scalar product. All of these factors are included in the Hamiltonian given.

Finally, we then use well-known mathematical procedures to determine the energy levels of the system and to describe the true wave functions in terms of linear combinations of the basis functions. As we shall see, this means that we set up the Hamiltonian as a matrix in terms of the spin product basis functions, then manipulate it into a form in which it is diagonal. The functions used to represent the Hamiltonian in diagonal form are, then, the true eigenfunctions, which are linear combinations of the basis functions.

Nuclear Spin Basis Functions

In Section 2.3 we described the properties of α and β, the basis functions for a spin $\frac{1}{2}$. The definitions given there and the relationships developed in that section are essential for the presentation in this chapter and might profitably be reviewed at this point.

For two or more noninteracting spins, the correct wave function is simply the product of the functions for the individual spins. For N spins, each with $I = \frac{1}{2}$, there are 2^N such product functions. For N spins that may interact with each other, it makes sense to start with the 2^N products of α and β as basis functions and to see how the spin interactions cause them to be modified.

The Spin Hamiltonian

The Hamiltonian for a single spin was given in Eq. 2.4. For an N-spin system, $\mathcal{H}$ may be thought of as containing two parts:

1. $\mathcal{H}^{(0)}$, which describes the interaction of each nuclear spin with the imposed magnetic field. $\mathcal{H}^{(0)}$ is the sum of N terms of the sort given in Eq. 2.4, with each

modified to include the effect of chemical shielding of each nucleus:

$$\mathcal{H}^{(0)} = -\sum_{i=1}^{N} \frac{\gamma_i}{2\pi} B_0 (1 - \sigma_i) I_{zi} \tag{6.3}$$

In this and subsequent expressions we drop Planck's constant h in order to convert to the more useful unit of hertz, rather than joules. To further simplify the notation, we substitute the Larmor relation, Eq. 2.14, into Eq. 6.3 to get

$$\mathcal{H}^{(0)} = -\sum_{i=1}^{N} \nu_i I_{zi} \tag{6.4}$$

2. $\mathcal{H}^{(1)}$, which describes the coupling between all possible pairs of nuclei. In accord with the discussion in Section 5.1, each such interaction is described by a scalar product of the spin operators, and the sum covers all possible pairwise interactions:

$$\mathcal{H}^{(1)} = \sum_{i=1}^{N} \sum_{j=i}^{N} J_{ij} \mathbf{I}_i \cdot \mathbf{I}_j$$

$$\mathcal{H}^{(1)} = \sum_{i=1}^{N} \sum_{j=i}^{N} J_{ij} [I_{xi} I_{xj} + I_{yi} I_{yj} + I_{zi} I_{zj}] \tag{6.5}$$

Equation 6.5 is just the expanded form of a vector dot product and is more useful for later computations. For N nuclei the sum includes $N(N - 1)$ terms. For example, for a system of four nuclei there are four terms in $\mathcal{H}^{(0)}$ and six terms in $\mathcal{H}^{(1)}$.

For molecules that are tumbling rapidly so that magnetic dipole interactions can be neglected, the sum of $\mathcal{H}^{(0)}$ and $\mathcal{H}^{(1)}$ is adequate as the complete Hamiltonian to determine energy levels for the nuclear spin system. However, as we noted in Section 5.5, additional terms must be added to account for external perturbations, such as strong rf fields. In this chapter we take the steady-state Hamiltonian as

$$\mathcal{H} = \mathcal{H}^{(0)} + \mathcal{H}^{(1)} \tag{6.6}$$

Energy Levels and Eigenfunctions

A nuclear spin system may be described by a steady-state wave function arising from the Hamiltonian operator $\mathcal{H}$ satisfying the time-independent Schrödinger equation

$$\mathcal{H}\psi = E\psi \tag{6.7}$$

E gives the energy of the system, which turns out to be quantized into discrete energy levels.

As shown in standard quantum mechanics texts, a straightforward means for solving the Schrödinger equation is to express $\mathcal{H}$ as a matrix in an appropriate basis — in this case the 2^N product functions described earlier — and then to diagonalize the matrix. The diagonal terms give the energy levels (eigenvalues), and the eigenfunctions that form the basis of the diagonalized $\mathcal{H}$ are simple linear combinations of the basic product functions. In practice, this process is equivalent to solving the secular equation, which is a determinantal equation of the form

$$\left| \mathcal{H}_{mn} - E\delta_{mn} \right| = 0 \tag{6.8}$$

where the $\mathcal{H}_{mn}$ are the matrix elements of the Hamiltonian

$$\mathcal{H}_{mn} = \left\langle \phi_m \right| \mathcal{H} \left| \phi_n \right\rangle \tag{6.9}$$

and the Kronecker delta is

$$\delta_{mn} = \begin{array}{ll} 1, & m = n \\ 0, & m \neq n \end{array} \tag{6.10}$$

In expanded form, the determinant is

$$\begin{vmatrix} \mathcal{H}_{11} - E & \mathcal{H}_{12} & \mathcal{H}_{13} & \cdots \\ \mathcal{H}_{21} & \mathcal{H}_{22} - E & \mathcal{H}_{23} & \cdots \\ \mathcal{H}_{31} & \mathcal{H}_{32} & \mathcal{H}_{33} - E & \cdots \\ \cdots & \cdots & \cdots & \cdots \end{vmatrix} = 0 \tag{6.11}$$

The quantum mechanical computations here are extremely simple and furnish an excellent example of such calculations. The simplicity arises from the facts that (1) we need utilize only spin coordinates, not spatial coordinates, that is, only the x, y, and z components of the nuclear spin I (or equivalently the nuclear magnetic moment μ, because Eqs. 2.1 and 2.2 show that μ is just a constant times I), and (2) for nuclear spin interactions the determinant is $2^N \times 2^N$, not infinitely large as in many other quantum mechanical problems.

6.5 THE TWO-SPIN SYSTEM WITHOUT COUPLING

We now apply the concepts developed in the preceding section to the system of just two nuclei, first considering the case in which there is no spin coupling between them. We digress from our usual notation to call the two spins A and B, rather than A and X, because we later wish to use some of the present results in treating the coupled AB system. The four product basis functions are given in Eq. 6.1. We now compute the matrix elements needed for the secular determinant. Because there are four basis functions, the determinant is 4×4 in size, with 16 matrix elements. Many of these will turn out to be zero. For $\mathcal{H}_{11}$ we have, from

Eqs. 6.1, 6.4, and 6.9,

$$\mathcal{H}_{11} = \langle \phi_1 | \mathcal{H}^{(0)} | \phi_1 \rangle$$
$$= -\langle \alpha_A \beta_B | \nu_A I_{zA} + \nu_B I_{zB} | \alpha_A \beta_B \rangle$$
$$= -\nu_A \langle \alpha_A \alpha_B | I_{zA} | \alpha_A \alpha_B \rangle - \nu_B \langle \alpha_A \alpha_B | I_{BA} | \alpha_A \alpha_B \rangle$$
$$= -\nu_A \langle \alpha_A | I_{zA} | \alpha_A \rangle \langle \alpha_B | \alpha_B \rangle - \nu_B \langle \alpha_A | \alpha_A \rangle \langle \alpha_B | I_{zB} | \alpha_B \rangle \tag{6.12}$$

Using the relations in Eqs. 2.6 and 2.9, we obtain

$$\mathcal{H}_{11} = -\nu_A(\tfrac{1}{2})(1) - \nu_B(1)(\tfrac{1}{2}) = -\tfrac{1}{2}(\nu_A + \nu_B) \tag{6.13}$$

By the same procedure, the other three *diagonal* matrix elements (those on the principal diagonal) may be evaluated as

$$\mathcal{H}_{22} = -\tfrac{1}{2}(\nu_A - \nu_B)$$
$$\mathcal{H}_{33} = -\tfrac{1}{2}(-\nu_A + \nu_B) \tag{6.14}$$
$$\mathcal{H}_{44} = \tfrac{1}{2}(\nu_A + \nu_B)$$

In the absence of spin coupling, all *off-diagonal* elements are zero by virtue of the orthogonality of α and β. This is a general theorem, not restricted to the case of two nuclei. We shall illustrate the result for $\mathcal{H}_{12}$:

$$\mathcal{H}_{12} = \langle \phi_1 | \mathcal{H}^{(0)} | \phi_2 \rangle$$
$$= -\langle \alpha_A \alpha_B | \nu_A I_{zA} + \nu_B I_{zB} | \alpha_A \beta_B \rangle$$
$$= -\nu_A \langle \alpha_A \alpha_B | I_{zA} | \alpha_A \beta_B \rangle - \nu_B \langle \alpha_A \alpha_B | I_{zB} | \alpha_A \beta_B \rangle$$
$$= -\nu_A \langle \alpha_A | I_{zA} | \alpha_A \rangle \langle \alpha_B | \beta_B \rangle - \nu_B \langle \alpha_A | \alpha_A \rangle \langle \alpha_B | I_{zB} | \beta_B \rangle$$
$$= -\nu_A(\tfrac{1}{2})(0) - \nu_B(1)(0) = 0 \tag{6.15}$$

With all off-diagonal elements equal to zero, the secular determinant becomes

$$\begin{vmatrix} \mathcal{H}_{11} - E & 0 & 0 & 0 \\ 0 & \mathcal{H}_{22} - E & 0 & 0 \\ 0 & 0 & \mathcal{H}_{33} - E & 0 \\ 0 & 0 & 0 & \mathcal{H}_{44} - E \end{vmatrix} = 0 \tag{6.16}$$

It is well known that a determinant in the form of blocks connected with each other only by zeros may be written as the product of factors. Equation 6.16 can thus be factored into four 1×1 blocks, solutions of which are

$$E_1 = \mathcal{H}_{11} = -\tfrac{1}{2}(\nu_A + \nu_B)$$
$$E_2 = \mathcal{H}_{22} = -\tfrac{1}{2}(\nu_A - \nu_B)$$
$$E_3 = \mathcal{H}_{33} = \tfrac{1}{2}(\nu_A - \nu_B) \tag{6.17}$$
$$E_4 = \mathcal{H}_{44} = \tfrac{1}{2}(\nu_A + \nu_B)$$

These energies are easily associated with the four energy levels in Fig. 6.2 in our qualitative discussion of the AX system. The four allowed transitions, which are also shown in Fig. 6.2, correspond to the selection rule $\Delta F_z = \pm 1$. (F_z was defined in Eq. 6.2.) The origin of this selection rule will be taken up later. It is clear from the expressions in Eq. 6.17 that each observed line results from two transitions with precisely the same frequency, just as depicted qualitatively in Fig. 6.2.

Because the Hamiltonian is diagonal in the original (ϕ_n) basis set, these functions are the true wave functions. This result is completely general for any case where there is no spin coupling. As we shall see, it is only the spin coupling interaction, not the chemical shift, that causes the basis functions to mix. However, as we shall see in the next section, even with coupling many of the basis functions do not mix.

6.6 FACTORING THE SECULAR EQUATION

Before extending our treatment to two *coupled* nuclei, it is helpful to consider some general conditions that cause zero elements to appear in the secular equation. With this knowledge, we can avoid the effort of calculating many of the elements specifically for each case we study; furthermore, the presence of zero elements usually results in the secular determinant being factored into several equations of much smaller order, the solution of which is simpler than that of a high order equation.

Suppose the basis functions used to construct the secular equation, that is, the ϕ_n, are eigenfunctions of some operator F. Consider two of these functions, ϕ_m and ϕ_n, with eigenvalues f_m and f_n, respectively. Then

$$F\phi_m = f_m\phi_m \qquad F\phi_n = f_n\phi_n \tag{6.18}$$

Suppose further that the operator F commutes with the Hamiltonian:

$$F\mathcal{H} - \mathcal{H}F = 0 \tag{6.19}$$

From these premises it is shown in standard texts on quantum mechanics that

$$(f_m - f_n)\langle\phi_m|\mathcal{H}|\phi_n\rangle = 0$$
$$(f_m - f_n)\mathcal{H}_{mn} = 0 \tag{6.20}$$

Thus if $f_m - f_n \neq 0$, $\mathcal{H}_{mn} = 0$; that is, if ϕ_m and ϕ_n have *different* eigenvalues of F, the matrix element connecting them in the secular equation must be zero. (If $f_m = f_n$, nothing can be said from Eq. 6.20 about the value of $\mathcal{H}_{mn}$.)

Two types of operators F are important in the treatment of nuclear spin systems. One is the class of operators describing the symmetry of many molecules.

We shall defer a discussion of symmetry until Section 6.11. The other operator is F_z, which was defined in Eq. 6.2. Because α and β are eigenfunctions of I_z, the product functions ϕ_n are eigenfunctions of F_z. By using the well-established commutation rules for angular momentum, it can be shown that F_z and $\mathcal{H}$ commute, so Eq. 6.20 is applicable. For the two-spin case, Eq. 6.1 shows that the four basis functions are classified according to $F_z = 1$, 0, or -1. Only ϕ_2 and ϕ_3, which have the same value of F_z, can mix. Thus only $\mathcal{H}_{23}$ and $\mathcal{H}_{32}$ might be nonzero; all 10 other off-diagonal elements of the secular equation are clearly zero and need not be computed.

6.7 TWO COUPLED SPINS

We are now in position to complete our calculation of the AB system in general, with no restrictions whatsoever regarding the magnitude of the coupling constant J_{AB}. By virtue of the factoring due to F_z, the secular equation is

$$\begin{vmatrix} \mathcal{H}_{11} - E & 0 & 0 & 0 \\ 0 & \mathcal{H}_{22} - E & \mathcal{H}_{23} & 0 \\ 0 & \mathcal{H}_{32} & \mathcal{H}_{33} - E & 0 \\ 0 & 0 & 0 & \mathcal{H}_{44} - E \end{vmatrix} = 0 \qquad (6.21)$$

The secular equation is always symmetric about the principal diagonal; hence $\mathcal{H}_{23} = \mathcal{H}_{32}$. We thus have five matrix elements to evaluate. Because $\mathcal{H} = \mathcal{H}^{(0)} + \mathcal{H}^{(1)}$, we need evaluate only the portion of the matrix elements arising from $\mathcal{H}^{(1)}$ and then add the portion evaluated in Section 6.5 from $\mathcal{H}^{(0)}$.

For the first matrix element we find, using Eqs. 6.5 and 6.9,

$$\mathcal{H}_{11}^{(1)} = \langle \phi_1 | \mathcal{H}^{(1)} | \phi_1 \rangle$$

$$= J_{AB}\langle \alpha_A \alpha_B | I_{xA}I_{xB} + I_{yA}I_{yB} + I_{zA}I_{zB} | \alpha_A \alpha_B \rangle$$

$$= J_{AB}[\langle \alpha_A | I_{xA} | \alpha_A \rangle \langle \alpha_B | I_{xB} | \alpha_B \rangle + \langle \alpha_A | I_{yA} | \alpha_A \rangle \langle \alpha_B | I_{yB} | \alpha_B \rangle$$

$$\quad + \langle \alpha_A | I_{zA} | \alpha_A \rangle \langle \alpha_B | I_{zB} | \alpha_B \rangle]$$

$$= J_{AB}[0 + 0 + (\tfrac{1}{2})(\tfrac{1}{2})]$$

$$= \tfrac{1}{4} J_{AB} \qquad (6.22)$$

Similar computations for the other diagonal elements give

$$\mathcal{H}_{22}^{(1)} = -\tfrac{1}{4} J_{AB}$$

$$\mathcal{H}_{33}^{(1)} = -\tfrac{1}{4} J_{AB} \qquad (6.23)$$

$$\mathcal{H}_{44}^{(1)} = \tfrac{1}{4} J_{AB}$$

The lone nonzero off-diagonal element may be evaluated in a similar manner:

$$
\begin{aligned}
\mathcal{H}_{23}{}^{(1)} &= \langle \phi_2 | \mathcal{H}^{(1)} | \phi_3 \rangle \\
&= J_{AB} \langle \alpha_A \beta_B | I_{xA} I_{xB} + I_{yA} I_{yB} + I_{zA} I_{zB} | \beta_A \alpha_B \rangle \\
&= J_{AB} [\langle \alpha_A | I_{xA} | \beta_A \rangle \langle \beta_B | I_{xB} | \alpha_B \rangle + \langle \alpha_A | I_{yA} | \beta_A \rangle \langle \beta_B | I_{yB} | \alpha_B \rangle \\
&\quad + \langle \alpha_A | I_{zA} | \beta_A \rangle \langle \beta_B | I_{zB} | \alpha_B \rangle] \\
&= J_{AB} [(\tfrac{1}{2})(\tfrac{1}{2}) + (\tfrac{1}{2}i)(-\tfrac{1}{2}i) + 0] \\
&= \tfrac{1}{2} J_{AB}
\end{aligned}
\tag{6.24}
$$

Adding the contributions from $\mathcal{H}^{(0)}$ gives the complete matrix elements:

$$
\begin{aligned}
\mathcal{H}_{11} &= -\tfrac{1}{2}(\nu_A + \nu_B) + \tfrac{1}{4} J_{AB} \\
\mathcal{H}_{22} &= -\tfrac{1}{2}(\nu_A - \nu_B) - \tfrac{1}{4} J_{AB} \\
\mathcal{H}_{33} &= \tfrac{1}{2}(\nu_A - \nu_B) - \tfrac{1}{4} J_{AB} \\
\mathcal{H}_{44} &= \tfrac{1}{2}(\nu_A + \nu_B) + \tfrac{1}{4} J_{AB} \\
\mathcal{H}_{23} &= \tfrac{1}{2} J_{AB}
\end{aligned}
\tag{6.25}
$$

When these values are inserted into Eq. 6.21, it factors into three blocks:

$$
[-\tfrac{1}{2}(\nu_A + \nu_B) + \tfrac{1}{4} J_{AB}] - E = 0
\tag{6.26}
$$

$$
\begin{vmatrix}
-\tfrac{1}{2}(\nu_A - \nu_B) - \tfrac{1}{4} J_{AB} - E & \tfrac{1}{2} J_{AB} \\
\tfrac{1}{2} J_{AB} & \tfrac{1}{2}(\nu_A - \nu_B) - \tfrac{1}{4} J_{AB} - E
\end{vmatrix} = 0
\tag{6.27}
$$

$$
[\tfrac{1}{2}(\nu_A + \nu_B) + \tfrac{1}{4} J_{AB}] - E = 0
\tag{6.28}
$$

Equations 6.26 and 6.28 immediately give the values of two energy levels, E_1 and E_4. Equation 6.27 is a quadratic equation that is readily solved to give the values of E_2 and E_3. The four energy levels are

$$
\begin{aligned}
E_1 &= -\tfrac{1}{2}[\nu_A + \nu_B] + \tfrac{1}{4} J_{AB} \\
E_2 &= -\tfrac{1}{2}[(\nu_A - \nu_B)^2 + J_{AB}^2]^{1/2} - \tfrac{1}{4} J_{AB} \\
E_3 &= \tfrac{1}{2}[(\nu_A - \nu_B)^2 + J_{AB}^2]^{1/2} - \tfrac{1}{4} J_{AB} \\
E_4 &= \tfrac{1}{2}[\nu_A + \nu_B] + \tfrac{1}{4} J_{AB}
\end{aligned}
\tag{6.29}
$$

One point regarding the expressions in Eq. 6.29 deserves further comment: If $(\nu_A - \nu_B) \gg J_{AB}$ (the general AX case), J_{AB}^2 is negligible compared with

$(\nu_A - \nu_B)^2$ and may be dropped in Eq. 6.29. The resultant expressions for E_2 and E_3 are then considerably simplified. Note that the same result could have been obtained when the secular equation was written by dropping the off-diagonal elements $\frac{1}{2}J_{AB}$, which are much smaller than the diagonal elements for an AX spin system. Dropping such small off-diagonal terms leads to a factorization of the secular equation beyond that given by the factorization according to F_z. This is a general and very important procedure, applicable to cases in which certain differences in chemical shifts (expressed in Hz) are large compared with the corresponding J's. This is tantamount to a classification of the basis functions not only according to the total F_z but also according to $F_z(G)$, where G refers to a set of strongly coupled nuclei. The definition of $F_z(G)$ and the application to the ABX case will be taken up in Section 6.13.

Wave Functions

Now that we have expressions for the four energy levels, we can examine the wave functions $\psi_1 - \psi_4$ corresponding to each energy level. For states (1) and (4) we have seen that the basis functions ϕ_i do not mix, hence $\psi_1 = \phi_1$ and $\psi_4 = \phi_4$. However, ψ_2 and ψ_3 are linear combinations (mixtures) of ϕ_2 and ϕ_3, the extent of mixing depending on the ratio $(\nu_A - \nu_B)/J_{AB}$, as shown in the following expressions:

$$\psi_1 = \phi_1$$

$$\psi_2 = \frac{1}{(1 + Q^2)^{1/2}} (\phi_2 + Q\phi_3)$$

$$\psi_3 = \frac{1}{(1 + Q^2)^{1/2}} (-Q\phi_2 + \phi_3) \qquad (6.30)$$

$$\psi_4 = \phi_4$$

In Eq. 6.30, Q is

$$Q = \frac{J_{AB}}{(\nu_A - \nu_B) + [(\nu_A - \nu_B)^2 + J_{AB}^2]^{1/2}} \qquad (6.31)$$

The calculation of the wave functions ψ from the Hamiltonian matrix uses standard matrix manipulations, which are not reproduced here. We shall discuss the spectrum arising from an AB system in detail in Section 6.8.

Selection Rules and Intensities

To predict the spectrum for a two-spin system we now need only one more piece of information—the requirements for transitions between two energy levels.

Equations 2.12 and 2.17 provided the results of using time-dependent perturbation theory to assess the interaction between the x component of magnetization (proportional to I_x) and the applied rf field B_1. The essential qualitative result for the single nucleus then being considered was the selection rule

$$\Delta m = \pm 1 \tag{6.32}$$

where m is an eigenvalue of I_z. For a set of N nuclei, the operator for the rf interaction is a summation of I_x for all nuclei. It is easy to show that for N noninteracting nuclei, the selection rule for a specific transition is $\Delta m_i = \pm 1$ for one nucleus and $\Delta m_j = 0$ for all others; that is, only one nuclear spin "flips" during a single transition. For coupled nuclei, where wave functions are mixed, as we have seen, the selection is expressed more generally:

$$\Delta f_z = \pm 1 \tag{6.33}$$

where f_z is an eigenvalue of the total spin operator F_z.

Equation 6.33 is completely general. For the two-spin system, it results in the transitions we identified in Fig. 6.2, while the *double quantum transition* between ϕ_1 and ϕ_4, and the *zero quantum transition* between ϕ_2 and ϕ_3 are forbidden. Note that this statement is true for this treatment, which employs stationary state wave functions and time-dependent perturbations, but as we shall see in Chapter 11, it is easy with suitable pulse sequences to elicit information on zero quantum and quantum double processes. For our present purposes in the remainder of this chapter we accept the validity of Eq. 6.33.

6.8 THE AB SPECTRUM

For the AB spin system we have seen that there are four allowed transitions: $\phi_3 \rightarrow \phi_1$, $\phi_4 \rightarrow \phi_2$, $\phi_4 \rightarrow \phi_3$, and $\phi_2 \rightarrow \phi_1$. From the expressions that we derived for energy levels we can easily determine the frequencies of these four transitions, and from the expressions for the wave functions and transition probabilities, we can compute the relative intensities of the four lines.

Frequencies of Lines

From the differences in the energy level expressions, Eqs. 6.26, 6.28, and 6.29, the frequencies on the lines $\nu_1 - \nu_4$ are given by

$$\nu_1 = E_4 - E_2 = \tfrac{1}{2}(\nu_A + \nu_B) + \tfrac{1}{2}[(\nu_A - \nu_B)^2 + J_{AB}^2]^{1/2} + \tfrac{1}{2}J_{AB}$$

$$\nu_2 = E_3 - E_1 = \tfrac{1}{2}(\nu_A + \nu_B) + \tfrac{1}{2}[(\nu_A - \nu_B)^2 + J_{AB}^2]^{1/2} - \tfrac{1}{2}J_{AB}$$

TABLE 6.1 Transitions, Frequencies, and Relative Intensities for the AB System

Line	Transition	Frequency (Hz)[a]	Relative intensity
1	$T_{2\rightarrow4}$	$C + \frac{1}{2}J$	$1 - \dfrac{J}{2C}$
2	$T_{1\rightarrow3}$	$C - \frac{1}{2}J$	$1 + \dfrac{J}{2C}$
3	$T_{3\rightarrow4}$	$-C + \frac{1}{2}J$	$1 + \dfrac{J}{2C}$
4	$T_{1\rightarrow2}$	$-C - \frac{1}{2}J$	$1 - \dfrac{J}{2C}$

[a]Referred to the center of the four-line pattern, $\frac{1}{2}(\nu_A + \nu_B)$.

$$\nu_3 = E_4 - E_3 = \tfrac{1}{2}(\nu_A + \nu_B) - \tfrac{1}{2}[(\nu_A - \nu_B)^2 + J_{AB}^2]^{1/2} + \tfrac{1}{2}J_{AB}$$
$$\nu_4 = E_2 - E_1 = \tfrac{1}{2}(\nu_A + \nu_B) - \tfrac{1}{2}[(\nu_A - \nu_B)^2 + J_{AB}^2]^{1/2} - \tfrac{1}{2}J_{AB}$$

$$(6.34)$$

The four lines described by Eq. 6.34 are symmetrically spaced about the average of the chemical shifts of A and B. Using that value as the origin, dropping the subscript on J because there is only one coupling constant here, and simplifying the expression with the substitution

$$C = \tfrac{1}{2}[(\nu_A - \nu_B)^2 + J^2]^{1/2} \qquad (6.35)$$

We obtain the values listed in Table 6.1, and the spectrum is illustrated in Fig. 6.3. It is apparent that lines 1 and 2 are always separated by J, as are lines 3 and 4. Thus the value of J may be extracted immediately from an AB spectrum.

In general, the calculation ν_A and ν_B for an AB spectrum requires slightly more effort. By squaring both sides of Eq. 6.35 and rearranging terms,

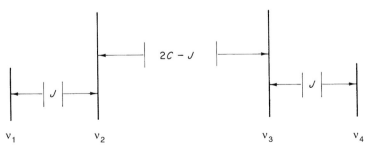

FIGURE 6.3 Schematic representation of an AB spectrum.

we obtain

$$(\nu_A - \nu_B)^2 = 4C^2 - J^2$$

$$|\nu_A - \nu_B| = [4C^2 - J^2]^{1/2}$$

$$= [(2C - J)(2C + J)]^{1/2}$$

$$= [(\nu_2 - \nu_3)(\nu_1 - \nu_4)]^{1/2} \qquad (6.36)$$

The absolute value sign is used, because the spectrum is symmetric and provides no information on whether ν_A or ν_B is larger. Note in Eq. 6.36 and Fig. 6.3 that $|\nu_A - \nu_B|$ is equal to the *geometric average* of the spacing of the outer and inner lines. It is *not* equal to the spacing between the average of the two outer "doublets" unless J^2 is negligible relative to $(\nu_A - \nu_B)^2$.

Intensities of Lines

To find the relative intensities of the four AB lines we use the results of time-dependent perturbation theory, as described in Section 2.3. For N spins, the probability P_{mn} of a transition between states m and n is given by a matrix element of the x component of magnetization or spin:

$$P_{mn} = \left\langle \psi_m \middle| \sum_{i=1}^{N} \gamma_i I_{xi} \middle| \psi_n \right\rangle \qquad (6.37)$$

For the AB system there are only two terms in the summation, and we can omit γ, as we are interested only in relative intensities. For the transition between states ψ_1 and ψ_2, we use the wave functions of Eq. 6.30 to obtain

$$P_{12} = \left\langle \alpha\alpha \middle| I_{xA} + I_{xB} \middle| \frac{1}{(1 + Q^2)^{1/2}} (\alpha\beta + Q\beta\alpha) \right\rangle \qquad (6.38)$$

Evaluation of P_{12} follows the same general procedure of expansion and simplification of terms that we showed in detail in Eqs. 6.15 and 6.22. For the four transition probabilities we obtain

$$P_{13} = P_{34} = \frac{1 + Q}{2(1 + Q^2)^{1/2}}$$

$$P_{12} = P_{24} = \frac{1 - Q}{2(1 + Q^2)^{1/2}} \qquad (6.39)$$

Line intensities are proportional to the square of the P's, and we can drop the normalizing factor in the denominator to show that the relative intensities depend on $(1 \pm Q)^2$. By substituting the definitions for Q (Eq. 6.31) and C (Eq. 6.35), we obtain the simple expressions given in Table 6.1. As shown in Table 6.1 and Fig. 6.3, the inner lines of the quartet are always more intense than the outer lines. In fact, the expressions in Table 6.1 show that for any AB spectrum, the ratio of the *intensities* of inner lines to outer lines is $(2C + J)/(2C - J)$, which is just the ratio of the *spacing* of the outer/inner lines.

6.9   AX, AB, AND A_2 Spectra

Let us examine the AB spectrum as the ratio $(\nu_A - \nu_B)/J$ varies. If the ratio is very large, we have an AX system, and as illustrated in Figure 6.4, the spectrum consists of two well-separated doublets, and $(\nu_A - \nu_X)$ can be measured directly from the midpoints of the doublets. As the ratio decreases somewhat, the spectrum is still essentially AX, but the intensities of the inner lines of the quartet are slightly greater than those of the outer lines. If $\nu_A - \nu_B \sim J$, we find the "classic" AB spectrum, as described also in Fig. 6.3. For $\nu_A - \nu_B < J$, the inner lines grow in intensity, while the outer lines may become almost unobservable in a noisy spectrum. Finally, when $\nu_A = \nu_B$ (i.e., an A_2 system), the spectrum degenerates to a single line. For this situation, where from Eq. 6.35, $2C = J$, Table 6.1 shows that we are unable to extract the value of J because the outer lines have zero intensity.

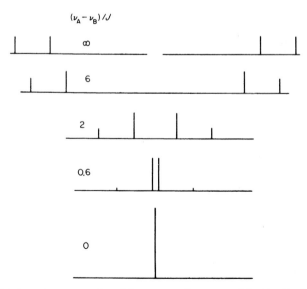

FIGURE 6.4 Schematic representation of the spectrum from two nuclei with a fixed coupling constant J and a varying value of $\nu_A - \nu_B$.

Magnetically Equivalent Nuclei

The A_2 spin system is just one illustration of the lack of splitting due to spin coupling in spectra where certain types of equivalence exist among the nuclei. We shall state without proof an extremely important NMR theorem and its corollary:

1. The spectrum arising from a system of nuclear spins that includes one or more sets of *magnetically* equivalent nuclei is independent of the spin coupling between nuclei within a magnetically equivalent set. For example, the spectrum of an A_3B system does not in any way depend on J_{AA}. However, the spectrum of an $AA'BB'$ spectrum is a function of $J_{AA'}$, because the two A nuclei are not magnetically equivalent. The proof of this theorem depends on commutation properties of angular momentum operators.

2. A molecule in which *all* coupled nuclei have the same chemical shift gives a spectrum consisting of a single line. For example, CH_4, which has four magnetically equivalent protons, gives a single proton resonance line; but CH_3D does not, because here one of the set of coupled nuclei clearly has a "chemical shift" different from the rest. This corollary does not depend on the reason for the chemical equivalence, which may result from molecular symmetry or from an accidental equivalence in shielding. When *all* coupled nuclei in a molecule are chemically equivalent, they are also magnetically equivalent, as they are equally coupled ($J = 0$) to all *other* nuclei.

6.10 "FIRST-ORDER" SPECTRA

In the early history of high resolution NMR, the theory was developed by use of perturbation theory. First-order perturbation theory was able to explain certain spectra, but second-order perturbation theory was needed for other cases, including the AB system. Spectra amenable to a first-order perturbation treatment give very simple spectral patterns ("first-order" spectra), as described in this section. More complex spectra are said to arise from "second-order effects."

Most introductory treatments of NMR describe coupling in the simple terms we used in Section 6.3 to discuss the AX system. That description can readily be extended to systems in which there are several magnetically equivalent nuclei. For example, consider acetaldehyde, CH_3CHO, in which equal coupling occurs (assuming rapid internal rotation) between each of the three CH_3 protons and the aldehyde proton. For this A_3X system, the molecules can be thought of as divided into two classes, depending on the spin orientation of X, just as in the AX system of Section 6.3, and the A resonance appears here also as a doublet. However, there are now four classes of A spin orientation—the molecules with all three A spins in the α state, that is, $\alpha_A\alpha_A\alpha_A$ ($f_z = {}^3/_2$), $\alpha_A\alpha_A\beta_A$ ($f_z = {}^1/_2$), $\alpha_A\beta_A\beta_A$ ($f_z = -{}^1/_2$), and $\beta_A\beta_A\beta_A$ ($f_z = -{}^3/_2$), with the $f_z = {}^1/_2$ and $-{}^1/_2$ states being statistically

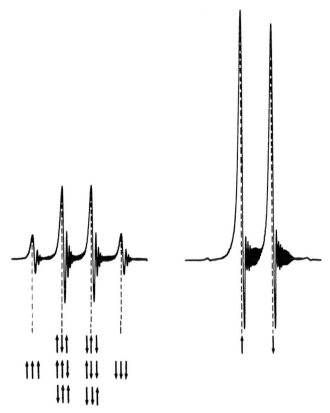

FIGURE 6.5 ^{1}H NMR spectrum of acetaldehyde. Right: CH_3 resonance, with the two spin states of the aldehyde proton indicated. Left, CHO resonance, with the four spin orientations of the methyl protons indicated.

three times as probable. Because each transition is equally probable, the X spectrum consists of a 1:3:3:1 quartet, as shown in Fig. 6.5.

From our discussion in the preceding sections, it is clear that this explanation cannot describe NMR spectra in general. However, it is correct and does account for the spectrum provided that two conditions are simultaneously satisfied:

1. *The chemical shift difference in hertz between nuclei (or groups of nuclei) must be much larger than the spin coupling between them.* (We give a quantitative sense to the phrase "much larger than" in the following section.) Because chemical shifts expressed in frequency units increase linearly with applied field (and radio frequency), spectra obtained at higher field strength are more likely to adhere to this relation than those obtained at lower field strength. At the operating frequencies of most modern NMR spectrometers (>200 MHz for ^{1}H) this condition is met

for a large majority of ^{1}H spin systems, but there are many instances in which the condition is not met. For example, the protons in CH_2 groups in alicyclic systems and in other asymmetric molecules often differ only slightly in chemical shift and display AB patterns, sometimes as part of a more complex spectrum.

2. *Coupling must involve groups of nuclei that are magnetically equivalent, not just chemically equivalent.* As this simple treatment deals with *the* coupling between A and X, each of the A nuclei is assumed to be equally coupled to X, hence the A nuclei are magnetically equivalent. However, in a number of instances a set of nuclei turns out to be only chemically equivalent, not magnetically equivalent, and the first-order analysis is inapplicable. Because high field strength does not in any way convert chemical equivalence to magnetic equivalence, we frequently encounter situations in which condition 2 is not satisfied. We see several examples in Section 6.16.

When first-order analysis is applicable, the number of components in a multiplet, their spacing, and their relative intensities can be determined easily from the following rules:

1. A nucleus or group of nuclei coupled to a set of n nuclei with spin I will have its resonance split into $2nI + 1$ lines. For the common case of $I = \frac{1}{2}$, there are then $n + 1$ lines.

2. The relative intensities of the $2nI + 1$ lines can be determined from the number of ways each spin state may be formed. For the case of $I = \frac{1}{2}$, the intensities of the $n + 1$ lines correspond to the coefficients of the binomial theorem, as indicated in Table 6.2.

3. The $2nI + 1$ lines are equally spaced, with the frequency separation between adjacent lines being equal to J, the coupling constant.

TABLE 6.2 Relative Intensities of First-Order Multiplets from Coupling with n Nuclei of Spin $\frac{1}{2}$

n	Relative intensity								
0					1				
1					1	1			
2				1	2	1			
3				1	3	3	1		
4			1	4	6	4	1		
5			1	5	10	10	5	1	
6		1	6	15	20	15	6	1	
7		1	7	21	35	35	21	7	1
8	1	8	28	56	70	56	28	8	1

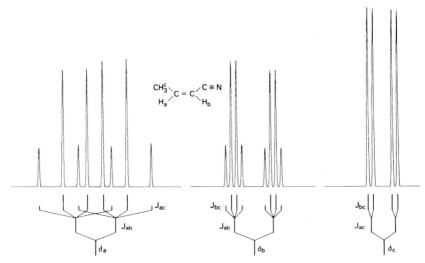

FIGURE 6.6 Repetitive first-order splitting analysis applied to a simulated spectrum of *cis*-CH₃CH═CHCN.

4. Coupling between nuclei within a magnetically equivalent set does not affect the spectrum.

For sets of magnetically equivalent nuclei, first-order analysis is usually considered applicable when $(\nu_A - \nu_B)/J_{AB} \gtrsim \sim 7$. [When $7 < (\nu_A - \nu_B)/J_{AB} < 20$, there is some distortion of intensities from the pattern given in Table 6.2, but the multiplet is still recognizable.] The deviation in intensities always occurs in the direction of making the lines near the center of the overall spectrum more intense and those toward the edges less intense.

When two or more couplings are present that may be treated by the first-order rules, a repetitive procedure can be used. For example, Fig. 6.6 gives an illustration of the repetitive application of first-order analysis. Usually it is convenient to consider the largest coupling first, but it is immaterial to the final result.

6.11 SYMMETRY OF SPIN WAVE FUNCTIONS

The presence of symmetry in a molecule imposes severe restrictions on many chemical and spectral properties. Often the existence of symmetry permits considerable simplification in the analysis of NMR spectra.

We speak of a *symmetry operation* $\Re$ as an operation that can be applied to a molecule to leave it in a configuration that is physically indistinguishable from the original configuration. With regard to wave functions, it is ψ^2 that corresponds to a physically measurable quantity, not ψ itself. Hence we require that

$$\Re\psi^2 = \psi^2 \tag{6.40}$$

For a nondegenerate wave function, which is to remain normalized, Equation 6.40 is valid *only* if the new ψ following the symmetry operation is either equal to the original ψ or is its negative:

$$\Re\psi = +\psi \quad \text{or} \quad -\psi \tag{6.41}$$

A function that is unchanged by the operation is said to be *symmetric*; one changed into its negative is *antisymmetric*. A function that is *asymmetric* (neither symmetric nor antisymmetric) cannot be a true wave function.

Let us now consider the application of this concept to a two-spin system in which the two spins are chemically equivalent (not necessarily magnetically equivalent) by virtue of their positions in a symmetric molecule. (Such nuclei are said to be *symmetrically equivalent*.) For clarity, we designate the spins by the subscripts a and b, but this does not imply any difference between them. Following the procedure of Section 6.4, we can write four basis functions for this system:

$$
\begin{aligned}
\phi_1 &= \alpha_a\alpha_b && \textit{symmetric} \\
\phi_2 &= \alpha_a\beta_b && \textit{asymmetric} \\
\phi_3 &= \beta_a\alpha_b && \textit{asymmetric} \\
\phi_4 &= \beta_a\beta_b && \textit{symmetric}
\end{aligned}
\tag{6.42}
$$

If we perform the symmetry operation of interchanging these two indentical nuclei, ϕ_1 and ϕ_4 will clearly be unchanged; hence we label them as symmetric functions. However, ϕ_2 and ϕ_3 are neither symmetric nor antisymmetric, because

$$\Re\phi_2 = \Re(\alpha_a\beta_b) = \beta_a\alpha_b = \phi_3 \quad (\text{nor } \phi_2 \text{ or } -\phi_2) \tag{6.43}$$

ϕ_2 and ϕ_3 are not in themselves acceptable wave functions. We may, if we wish, use them as basis functions in a calculation involving the secular equation, and we would ultimately obtain the correct solution to the problem. However, the calculations are often simplified by selecting in place of ϕ_2 and ϕ_3 two functions that have the proper symmetry and yet maintain the desirable properties of product basis functions. For the two-spin system this can be accomplished easily by using instead of ϕ_2 and ϕ_3 themselves functions that are the sum and difference of ϕ_2 and ϕ_3:

$$
\begin{aligned}
\phi_2' &= \frac{1}{\sqrt{2}}(\alpha_a\beta_b + \beta_a\alpha_b) && \textit{symmetric} \\
\phi_3' &= \frac{1}{\sqrt{2}}(\alpha_a\beta_b - \beta_a\alpha_b) && \textit{antisymmetric}
\end{aligned}
\tag{6.44}
$$

The factor $1/\sqrt{2}$ maintains the normalization condition.

We know from Section 6.7 that the true wave function $\psi_1, \ldots, \psi_4$ are linear combinations of the basis functions. If we begin with symmetrized functions, such as ϕ_2' and ϕ_3', then each of the ψ's can be formed exclusively from symmetric functions or exclusively from antisymmetric functions. Stated another way, *functions of different symmetry do not mix.* The result is that, like the situation with F_z, many off-diagonal elements of the secular equation must be zero, and the equation factors into several equations of lower order. We shall study an example of this factoring in Section 6.13, when we consider the A_2B system.

Simplification of the spectrum itself also results from the presence of symmetry, because transitions are permitted *only* between two symmetric or two antisymmetric states. We see in Section 6.13 that there is often a considerable reduction in the number of NMR lines.

For the two-spin system the only symmetry operation is the interchange of the two nuclei, and the correct linear combinations, ϕ_2' and ϕ_3', could be constructed by inspection. When three or more symmetrically equivalent nuclei are present, the symmetry operations consist of various permutations of the nuclei. The correct symmetrized functions can be determined systematically only by application of results from group theory. We shall not present the details of this procedure.

6.12 GENERAL PROCEDURES FOR SIMULATING SPECTRA

In the preceding sections we derived in considerable detail the expressions for the secular equation and the transitions of the AB system. It is apparent that the calculation of each element in the secular equation would become very tedious for systems of three or more nuclei. Fortunately, general algebraic expressions have been derived that are readily amenable to implementation in digital computers. Most NMR spectrometers include in their software packages simple programs for simulating the spectra of multispin systems from a set of assumed chemical shifts and coupling constants.

The inverse problem of analyzing an observed spectrum to obtain the chemical shifts and coupling constants is more difficult, because the secular equation is not easily solved in this sense. However, computer programs permit the simulation of trial spectra from assumed parameters (chemical shifts and coupling constants) and their systematic variation in an iterative process that converges on a spectrum ideally identical to that observed. Thus, in general, the problem of analyzing almost any specific complex spectrum can be solved by digital techniques.

Nevertheless, it is valuable to investigate briefly a few spin systems beyond the AB system to observe their general spectral patterns and to derive some algebraic (as contrasted with digital) expressions that provide insight into the overall behavior of coupled spin systems.

6.13 THREE-SPIN SYSTEMS

Let us consider first the general three-spin system with no restrictions regarding relative sizes of chemical shifts and coupling constants and then discuss three special (but very common) cases.

ABC

For the three-spin system there are 2^3 basis functions that can be formed as products without regard to any symmetry considerations. These can be classified into four sets according to the values of f_z, as indicated in Table 6.3. Application of the selection rule $\Delta f_z = \pm 1$ shows that there are 15 allowed transitions.

Because the functions for $f_z = \frac{1}{2}$ and $-\frac{1}{2}$ lead to cubic equations, it is not possible to express the transition energies in simple algebraic form as functions of the six parameters ν_A, ν_B, ν_C, J_{AB}, J_{AC}, and J_{BC}. Hence, analysis of an ABC spectrum must be carried out for each case individually, using an iterative procedure such as that mentioned in the preceding section. Most such computations seek the "best" agreement between calculated and observed frequencies, but the parameters that emerge for the best fit may not be unique. Only by taking relative intensities into account can the correct solution be ensured. At high magnetic fields true ABC spectra are rather uncommon.

A$_2$B

When two of three strongly coupled nuclei are magnetically equivalent, the presence of symmetry results in considerable simplification. In the first place, the spectrum is now determined by only two chemical shifts, ν_A and ν_B, and one coupling constant, J_{AB}, because as we saw in Section 6.9, the appearance of the spectrum does not depend on J_{AA}. Each of the eight basis functions is now formed as the product of the spin function of B (α or β) with one of the

TABLE 6.3 Basis Functions for the ABC System

$\phi_1 = \alpha\alpha\alpha,$			$f_z = \frac{3}{2};$
$\phi_2 = \alpha\alpha\beta,$	$\phi_3 = \alpha\beta\alpha,$	$\phi_4 = \beta\alpha\alpha,$	$f_z = \frac{1}{2};$
$\phi_5 = \beta\beta\alpha,$	$\phi_6 = \beta\alpha\beta,$	$\phi_7 = \alpha\beta\beta,$	$f_z = -\frac{1}{2};$
$\phi_8 = \beta\beta\beta,$			$f_z = -\frac{3}{2};$

TABLE 6.4 Base Functions for the A_2B System

Function	A_2	B	f_z	Symmetry
ϕ_1	$\alpha\alpha$	α	$\frac{3}{2}$	s
ϕ_2	$(1/\sqrt{2})(\alpha\beta + \beta\alpha)$	α	$\frac{1}{2}$	s
ϕ_3	$(1/\sqrt{2})(\alpha\beta - \beta\alpha)$	α	$\frac{1}{2}$	a
ϕ_4	$\beta\beta$	α	$-\frac{1}{2}$	s
ϕ_5	$\alpha\alpha$	β	$\frac{1}{2}$	s
ϕ_6	$(1/\sqrt{2})(\alpha\beta + \beta\alpha)$	β	$-\frac{1}{2}$	s
ϕ_7	$(1/\sqrt{2})(\alpha\beta - \beta\alpha)$	β	$-\frac{1}{2}$	a
ϕ_8	$\beta\beta$	β	$-\frac{3}{2}$	s

symmetrized functions given in Eqs. 6.42 and 6.44. These basis functions are given in Table 6.4.

As in the ABC case, the basis functions divide into four sets according to f_z with 1, 3, 3, and 1 functions in each set. However, of the three functions in the set with $f_z = \frac{1}{2}$ or $-\frac{1}{2}$, two are symmetric and one antisymmetric. Hence each of the two 3 × 3 blocks of the secular equation factors into a 2 × 2 block and 1 × 1 block. Algebraic solutions are thus possible. Furthermore, the presence of symmetry reduces the number of allowed transitions from 15 to 9, because no transitions are allowed between states of different symmetry. (One of the nine is of extremely low intensity and is not observed.) Thus the A_2B system provides a good example of the importance of symmetry in determining the structure of NMR spectra.

Some examples of simulated and experimental A_2B spectra are given in Fig. 6.7.

ABX

A second variant of the ABC system occurs when the chemical shift of one nucleus differs substantially from that of the other two—an ABX system. The presence of one nucleus only weakly coupled to the others permits factoring of the secular equation so that algebraic solutions are possible. The basis functions for the ABX system are just those shown in Table 6.3 for the general three-spin system. However, because $(\nu_A - \nu_X)$ and $(\nu_B - \nu_X)$ are much larger than J_{AX} and J_{BX}, we can define an F_z for the AB nuclei separately from F_z for the

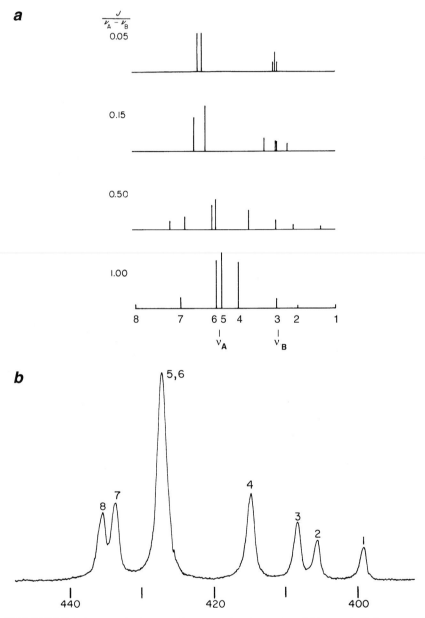

FIGURE 6.7 A$_2$B spectra. (a) Computed spectra with various ratios of $J/(\nu_A - \nu_B)$. (b) Spectrum of the aromatic protons in 2,6-di-*tert*-butylphenol in CDCl$_3$ at 60 MHz.

TABLE 6.5 Basis Functions for the ABX System

Function	AB	X	$f_z(AB)$	$f_z(X)$	f_z
ϕ_1	$\alpha\alpha$	α	1	$\frac{1}{2}$	$\frac{3}{2}$
ϕ_2	$\alpha\alpha$	β	1	$-\frac{1}{2}$	$\frac{1}{2}$
ϕ_3	$\alpha\beta$	α	0	$\frac{1}{2}$	$\frac{1}{2}$
ϕ_4	$\beta\alpha$	α	0	$\frac{1}{2}$	$\frac{1}{2}$
ϕ_5	$\beta\beta$	α	-1	$\frac{1}{2}$	$-\frac{1}{2}$
ϕ_6	$\beta\alpha$	β	0	$-\frac{1}{2}$	$-\frac{1}{2}$
ϕ_7	$\alpha\beta$	β	0	$-\frac{1}{2}$	$-\frac{1}{2}$
ϕ_8	$\beta\beta$	β	-1	$-\frac{1}{2}$	$-\frac{3}{2}$

X nucleus (see Section 6.7). The basis functions classified in this way are given in Table 6.5. Of the three functions with $f_z = \frac{1}{2}$, ϕ_3 and ϕ_4 have the same values of $f_z(AB)$ and $f_z(X)$, but ϕ_2 is in a separate class and does not mix with ϕ_3 and ϕ_4. Thus the 3 × 3 block of the secular equation factors into a 2 × 2 block and a 1 × 1 block. Analogous factoring occurs for the 3 × 3 block arising from the three functions with $f_z = -\frac{1}{2}$.

Appendix B provides details of the computation of the matrix elements, the solution of the factors of the secular equation, and the calculation of the transition frequencies and intensities, which are readily carried out. The ABX spectrum consists of eight lines in the AB region and six in the X region, of which two may be either very weak or quite strong, depending on the individual parameters. (The 15th line of the general ABC spectrum is forbidden here.) A typical ABX spectrum is shown in Fig. 6.8. The AB portion consists of two overlapping AB-type quartets (weak, strong, strong, weak lines), except that C of Eq. 6.35 is replaced by D_+ and D_-:

$$D_+ = \tfrac{1}{2}\{[(\nu_A - \nu_B) + \tfrac{1}{2}(J_{AX} - J_{BX})]^2 + J_{AB}^2\}^{1/2}$$
$$D_- = \tfrac{1}{2}\{[(\nu_A - \nu_B) - \tfrac{1}{2}(J_{AX} - J_{BX})]^2 + J_{AB}^2\}^{1/2}$$

(6.45)

The centers of these two quartets—designated $(ab)_+$ and $(ab)_-$—are separated by $\frac{1}{2}|J_{AX} + J_{BX}|$. The X portion of the spectrum consists of three pairs of lines symmetrically placed around ν_X. The two strongest lines are separated by $|J_{AX} + J_{BX}|$, which is just twice the separation of the centers of the $(ab)_+$ and $(ab)_-$ quartets. The spacings and relative intensities of the other lines depend on the chemical shifts and coupling constants, as described in Appendix B.

ABX spectra occur even at high magnetic fields when two nuclei are accidentally nearly equivalent in chemical shift. Also, we shall find in Sections 6.15 and

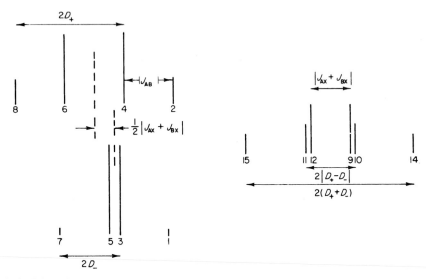

FIGURE 6.8 Schematic representation of a typical ABX spectrum, showing the AB portion separated into its constituent $(ab)_+$ and $(ab)_-$ components.

6.16 that the ABX system provides a useful model for exploring the behavior in systems in which two nuclei accidentally have the same chemical shift, an AA′X system.

AMX

In the limit of large chemical shift differences, the three-spin system displays a first-order spectrum, consisting of 12 lines, the other three being forbidden. The spectral line positions can be derived from the ABX spectrum or simply by application of repetitive first-order splittings.

6.14 RELATIVE SIGNS OF COUPLING CONSTANTS

There is a tendency to focus on the magnitude of a coupling constant, which can often be obtained directly from splitting of a resonance, and to ignore the fact that J has both sign and magnitude. Signs of coupling constants are important in basic theory and in developing correlations between J and molecular structure, as discussed in Chapter 5. Also, the appearance of many spectra depends on the sum and/or difference of coupling constants (e.g., $J_{AX} + J_{BX}$, as seen earlier), where a combination of two coupling constants with opposite signs may cause unexpected results.

The AMX system is useful in describing the effect of changes in signs of coupling constants. In Section 6.3 we noted that a change in sign of J in an AB or AX system results in the interchange of pairs of spectral lines but produces no observable change in the spectrum. This is also true in an AMX spectrum. However, here it is possible to carry out decoupling experiments (see Section 5.5) in such a way as to cause selective collapse of doublets to singlets. Which lines are altered depends on the relative signs of the coupling constants.

From a simple consideration of the origin of the lines in an AMX spectrum we can write the following expressions:

$$\nu(A_i) = \nu_A - J_{AM}m_M - J_{AX}m_X$$

$$\nu(M_i) = \nu_M - J_{AM}m_A - J_{MX}m_X \qquad (6.46)$$

$$\nu(X_i) = \nu_X - J_{AX}m_A - J_{MX}m_M$$

The small m's can independently assume the value $+\frac{1}{2}$ or $-\frac{1}{2}$ to account for all 12 lines. The index i runs from 1 to 4 and denotes the four A, four M, and four X lines in order of decreasing frequency, that is, in the order in which they appear from left to right in the spectrum. For example, if both J_{AM} and J_{AX} are positive (and if all values of γ are positive), line A_1 arises from a "flip" of the A spin, while spins M and X retain their orientations (i.e., $m_M = m_X = -\frac{1}{2}$), as illustrated in Fig. 6.9a.

Consider, for example, the effect of the relative signs of J_{AX} and J_{MX}. As indicated in Fig. 6.9a, lines A_1 and A_2 differ only in the value of m_M assigned to them; they have the same value of m_X and may be said to correspond to the same X spin state. Identical statements may be made about the pairs of lines A_3 and A_4, M_1 and M_2, and M_3 and M_4. If J_{AX} and J_{MX} are both positive, then the high frequency pair of A lines, A_1 and A_2, arise from the $-\frac{1}{2}$ X state, and the high frequency pair of M lines, M_1 and M_2, also arise from the $-\frac{1}{2}$ X state. The low-frequency pairs in each case then, of course, arise from the $+\frac{1}{2}$ X state. If, however, J_{AX} and J_{MX} have opposite signs, then the $+\frac{1}{2}$ X state is responsible for the low frequency pair in one case and the high frequency pair in the other.

An easy way to determine which combination of signs is correct is to *selectively decouple* only half the molecules—those with a specific X state, either $+\frac{1}{2}$ or $-\frac{1}{2}$—by carefully adjusting the strength and placement of the irradiating rf field. Figure 6.9b gives the spectrum of an AMX system that can be used to demonstrate this result. Figure. 6.9c shows that irradiating at the center of the M_1-M_2 doublet causes the A_1-A_2 doublet to collapse, thus demonstrating that J_{AX} and J_{MX} have the same signs in this substituted pyridine, and Fig. 6.9d and e confirm this result. Other selective decoupling experiments show that all three coupling constants have the same sign in this molecule.

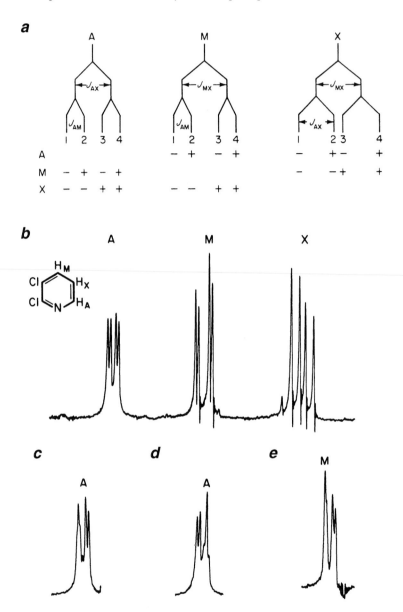

FIGURE 6.9 (*a*) Schematic representation of an AMX spectrum, showing the spin states associated with each spectral line. The situation depicted is that for all three *J*'s positive. The signs of *m* are indicated. (*b*) Spectrum of 2,3-dichloropyridine at 60 MHz, an example of an AMX spin system. (*c–e*) Results of selective decoupling in 2,3-dichloropyridine, with decoupling frequencies centered between the following lines: (*c*) M_1 and M_2, (*d*) M_3 and M_4, (*e*) A_1 and A_2.

6.15 SOME CONSEQUENCES OF STRONG COUPLING AND CHEMICAL EQUIVALENCE

The ABX system provides an excellent framework for describing features that occur all too frequently in NMR spectra, even at very high field strengths where we might expect to find simple first-order patterns. These features arise from strong coupling between two or more nuclei and are particularly pronounced when the nuclei have the same or nearly the same chemical shift but are not magnetically equivalent. There is nothing conceptually new in this section; we merely point out by means of several examples how easy it is to be deceived by the appearance of certain spectra and to draw erroneous conclusions.

Our description of an ABX spectrum in Section 6.13 pointed out that eight lines are normally observed in the AB region and either four or six lines in the X region. Frequently, however, some of the lines coincide, creating a spectrum whose appearance is not that of a typical ABX pattern. Consider, for example, an ABX system with the parameters shown in Fig. 6.10, which depicts calculated ABX spectra as a function of only one changing parameter, ν_B. The spectrum in (a) is readily recognized as an ABX pattern. However, as ν_B approaches ν_A in (b), the appearance changes drastically, and when $\nu_B = \nu_A$ in (c), the AB region appears to be simply a doublet, while the X region is a 1:2:1 triplet.

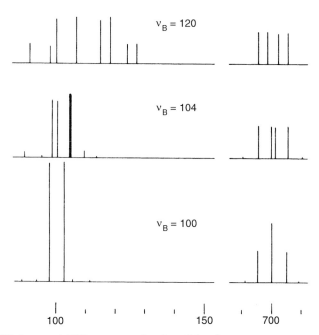

FIGURE 6.10 Computed ABX spectra as a function of changing one parameter, ν_B, while other parameters remain fixed at $\nu_A = 100$, $J_{AB} = 9$, $J_{AX} = 7$, and $J_{BX} = 3$ Hz.

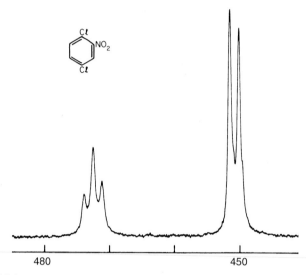

FIGURE 6.11 ^{1}H NMR spectrum (60 MHz) of 2,5-dichloronitrobenzene in CDCl$_3$. The triplet is due to H$_6$ and the doublet to H$_3$ and H$_4$, which are accidentally chemically equivalent. The observed splitting of 1.6 Hz is what would be expected as the average of J_{meta} (~3 Hz) and J_{para} (close to zero).

This spectrum is an excellent example of what have been termed "deceptively simple spectra." If one were confronted with such a spectrum and did not realize that it is a special case of ABX (actually AA'X, because the chemical shifts of A and B are equal), the doublet and triplet might mistakenly be interpreted as the components of a first-order A$_2$X spectrum, with $J_{AX} = 5$ Hz. Actually, the observed splitting is the *average* of J_{AX} and $J_{A'X}$, rather than any coupling constant in the molecule. An experimental example is shown in Fig. 6.11.

Deceptively simple spectra are not limited to ABX systems; they may occur for specific combinations of parameters in any spin system. Caution should be exercised in interpreting spectra that contain accidental chemical shift equivalences and show apparent equivalences of coupling constants that one might intuitively think would not be equal.

The ABX system also provides a convenient framework for exploring the consequences of strong coupling from a slightly different perspective. Consider the ABX system in the molecular fragment **VI**.

$$
\begin{array}{ccc}
\text{R} & \text{R} & \text{R} \\
| & | & | \\
\text{R—C—C—C—OR} \\
| & | & | \\
\text{H}_\text{A} & \text{H}_\text{B} & \text{H}_\text{X}
\end{array}
$$

VI

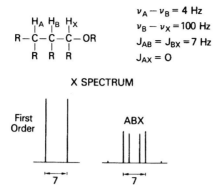

FIGURE 6.12 Simulated spectra for H_X in the molecular fragment shown. The spectrum labeled "first order" is what might be predicted on the basis of a coupling between H_X and H_B. The spectrum labeled "ABX" is the X portion of an ABX spectrum calculated according to the procedures in Section 6.13.

The exact nature of the substituents is not important; for this illustration they are taken to have no magnetic nuclei but to result in the chemical shifts (in Hz) given in Fig. 6.12. For free rotation about the C—C bonds, the couplings J_{AB} and J_{BX} should be nearly equal and each of the order of 7 Hz, as indicated in Table 5.1. On the other hand, J_{AX} (through four single bonds) should be nearly zero. If we direct our attention only to the X portion of the spectrum (as might often happen in a complex molecule where the A and B portions would be overlapped by other aliphatic protons), we might well be tempted to treat the X portion of the spectrum by simple first-order analysis because $(\nu_B - \nu_X) \gg J_{BX}$. First-order analysis would then predict a simple doublet of approximately equal intensities with a splitting of 7 Hz. However, the correctly computed ABX spectrum, given in Fig. 6.12, shows that the X portion has six lines. The two most intense lines are again separated by 7 Hz, but the two central lines are almost as strong.

The first-order calculation fails because the proton in question (X) is coupled to one of a set of strongly coupled nuclei (i.e., a set in which J is greater than the chemical shift difference in Hz). As we saw in Section 6.7, the spin wave functions for such nuclei mix; hence, it is invalid to treat one of the nuclei in the strongly coupled set as though the others were not there. Such cases often occur in symmetric molecules where one might confuse chemical and magnetic equivalence.

An example of a system other than ABX may help emphasize the types of circumstances in which such effects occur in high field NMR spectra. Figure 6.13a shows the spectrum of 1,4-dibromobutane in which the protons attached to carbon 2 are chemically but not magnetically equivalent to those on carbon 3. They are coupled to each other with a coupling constant $J_{23} \approx 7$ Hz (a typical average value for such vicinal coupling). Since $\nu_2 - \nu_3 = 0$, the four protons on C_2 and C_3 behave as a strongly coupled group, $J_{23}/(\nu_2 - \nu_3) = \infty$. The result, as shown in Fig. 6.13a, is that the protons on C_1 and C_4 give rise not to a simple

a

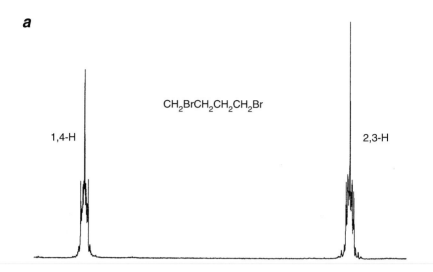

CH$_2$BrCH$_2$CH$_2$CH$_2$Br

1,4-H

2,3-H

b

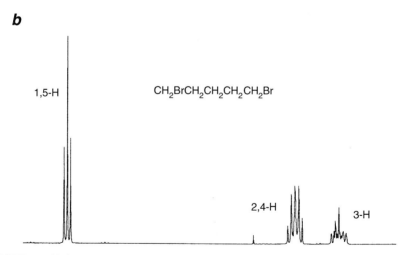

1,5-H

CH$_2$BrCH$_2$CH$_2$CH$_2$CH$_2$Br

2,4-H

3-H

FIGURE 6.13 ^{1}H NMR spectra at 300 MHz of (*a*) 1,4-dibromobutane and (*b*) 1,5-dibromopentane. Strong coupling effects are seen in (*a*) but not in (*b*). Spectra courtesy of Herman J. C. Yeh (National Institutes of Health).

triplet, as might be expected from first-order analysis, but to the complex multiplet shown. The spectrum for the protons on C$_2$ and C$_3$ is likewise complex.

This situation can be contrasted with that in 1,5-dibromopentane (Fig. 6.13*b*). In this case the chemical shifts of the protons on C$_2$ and C$_3$ are quite different;

for $J \approx 7$ Hz, $J_{23}/(\nu_2 - \nu_3) \approx 0.08$, so the C_2 and C_3 protons are not strongly coupled. The C_1 and C_5 protons thus display the 1:2:1 triplet that might be expected from a first-order analysis.

There is a common misconception that strong coupling effects need not be considered at higher field, because accidental chemical equivalences or near equivalences are much less likely to occur than at lower magnetic fields. Of course, that is true. However, symmetric, not accidental, chemical equivalence is the basis for the effect shown in Fig. 6.13a.

6.16 "SATELLITES" FROM CARBON-13 AND OTHER NUCLIDES

Proton NMR spectra in organic molecules can be interpreted without regard to the structural carbon framework because the predominant ^{12}C has no nuclear spin. However, ^{13}C has a spin of $\frac{1}{2}$, which not only permits its direct observation but also provides features in the ^1H spectrum from the ^{13}C that is present at a natural abundance of 1.1%. As we saw in Chapter 5, $^1J(^{13}C-H)$ is normally 100–200 Hz, whereas two- and three-bond coupling constants often run 5–10 Hz. Hence a resonance line from a proton attached to a ^{12}C atom is accompanied by weak ^{13}C "satellites" separated by $^1J(^{13}C-H)$ and placed almost symmetrically about the main line. (The departure from precisely symmetrical disposition arises from the $^{13}C/^{12}C$ isotope effect on the ^1H chemical shift, as described in Section 4.8.) For example, the proton resonance of chloroform in Fig. 6.14a shows ^{13}C satellites.

When the molecule in question contains more than one carbon atom, the ^{13}C satellites often become more complex. Consider, for example, the molecule $CHCl_2CHCl_2$, the proton resonance of which is shown in Fig. 6.14b. The ordinary spectrum is a single line because of the magnetic equivalence of the two protons. On the other hand, the approximately 2.2% of the molecules that contain one ^{13}C and one ^{12}C have protons that are not magnetically equivalent. The proton resonance spectrum of these molecules is an ABX spectrum in which $\nu_A \approx \nu_B$. A simulation of the ABX spectrum is also shown in Fig. 6.14. The value of $^3J_{HH'}$ (i.e., J_{AB} in the ABX spectrum) can readily be observed, even though it is not obtainable from the spectrum of the fully ^{12}C molecule, which constitutes an A_2 spin system.

Satellites are not restricted to ^{13}C, but may be seen with other magnetic nuclei that are present at low abundance when the principal isotope has $I = 0$. Among the best known are ^{29}Si (8.5%), ^{111}Cd and ^{113}Cd (each about 9%), ^{199}Hg (7.6%), and ^{207}Pb (8.9%). Other nuclides, such as ^{15}N, have a natural abundance so low that satellite signals are rarely observed in normal one-dimensional NMR spectra, but with polarization transfer methods described in Chapters 9 and 10, the existence of these weak satellites often permits observation of the less sensitive, low abundance nuclide by indirect detection.

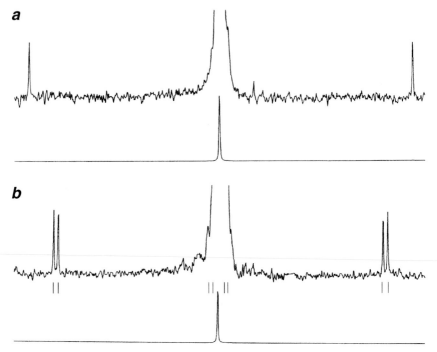

FIGURE 6.14 ¹H NMR spectrum (300 MHz) showing ¹³C satellites. (*a*) CHCl₃; (*b*) CHCl₂CHCl₂, with a simulation of the AB portion of an ABX spectrum.

6.17 THE AA′BB′ AND AA′XX′ SYSTEMS

These four-spin systems are characterized by two chemical shifts and four coupling constants, $J_{AA'}$, $J_{BB'}$, J_{AB}, and $J_{AB'}$. The last two are not equal, leading to magnetic nonequivalence, which occurs for both AA′BB′ and AA′XX′. A good example of an AA′XX′ system was given in Fig. 6.1. We provide a brief discussion of these spin systems, because they occur frequently.

The calculation of the energy levels and transitions is considerably simplified by inclusion of symmetry, as described in Section 6.11. For the AA′XX′ system, this results in factoring of the secular equation so that only linear and quadratic equations need to be solved. However, for AA′BB′ there remains a 4 × 4 factor, which precludes a general algebraic solution and requires computer simulation for specific parameters. There are 24 allowed transitions in an AA′BB′ spectrum. For AA′XX′, the A and X portions of the spectra are identical, each consisting of 10 lines (two pairs of transitions coincide), as illustrated in Fig. 6.15. In addition to the two very intense lines, the other lines are grouped into two AB-type subspectra (weak, strong, strong, weak lines in a symmetric pattern). The quantities

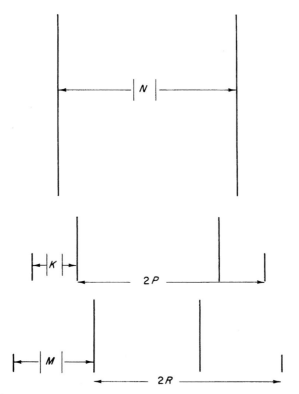

FIGURE 6.15 Schematic representation of the A portion of an AA′XX′ spectrum. The X portion is identical but centered at ν_X rather than ν_A.

K, M, N, P, and R shown in Fig. 6.15 are related to sums and differences of the four coupling constants. From the indicated spacings of lines, it is easy to determine the values of all coupling constants, as described in Appendix B.

For AA′BB′, the spectrum is symmetrical about the average $\frac{1}{2}(\nu_A + \nu_B)$, and in principle all 24 lines may be observed. Even at high field, some *ortho*-disubstituted benzenes, for example, may display an AA′BB′ spectra, rather than an AA′XX′ spectrum.

6.18 ADDITIONAL READING AND RESOURCES

The theory described here was fully developed during the early years of high resolution NMR investigations and is now "standard." The classic work is included in the *High Resolution NMR* by Pople, Schneider, and Bernstein,[34] and the book *Structure of High Resolution NMR Spectra* by P. L. Corio[79] covers some aspects of the theory in greater depth. Pertinent articles in the *Encyclopedia of NMR* are *Analysis of High Resolution Solution State Spectra*[80] and *Analysis of Spectra: Automatic Methods.*[81]

Software for simulating spectra from assumed values of chemical shifts and coupling constants is widely available — usually supplied with NMR spectrometers and with many data processing programs. Many programs also use the simulated spectrum as the first step in an iterative process to fit the frequencies observed in an experimentally obtained spectrum.

6.19 PROBLEMS

6.1 Describe the following spin systems as AB, etc.: $CH_2{=}CHF$; PF_3; cubane; $CH_3CHOHCH_3$; H_2; chlorobenzene; *n*-propane. Assume free rotation about single bonds.

6.2 Show that for N coupled nuclei there are $\frac{1}{2}N(N-1)$ coupling constants.

6.3 Verify that ψ_2 in Eq. 6.30 gives the value of E_2 in Eq. 6.29.

6.4 Which of the following spectra result from AB systems? Spectra are given as frequency (relative intensity).
   (a) 100 (1), 108 (2.3), 120 (2.3), 128 (1)
   (b) 100 (1), 104 (2.4), 113 (2.4), 120 (1)
   (c) 100 (1), 110 (4), 114 (4), 124 (1)
   (d) 100 (1), 107 (15), 108 (15), 115 (1)

6.5 A molecule is known to contain carbon, hydrogen, fluorine and oxygen. A *portion* of the 400 MHz proton NMR spectrum consists of four lines at 1791, 1800, 1809, and 1818 Hz relative to the reference TMS. The relative intensities of the four lines are 1:3:3:1, and the total area under the four lines corresponds to two protons.
   (a) Give two completely different explanations for this set of four lines.
   (b) For *each* of the two explanations, determine the chemical shift of the two protons in ppm.
   (c) For *each* of the two explanations, give the frequencies and relative intensities that would be observed at 600 MHz.

6.6 Find ν_A and ν_B from each of the following AB spectra: (a) 117, 123, 142, 148; (b) 206, 215, 217, 226.

6.7 Verify the nonmixing of basis functions of different symmetry for ϕ_2' and ϕ_3' in Eq. 6.44 by calculating $\mathcal{H}_{2'3'}$.

6.8 In which of the following molecules are strong coupling effects of the type illustrated in Fig. 6.13 likely to *appear* at high magnetic field?

(a) $CH_3CH_2OCH_2CH_3$; (b)

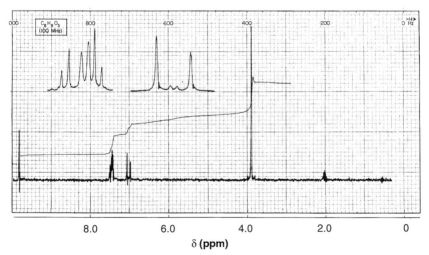

(c)

6.9   Determine the structural formula of the compound giving the ^1H spectrum in Fig. 6.16. Analyze the ABX portion of the spectrum by the procedure given in Table B.2 (Appendix B) or with the aid of a suitable computer simulation.

6.10 Analyze the AA'XX' spectra in Figs. 6.17 and 6.18 by the procedure described in Appendix B or with the aid of a suitable computer simulation. From the values of the coupling constants, deduce the correct isomeric structure of each compound. (See also Fig. 6.1.)

6.11 Give the type of spin system (e.g., AB_2X_2) expected for each of the following ethane derivatives, where Q, R, and S are substituents that do not spin couple with the protons. Assume that rotation about the C—C bond is rapid and that there is a large chemical shift difference between protons on different carbon atoms. (a) CH_3—CH_2R; (b) CH_2Q—CH_2R; (c) CH_3—CQRS; (d) CH_2Q—CHRS; (e) CHQ_2—CHRS.

FIGURE 6.16 ^1H NMR spectrum at 100 MHz of a sample with the molecular formula $C_8H_8O_3$. The multiplet near $\delta = 2.0$ ppm is from acetone-d_5. Frequencies of lines in inset: 753.3, 751.4, 748.2, 746.4, 744.8, 743.0, 708.1, 704.5, 702.8, 699.4 Hz from TMS.

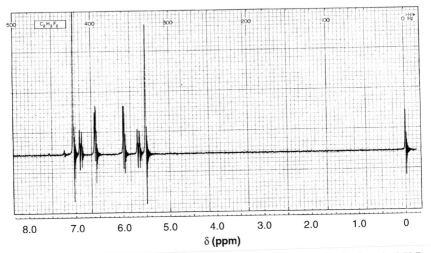

FIGURE 6.17 ^1H NMR spectrum at 60 MHz of a sample with the molecular formula $C_2H_2F_2$. Frequencies of lines: 422.6, 414.8, 412.1, 395.1, 393.7, 359.1, 357.6, 340.6, 337.9, 330.1 Hz from TMS.

6.12 Repeat Problem 6.11 for slow rotation about the C—C bond.

6.13 Prepare a table similar to Table 6.2 for the relative intensities of lines in a multiplet for a spin coupled to N spins with $I = 1$ for $N = 0, 1, 2,$ and 3.

6.14 Predict the ^1H NMR spectrum of dimethylmercury, $Hg(CH_3)_2$.

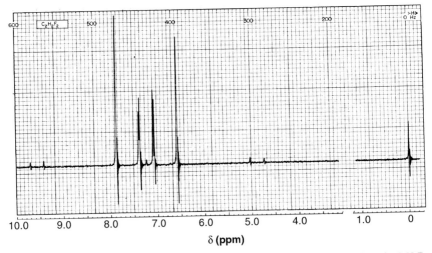

FIGURE 6.18 ^1H NMR spectrum at 60 MHz of a sample with the molecular formula $C_2H_2F_2$. Frequencies of lines: 583.9, 565.9, 472.8, 443.9, 442.2, 424.9, 423.3, 394.4, 301.0, 283.1 Hz from TMS.

6.15 Diborane has the structure

$$
\begin{array}{ccc}
H_a \backslash \diagup H_b \backslash \diagup H_a \\
B \quad B \\
H_a \diagup \backslash H_b \diagup \backslash H_a
\end{array}
$$

(a) Sketch the spectra ("first order" approximation is applicable) of both the terminal hydrogens and the bridgehead (central) hydrogens assuming the following: (i) only ^{11}B is present in the molecule; (ii) $^1J(^{11}B-H_a) = 137$ Hz; (iii) $^1J(^{11}B-H_b) = 48$ Hz; and (iv) $^2J(H_a-H_b) = 0$. (b) In the actual ^1H spectrum of $^{11}B_2H_6$ the lines are rather broad. Why? (c) What is the value of $^1J(^{10}B-H_a)$?

Spectra of Solids

In previous chapters we have seen that the Hamiltonian describing a nuclear spin system is considerably simplified when molecules tumble rapidly and randomly, as in the liquid state. However, that simplicity masks some fundamental properties of spins that help us to understand their behavior and that can be applied to problems of chemical interest. We turn now to the solid state, where these properties often dominate the appearance of the spectra. Our treatment is limited to substances such as molecular crystals, polymers, and glasses, that is, solids in which there are well-defined individual molecules. We do not treat metals, ionic crystals, semiconductors, superconductors, or other systems in which delocalization of electrons is of critical importance.

Our aims in this chapter are, first, to describe the important interactions in sufficient detail to explain the principal features in the NMR spectra of solids and, second, to discuss methods that have been developed to overcome selectively the effect of some of these interactions and thus to produce in solids high resolution spectra that are reminiscent of those found in liquids, albeit with line widths that do not quite achieve the very small values often found in liquids. In addition, we consider briefly the features of spectra arising from molecules that are partially oriented in liquid crystals or in very high magnetic fields.

7.1 SPIN INTERACTIONS IN SOLIDS

For spin $\frac{1}{2}$ nuclei, the major differences between liquids and solids involve magnetic dipolar interactions (which average to zero in liquids) and the anisotropy in the chemical shielding (which averages to a single value in liquids). In addition, electron-coupled spin–spin interactions demonstrate an anisotropy in J, but this anisotropy is observed only infrequently. For nuclei with $I > \frac{1}{2}$, the interaction between the nuclear electric quadrupole moment and the gradient in the electric field caused by surrounding electrons and ions is an important factor in solids. Usually it dominates chemical shift and magnetic dipolar interactions, and in some instances it competes with the Zeeman interaction between the nuclear magnetic moment and the applied magnetic field.

The Hamiltonian for each of these four interactions has the same form:

$$\mathscr{H} = c\boldsymbol{I}\cdot\mathbb{R}\cdot\boldsymbol{S} \tag{7.1}$$

where c is a collection of constants; $\mathbb{R}$ is a second-rank tensor that describes the interaction; and $\boldsymbol{I}$ and $\boldsymbol{S}$ are the spin vectors for the interacting pair of spins (except for the shielding, where $\boldsymbol{S}$ represents the static magnetic field $\mathbf{B}_0$). $\mathbb{R} = \mathbf{D}$, the direct dipolar tensor; $\boldsymbol{J}$, the indirect coupling tensor; σ, the shielding tensor; or $\mathbf{V}$, the electric field gradient tensor. For some purposes it is convenient to express the Hamiltonian in terms of Cartesian coordinates or spherical polar coordinates in the laboratory frame of reference; in other cases a coordinate system fixed in the molecule may be preferable, as we see subsequently.

Except for some quadrupolar effects, all the interactions mentioned are small compared with the Zeeman interaction between the nuclear spin and the applied magnetic field, which was discussed in detail in Chapter 2. Under these circumstances, the interaction may be treated as a perturbation, and the first-order modifications to energy levels then arise only from terms in the Hamiltonian that commute with the Zeeman Hamiltonian. This portion of the interaction Hamiltonian is often called the *secular* part of the Hamiltonian, and the Hamiltonian is said to be *truncated* when nonsecular terms are dropped. This secular approximation often simplifies calculations and is an excellent approximation except for large quadrupolar interactions, where second-order terms become important.

7.2 DIPOLAR INTERACTIONS

As we pointed out in Section 5.1, the magnetic moment of a proton gives rise to a local magnetic field that is about 1 gauss at a distance of 2 Å, that is, a distance typical of a chemical bond length or a close intermolecular approach. Depending on the orientation of the nuclear magnet, this local field can add to or subtract from the applied field. Thus, two magnetic moments $\boldsymbol{\mu}_1$ and $\boldsymbol{\mu}_2$ can interact with each other via a classical dipolar field. From electromagnetic theory, the energy of this interaction, which becomes the quantum mechanical Hamiltonian, is

$$\mathcal{H}_D = \frac{\boldsymbol{\mu}_1 \cdot \boldsymbol{\mu}_2}{r^3} - \frac{3(\boldsymbol{\mu}_1 \cdot \mathbf{r})(\boldsymbol{\mu}_2 \cdot \mathbf{r})}{r^3} \tag{7.2}$$

where $\mathbf{r}$ is the vector from $\boldsymbol{\mu}_1$ to $\boldsymbol{\mu}_2$, and the operator $\boldsymbol{\mu} = \gamma \hbar \mathbf{I}$. For an ensemble of N such nuclei, the interactions are summed over all possible pairs to give

$$\mathcal{H}_D = \sum_{i<j=1}^{N} \left[\frac{\boldsymbol{\mu}_i \cdot \boldsymbol{\mu}_j}{r_{ij}^3} - \frac{3(\boldsymbol{\mu}_i \cdot \mathbf{r}_{ij})(\boldsymbol{\mu}_j \cdot \mathbf{r}_{ij})}{r_{ij}^3} \right] \tag{7.3}$$

Consider the two-spin system, with energy levels shown in Fig. 7.1. We can derive a simple and useful expression for a pair of spins by expanding the scalar product of Eq. 7.2 in terms of the x, y, and z spin components and using spherical polar coordinates (r, θ, and ϕ) for the spatial coordinates. This gives the expression

$$\mathcal{H}_D = -\frac{\gamma_1 \gamma_2 \hbar^2}{r^3} [A + B + C + D + E + F] \tag{7.4}$$

where

$$A = I_{1z} I_{2z} (3 \cos^2 \theta - 1)$$

$$B = -\frac{1}{2} [I_{1x} I_{2x} + I_{1y} I_{2y}](3 \cos^2 \theta - 1)$$

$$C = \frac{3}{2} [I_1^+ I_{2z} + I_{1z} I_2^+] \sin \theta \cos \theta \, e^{-i\phi} \tag{7.5}$$

$$D = \frac{3}{2} [I_1^- I_{2z} + I_{1z} I_2^-] \sin \theta \cos \theta \, e^{i\phi}$$

$$E = \frac{3}{4} I_1^+ I_2^+ \sin^2 \theta \, e^{-2i\phi}$$

$$F = \frac{3}{4} I_1^- I_2^- \sin^2 \theta \, e^{2i\phi}$$

Terms $C-F$ can be shown not to commute with the Zeeman Hamiltonian, hence to contribute negligibly to the energy levels, but they are important in relaxation processes, as we see in Chapter 8. For this chapter we shall use the truncated Hamiltonian with terms A and B. The spin portion of this truncated Hamiltonian may look more familiar with some rearrangement of terms

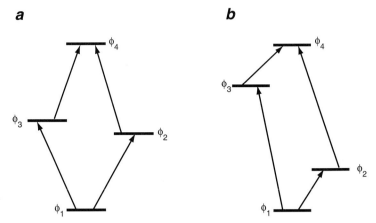

FIGURE 7.1 Energy levels for a two-spin system. (*a*) Homonuclear; (*b*) heteronuclear.

to give

$$\mathcal{H}_D = -\frac{\gamma_1\gamma_2\hbar^2}{r^3}\left[\frac{3}{2}I_{1z}I_{2z} - \frac{1}{2}(I_{1x}I_{2x} + I_{1y}I_{2y} + I_{1z}I_{2z})\right](3\cos^2\theta - 1)$$

$$\mathcal{H}_D = -\frac{\gamma_1\gamma_2\hbar^2}{r^3}\left[\frac{3}{2}I_{1z}I_{2z} - \frac{1}{2}\mathbf{I}_1\cdot\mathbf{I}_2\right](3\cos^2\theta - 1) \tag{7.6}$$

which includes the same scalar product spin terms as the Hamiltonian for indirect spin−spin coupling, Eq. 6.5. However, here the initial factor is the direct dipolar coupling

$$D_{12} = \frac{\gamma_1\gamma_2\hbar^2}{r^3} \tag{7.7}$$

rather than the indirect spin−spin coupling constant J_{12}.

Much of the treatment of the AB system in Section 6.7 is applicable here. Functions ϕ_1 and ϕ_4 do not mix with the others (but they would if the terms $C-F$ had been retained), while ϕ_2 and ϕ_3 may mix. However, as we saw previously (Eqs. 6.29 and 6.31), the extent of mixing depends on the magnitude of the off-diagonal element $\mathcal{H}_{23}$, which in turn depended there on the ratio $J/(\nu_A - \nu_B)$, while in the present case we must consider the ratio $D/(\nu_A - \nu_B)$. If the two dipole-coupled nuclei are of the same species (e.g., two protons), ν_A and ν_B differ only by a chemical shift, which is usually of the same magnitude as or smaller than the dipolar interaction D; hence the functions mix, and we must use $\mathcal{H}_D$ as given in Eq. 7.6. However, if the nuclei are of different species, $\nu_A - \nu_B \gg D$, and we can ignore the off-diagonal elements. This is tantamount to removing the terms in I_x and I_y, which leaves

$$\mathcal{H}_D = -\frac{\gamma_1 \gamma_2 \hbar^2}{r^3} I_{1z} I_{2z} (3 \cos^2 \theta - 1) \tag{7.8}$$

for the heteronuclear two-spin system.

We shall treat the heteronuclear and homonuclear systems separately. Although there are many features in common in the (wide line) spectra produced, the differences in the form of the Hamiltonian suggest quite different methods for manipulating the spin systems so as to produce narrow lines and "high resolution" spectra.

7.3 "SCALAR COUPLING"

Before examining the two-spin system in more detail, we should note that the indirect, electron-coupled spin interactions discussed in Chapter 5 are properties of the molecule, hence are also present in solids. These couplings are usually much smaller than direct dipolar interactions, but there are instances involving heavy nuclei in which the isotropic splittings may be over 30 kHz. Because both direct dipolar coupling and indirect spin coupling have the form of scalar products, the two cannot easily be distinguished in a solid. Thus, when we measure splittings attributed to dipolar interactions, the indirect coupling may also contribute. This consideration also applies in liquid crystal mesophases discussed in Section 7.13. Likewise, decoupling of dipolar interactions (Section 7.5) also eliminates J splittings.

As we noted previously, J_{ij} is a tensor, not a scalar, but unlike D_{ij} it has a nonvanishing trace, hence it is found in molecules tumbling in liquids. In solids and in liquid crystals it is possible in principle to investigate the off-diagonal elements of the J tensor, but measurements are difficult, and few instances of measurable anisotropy in J have been reported. However, one recent result shows that in benzene the anisotropies for one-, two-, and three-bond $C-C$ couplings are approximately $+17$, -4, and $+9$ Hz, respectively, as compared with the isotropic values, $^1J_{CC} = 56$, $^2J_{CC} = -2.5$, and $^3J_{CC} = 10$ Hz.[82]

7.4 THE HETERONUCLEAR TWO-SPIN SYSTEM

The spin part of the Hamiltonian in the solid is identical to that of the AX system, and the solution is the same — a pair of lines for A and a pair for X, as in Fig. 6.2. However, the separation is dependent on orientation of the vector r joining the spins relative to $\mathbf{B}_0$, because the effective coupling is $D(3 \cos^2 \theta - 1)$, rather than D alone. As indicated in Fig. 7.2, the lines cross as θ is varied, becoming coincident when $\theta = 54.7°$, the angle for which the term $(3 \cos^2 \theta - 1) = 0$. This angle that will appear frequently in our later discussion.

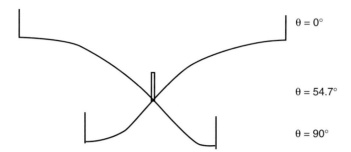

FIGURE 7.2 Illustration of the variation of the dipolar splitting of the two A lines as the internuclear vector **r** varies in direction from 0° (parallel to **B**$_0$) to 90°.

A single crystal, in which all molecules have the same orientation, produces a sharp doublet spectrum for the A nuclei for each orientation when the crystal is rotated through the angle θ. However, a random collection of crystallites produces a *powder pattern*, which is the superposition of spectra from all possible values of θ, each weighted according to the probability of its occurrence. As suggested in Fig. 7.3, the probability of the vector **r** lying at or near 90° from **B**$_0$ is far greater than its lying close to 0°. When this statistical factor is considered along with the basic $(3 \cos^2 \theta - 1)$ dependence in the energy term, the intensity distribution for two magnetic moments oriented parallel to each other is given by the solid curve in Fig. 7.4, and that for two moments oriented antiparallel to each other is given by the dashed curve. Each branch covers a frequency range from $\pm D/2$ to $\mp D$, with functional dependence of $D/2(1 + \nu)^{1/2}$.

For an X nucleus with a spin of $\frac{1}{2}$, the X spectrum is, of course, identical to that of A but centered at ν_X. As we might expect, for X with spin $> \frac{1}{2}$, the treatment of A is qualitatively similar, but the spectrum has more branches, and the existence of a quadrupole moment in X may have significant ramifications,

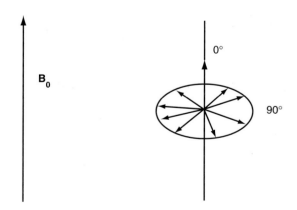

FIGURE 7.3 Probability of vector **r** orientation. The illustration shows that there is only one orientation at 0° but many orientations at 90°.

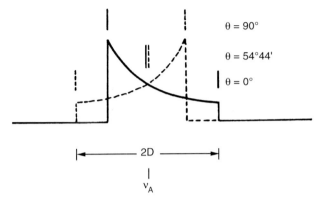

FIGURE 7.4 Schematic representation of the A portion of a powder pattern for an AX system. The solid line refers to nuclei with parallel spin orientation, the dotted line to nuclei with antiparallel spin orientation. Each branch covers a range from 0° to 90°, as indicated, and the two branches cross at 54.7°.

particularly from the large electric quadrupole interaction (see Section 7.11), which can invalidate the perturbation approach used here. We shall not consider such interactions here.

7.5   DIPOLAR DECOUPLING

For applications to most chemical problems, we need spectra that distinguish between chemically shifted nuclei, but the dipolar broadening usually obscures chemical shifts. The question is, can such broadening be removed? Fortunately, the form of Eq. 7.8, which involves the product of only the z components of the spin operator, suggests a simple method—decoupling. As discussed briefly in Section 5.5, irradiation with adequate rf power at ν_x causes A to appear to be decoupled from X. We need to make $\gamma B_2 / 2\pi \gg D$, which is about 23 kHz for the one-bond C–H interaction, but it is feasible to obtain large enough values of B_2 with modern spectrometers and probes designed specifically for study of solids. Thus, broadening from heteronuclear dipolar coupling can be readily suppressed by applying B_2 at ν_x while observing the FID of nucleus A. The effectiveness of this method in providing sufficient line narrowing depends on whether the heteronuclear interaction is the dominant contribution to line broadening. If the observed nucleus A is ^{19}F or some other nucleus with a large magnetic moment and large natural abundance, decoupling protons may have little effect on the width of the overall spectrum, but if A is ^{13}C or another nucleus at low abundance, where homonuclear interactions are very weak, dipolar decoupling can be very effective.

However, ^{13}C, ^{15}N, and many other low abundance nuclei give only weak signals because of their small magnetic moments, their low abundance, and in most instances their long longitudinal relaxation times in solids. Thus the advantage of line narrowing in insufficient for practical study of such nuclei, and another ingenious technique must be brought to bear along with dipolar decoupling.

7.6 CROSS POLARIZATION

We saw in Chapter 2 that nuclei of one species (e.g., ^{1}H) readily exchange energy with each other to develop an equilibrium magnetization because they precess at the same frequency (or approximately the same frequency if chemical shifts are taken into account). However, because of the wide disparity in precession frequencies at the magnetic fields normally employed for NMR, each type of nucleus represents to a high degree of approximation a thermodynamically closed system. In 1962, Hartmann and Hahn[83] found a way to bridge the gap between such spin systems. They realized that by simultaneously applying continuous wave rf power at both ν_A and ν_X we define *two* rotating frames and can consider the precessions of A and X in their respective rotating frames, which are

$$\omega_{1A} = \gamma_A B_{1A}$$
$$\omega_{1X} = \gamma_X B_{1X}$$
(7.9)

We now have an additional degree of freedom—the relative values of B_1. By choosing

$$\frac{B_{1X}}{B_{1A}} = \frac{\gamma_A}{\gamma_X}$$
(7.10)

the "Hartmann–Hahn condition," we can make $\omega_{1A} = \omega_{1X}$ and, as in any quantum mechanical system, facilitate exchange of energy between A and X.

This principle can be applied in a technique, usually called *cross polarization*, that uses the large magnetization reservoir of protons to observe nuclei such as ^{13}C, as illustrated in Fig. 7.5. The proton has a magnetic moment four times as large as that of ^{13}C, and proton spins are virtually 100% abundant. By applying rf simultaneously to ^{1}H and ^{13}C in accord with Eq. 7.10 until equilibrium is attained (usually 1–5 ms), the ^{13}C polarization is enhanced by a factor of 4 (the ratio γ_H/γ_C, a result that can be derived from thermodynamic arguments not reproduced here). Meanwhile, the proton reservoir is depleted by only 1% of its value. The ^{13}C rf is now turned off, and a ^{13}C FID is observed that is much larger than its normal value, ideally by a factor of 4 but in practice often by a factor of 2–3. The ^{1}H rf remains on to supply dipolar decoupling during the FID. At the conclusion of the FID, the Hartmann–Hahn contact between ^{1}H and ^{13}C can be reestablished and the process repeated if the proton magnetization has not dephased. The rate of relaxation might appear to be determined by the ^{1}H T_2, but as we discuss in Chapter 9, the presence of the continuous ^{1}H rf field causes the

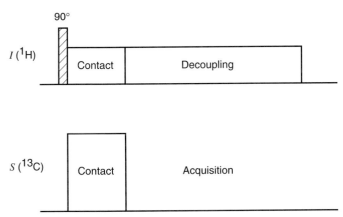

FIGURE 7.5 Pulse sequence for cross polarization. After a 90° pulse, the *I* spins transfer polarization to *S* spins during the contact time as described in the text. The *S* signal is then detected while the *I* spins are decoupled.

magnetization to be *spin locked* along x', where it relaxes with a considerably longer time, $T_{1\rho}$, the spin–lattice relaxation time in the rotating frame. Nevertheless, in many instances $T_{1\rho}$ is sufficiently short that the ^1H magnetization has been depleted and a further Hartmann–Hahn contact is not worthwhile. However, the entire pulse sequence can be repeated after longitudinal relaxation *of the proton magnetization* restores equilibrium. In solids T_1 for protons is usually much less than T_1 for ^{13}C, so that the entire process produces a much larger ^{13}C signal in a given time than would be obtained by a simple series of ^{13}C pulses.

Overall, the combination of cross polarization and dipolar decoupling has made feasible the solid-state study of ^{13}C, ^{15}N, ^{29}Si, and many other nuclides of low abundance. Cross polarization can also be applied in situations in which abundant spins can be diluted by isotopic substitution or by dispersing molecules in an inert solid state matrix.

7.7 THE HOMONUCLEAR TWO-SPIN SYSTEM

For a pair of magnetically equivalent nuclei (e.g., two protons), symmetrized wave functions must be used, as in Eqs. 6.42 and 6.44, and the resulting calculation gives eigenvalues and transitions as described in Section 6.11, except that the adjacent symmetric energy levels are separated by

$$\Delta \nu = \frac{3}{4} D_{AA}(3 \cos^2 \theta - 1) \tag{7.11}$$

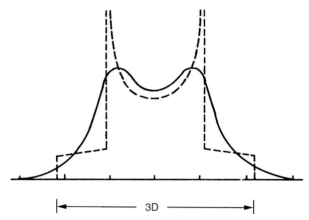

FIGURE 7.6 Powder patterns for homonuclear dipole coupling. Dashed line represents a two-spin system. Solid line shows broadening from other nearby nuclei in a multispin system.

rather than ν_A. The angular dependence is the same as in the heteronuclear case, and for a powder pattern (often called a *Pake pattern* after its discoverer[84]) the shape is of the same form as for the A portion of the AX pattern. However, note from Fig. 7.6, which shows a typical homonuclear powder pattern, that the separations are $^3/_2$ as large as for the AX case, as a result of both terms A and B of Eq. 7.5 coming into play.

If the nuclei are of the same species but are not magnetically equivalent (e.g., because of chemical shift differences), the four basis functions of Eq. 6.1 must be used. However, because the dipolar coupling is normally much larger than chemical shift differences, the resulting spectra and powder pattern are little changed from those in Fig. 7.6.

7.8 LINE NARROWING BY MULTIPLE PULSE METHODS

For the heteronuclear two-spin system we found that we could create an "average" of the two X spin states by decoupling. Clearly such heteronuclear decoupling is inapplicable to the homonuclear two-spin system, but examination of the Hamiltonian of Eq. 7.6 suggests another approach—manipulate the spin system to make the two spin terms equal *on the average* so that $\mathcal{H}_D = 0$. This can be accomplished by use of appropriate pulse sequences. Prior to our discussion of pulse sequences in Chapters 9–11, we are not prepared to treat this rather complex process in detail, but a simple pictorial presentation gives the essential features.

Consider the sequence of four pulses applied in the rotating frame to the two dipolar-coupled nuclei:

$$90^{\circ}_{x'},\ 2\tau,\ 90^{\circ}_{-x'},\ \tau,\ 90^{\circ}_{y'},\ 2\tau,\ 90^{\circ}_{-y'},\ \tau$$

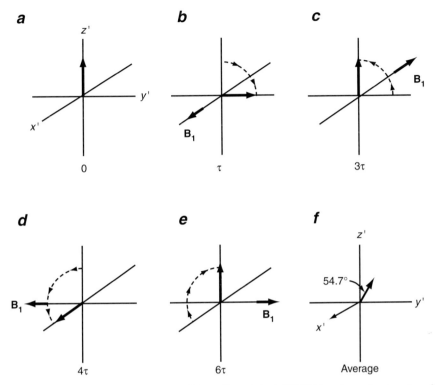

FIGURE 7.7 Pictorial representation of the effect of the WAHUHA pulse cycle. A series of 90° pulses applied as indicated causes **M** to spend a period 2τ along each of the Cartesian axes in the rotating frame. The *average* orientation of M during the period 6τ is thus along the diagonal at 54.7° from the z axis, as shown.

The effect of the pulses, as illustrated in Fig. 7.7, is to cause the magnetization to spend one third of its time along each of the three Cartesian axes and to end where it started, along z'. Such a sequence that brings **M** back to its initial position is called a pulse *cycle*. If the cycle is carried out rapidly, that is, if $6\tau \ll T_2$, we can treat only the *average* Hamiltonian, just as we did in the heteronuclear system under dipolar decoupling. In Eq. 7.6, the term $\mathbf{I}_1 \cdot \mathbf{I}_2$ is unaffected because it involves only the relative orientations of the two spins, which the pulse cycle does not alter. However, in the rotating frame the term $I_{1z}I_{2z}$ behaves as though it has been converted during each of the three 2τ periods:

$$\frac{3}{2} I_{1z}I_{2z} \Rightarrow \frac{1}{2} I_{1x}I_{2x} + \frac{1}{2} I_{1y}I_{2y} + \frac{1}{2} I_{1z}I_{2z} \tag{7.12}$$

The sum of these three terms cancels that from the scalar product and makes $\mathcal{H}_D = 0$. As shown in Fig. 7.7, the *average* orientation of **M** during the period of the pulse cycle is at an angle of 54.7° from the z axis, the angle at which we noted in Section 7.4 that dipolar interactions vanished.

A rigorous quantum mechanical treatment shows that the process is more complex, particularly when pulse imperfections are taken into account. The basic four-pulse cycle just given (called WAHUHA after its inventors)[85] has been supplanted in practice by cycles that use 8–24 pulses (e.g., MREV-8, MREV-16, BLEW-24). The cycle is repeated many times during the period T_2, and observation of the magnetization is made after each cycle during one of the τ periods. These *multiple pulse* cycles are difficult to apply but are quite effective in narrowing lines.

7.9 ANISOTROPY OF THE CHEMICAL SHIELDING

In Chapter 4 we saw that chemical shielding arises primarily from induced currents in the vicinity of the nucleus but depends on the distribution of electrons within the molecule. Except for highly symmetric molecules the induced current varies with relative orientation of the molecule and the magnetic field $\mathbf{B}_0$ that induces the currents. As we saw in Section 4.1, the shielding σ is clearly a tensor, which can be expressed in terms of its principal components σ_{11}, σ_{22}, and σ_{33} along three mutually orthogonal coordinates in the molecules. The normal convention is to take

$$\sigma_{11} \leq \sigma_{22} \leq \sigma_{33} \qquad (7.13)$$

with axially symmetric tensors designated by $\sigma_\parallel$ and $\sigma_\perp$, as indicated in Section 4.1. In a liquid with rapid, random molecular tumbling, we saw in Eq. 4.2 that the isotropic value of the shielding, σ_{iso}, is the average of the three principal components.

If the molecule is in a solid, σ can also be described in terms of Cartesian space-fixed (laboratory) coordinates in which $\mathbf{B}_0$ is normally taken along the z axis, and it is the shielding along this axis that alters the resonance frequency. Within the secular approximation, it is only σ_{zz} that contributes, and it may be related to the three principal components via their direction cosines relative to $\mathbf{B}_0$:

$$\sigma_{zz} = \sigma_1 \cos^2 \theta_1 + \sigma_2 \cos^2 \theta_2 + \sigma_3 \cos^2 \theta_3 \qquad (7.14)$$

Equation 7.14 may readily be rewritten in terms of $\sigma_{iso} = \frac{1}{3}(\sigma_{11} + \sigma_{22} + \sigma_{33})$:

$$\sigma_{zz} = \sigma_{iso} + \frac{1}{3} \sum_{i=1}^{3} (3 \cos^2 \theta_i - 1)\sigma_i \qquad (7.15)$$

For the commonly encountered axial symmetry, where $\sigma_1 = \sigma_2 = \sigma_\perp$ and $\sigma_3 = \sigma_\parallel$, we can use the relation

$$\cos^2 \theta_1 + \cos^2 \theta_2 + \cos^2 \theta_3 = 1 \qquad (7.16)$$

a

b

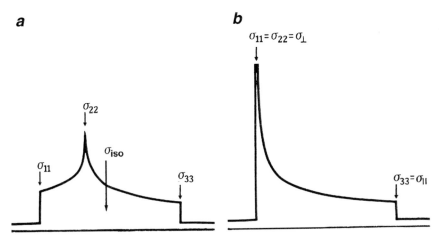

FIGURE 7.8 Schematic representation of a powder pattern for anisotropic chemical shielding. (a) Typical pattern for $\sigma_{11} \neq \sigma_{22} \neq \sigma_{33}$. ($b$) Typical pattern for axial symmetry where $\sigma_{11} = \sigma_{22} = \sigma_\perp$ and $\sigma_{33} = \sigma_\parallel$.

to show that

$$\sigma_{zz} = \sigma_{iso} + \tfrac{1}{3}[3 \cos^2 \theta_\parallel - 1](\sigma_\parallel - \sigma_\perp) \tag{7.17}$$

This expression shows that the chemical shielding anisotropy (CSA) leads to a variation of the resonance frequency with orientation according to the familiar $(3 \cos^2 \theta - 1)$ dependence and thus produces typical powder patterns, as illustrated in Fig. 7.8 for axial and nonaxial symmetries.

An individual powder pattern is far more informative than the isotropic value of σ as deduced from a liquid state spectrum, but overlap of a number of powder patterns for a complex molecule may render a spectrum uninterpretable. Thus, after carrying out a cross polarization experiment with dipolar coupling, we may be left with a very confusing spectrum. Fortunately, there is a simple sample rotation technique for removing the effect of the CSA to give high resolution spectra in solids that resemble the spin-decoupled spectra in liquids. Moreover, there are other sample rotation procedures for obtaining the CSA information while still providing relatively narrow lines.

7.10   MAGIC ANGLE SPINNING

In 1958, long before dipolar decoupling and multiple pulse methods were developed, a very different approach was introduced in an attempt to narrow magnetic dipolar broadened NMR lines in solids. This method focuses on the geometric part of the Hamiltonians (Eqs. 7.6, 7.8, and 7.17), namely the term $(3 \cos^2 \theta - 1)$.

As we have seen, this term vanishes if θ is averaged over all space, but clearly it also vanishes if $\cos^2 \theta = \frac{1}{3}$, which implies that $\theta = 57.4°$. By transforming from molecular coordinates to those that describe the orientation of a macroscopic cylindrical sample, one can show that rapid rotation of the macroscopic sample at an angle of 54.7° to $\mathbf{B}_0$ eliminates the θ dependence, hence all dipolar interactions and CSA effects.

The initial impetus for such sample spinning was to remove dipolar interactions, and in instances in which such interactions are small (e.g., between two ^{31}P nuclei with relatively small magnetic moments) the effect was demonstrated, prompting the comment from a pioneer in NMR that the result was "magic" and giving rise to the term *magic angle spinning* (MAS). However, for homonuclear dipolar coupling, the spinning rate for significant line narrowing must be much

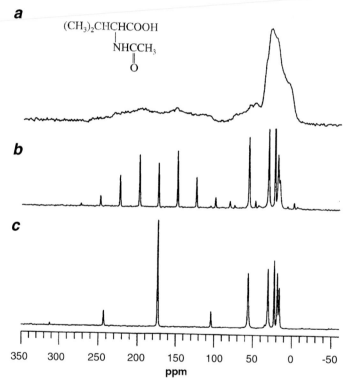

FIGURE 7.9 ^{13}C NMR spectra of N-acetyl-L-valine at 100 MHz (9.4 T) with ^1H cross polarization. (*a*) Static spectrum, showing a powder pattern from chemical shift anisotropy. (*b*) Magic angle spinning at 2.5 kHz is nearly sufficient to average CSA for the five aliphatic carbon nuclei, but the spinning sidebands follow the powder pattern contour for the carbonyl carbons. (*c*) Spinning at 7 kHz eliminates most spinning sidebands and shows sharp peaks for all nuclei. Spectra courtesy of Robert Tycko (National Institutes of Health).

greater than the magnitude of the interaction (in Hz), and multiple pulse methods (WAHUHA, etc.) are much more effective. For heteronuclear dipolar coupling MAS has effects at low spinning speeds (see Section 7.12), but dipolar decoupling is generally more effective in providing narrow lines.

MAS is normally applied concurrently with the dipole line-narrowing methods in order to eliminate the effects of CSA and provide true "high resolution" NMR spectra in the solid phase. For heteronuclear systems the combined method is usually referred to by the initials CP-MAS, and for homonuclear systems the acronym CRAMPS (combined rotation and multiple pulse spectroscopy) has been coined.

An illustration of CP-MAS is given in Fig. 7.9. The sharp lines that are obtained permit CP-MAS to be used for structure elucidation of organic and inorganic compounds, in much the same way as liquid state spectra are used. As we see in Chapter 10, two-dimensional NMR methods are applicable to solids and form part of the analytical capability of CP-MAS and CRAMPS. One important application is in combinatorial chemistry, where molecules are synthesized on resin beads. It is feasible to obtain good ^1H and ^{13}C spectra from a single bead.

Spinning Sidebands and Recoupling Techniques

When the spinning speed is much less than the magnitude of the CSA, spinning sidebands are generated at multiples of the spinning rate, reminiscent of the spinning sidebands for liquid samples discussed in Section 3.3. The envelope of the sidebands approximates the shape of the CSA-broadened line in the absence of spinning, and analysis of the intensity pattern provides the values of σ_1, σ_2, and σ_3.

Although cross polarization and dipolar decoupling remove the effect of proton interactions, other dipolar interactions, such as those between $^{13}C-^{13}C$ or $^{13}C-^{15}N$ in isotopically enriched samples, persist. However, as such couplings are relatively small because of the small magnetogyric ratios, MAS causes them to average out. Where single sharp lines are desired for study of complex molecules, such simplification is desirble, but in other molecules this dipolar coupling information can be useful for estimating internuclear distances and attacking molecular structure problems in solids. As a result, techniques have been developed that cause *recoupling* of these homonuclear or heteronuclear dipolar interactions while maintaining the resolution improvement from MAS.

A number of recoupling techniques are utilized for specific purposes, but they all rely on synchronizing measurements, and, in most instances, pulse sequences, with the rotation period. For homonuclear recoupling (e.g., $^{13}C-^{13}C$ or $^{31}P-^{31}P$), rotational resonance (R^2) is based on magnetization exchange when the spinning rate ω_r is adjusted so that a multiple of it equals the chemical shift difference between the two dipolar coupled nuclei, and DRAMA (*d*ipolar *r*ecovery *a*t the *m*agic *a*ngle) uses a multiple pulse sequence to interfere with spatial averaging imposed by MAS. Heteronuclear recoupling is most frequently carried out by *r*otational *e*cho *d*ouble *r*esonance (REDOR) or one of its offshoots. For an $^1H-^{13}C-^{15}N$ system, for example, a normal ^{13}C CP-MAS experiment is carried out, but $180°$ ^{15}N pulses are applied at $2\omega_r$. These pulses invert the ^{15}N spins, thus modulating the $I_z S_z$ dipolar coupling term and interfering with the averaging achieved by the MAS. The resulting signals can be analyzed to obtain the value of the dipolar coupling.

7.11 QUADRUPOLE INTERACTIONS AND LINE-NARROWING METHODS

For nuclei with $I \geq 1$, the nuclear electric quadrupole moment interacts with the electric field gradient in a nonsymmetric environment of electrons around the nucleus. The electric field gradient is described by a tensor that can be expressed in diagonal form (components V_X, V_Y, and V_Z) in a principal axis system fixed in the molecule. By Laplace's equation the sum of the three components is zero, so there are two independent parameters, usually taken as the largest of the three components, $V_Z = \partial^2 V/\partial z^2$, and the asymmetry parameter $\eta = (V_X - V_Y)/V_Z$, with the convention that $|V_Z| \geq |V_Y| \geq |V_X|$. The electrical interaction leads to quantized energy levels, which can be studied by nuclear quadrupole resonance (NQR), in which rf is applied to induce transitions *without* application of a static magnetic field. NQR is applicable only in solids, where molecules are fixed in space, hence fixed relative to the rf coil. We shall not explore NQR here.

With application of $\mathbf{B}_0$, it is possible to study NMR of quadrupolar nuclei. Because the principal axis of $\mathbf{V}$ does not in general correspond to the direction of $\mathbf{B}_0$ in the laboratory frame, geometric factors, including the ubiquitous $(3 \cos^2 \theta - 1)$ must be considered, but the situation is more complex than we have seen previously. In many instances we cannot treat the quadrupolar interaction as merely a small perturbation on the Zeeman energy levels, as we did with dipolar interactions and CSA, because quadrupole energies can often be of the order of MHz (except for ^2H). Moreover, with a quadrupolar nucleus there are more than two energy levels, which respond differently to quadrupole interactions, as do the transitions between them. Within the scope of this book, we do not attempt to provide general equations but limit ourselves to describing only a few special cases and indicating approaches to line narrowing in those cases.

When the Zeeman interaction is much larger than the quadrupole interaction, nonsecular terms may be discarded to give a relatively simple expression for the quadrupole Hamiltonian:

$$\mathscr{H}_Q = \frac{\chi}{8I(2I - 1)} [3 \cos^2 \theta - 1 + \eta \sin^2 \theta \cos 2\phi][3I_z^2 - I^2] \qquad (7.18)$$

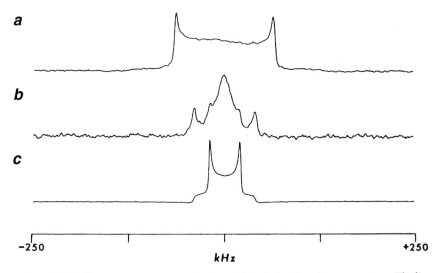

FIGURE 7.10 ^{2}H (D) NMR spectrum of polycrystalline alanine-d_3 at three temperatures. The line shape variation results from reorientation of the CD$_3$ group. (*a*) At 123 K the powder pattern represents a nearly static CD$_3$. (*b*) At 177 K, the line shape is distorted by motional averaging. (*c*) At 293 K, motion is fast enough to produce an undistorted axially symmetric averaged powder pattern. Spectra courtesy of Dennis A. Torchia (National Institutes of Health).

The angles θ and ϕ define the relative orientations of $\mathbf{B}_0$ and the molecule-based principal axis system of the electric field gradient. The *quadrupole coupling constant* $\chi = eQV_z/h$ describes the magnitude of the interaction between the nuclear quadrupole moment eQ (where e is the charge on the electron) and the electric field gradient, whose maximum component V_z is also often given the symbol eq, making $\chi = e^2Qq/h$. $\mathcal{H}_Q$ may be used in a first-order perturbation treatment to obtain energy levels.

The treatment of ^{2}H is simplified because $I = 1$, so there are only three energy levels, and because the quadrupole interactions range only as high as about 200 kHz. The frequencies of the two transitions vary with $(3 \cos^2 \theta - 1)$ and produce a powder pattern of the sort shown in Fig. 7.10. As indicated in the figure, the full line shape can sometimes provide valuable information on dynamic processes, but in other instances we would like to obtain only a single narrow line. In principle, MAS could collapse the powder pattern. However, even for small quadrupole interactions the spinning speed would normally be too high to produce a single line, even though a series of sharp spinning sidebands can be obtained, just as in the heteronuclear dipolar example. An alternative approach is suggested by the fact that, although the frequencies of the individual transitions vary with θ, their sum is invariant because the θ dependence of the energies of the $m = +1$ and $m = -1$ levels is the same, as indicated in Fig. 7.11. Thus, a *double quantum*

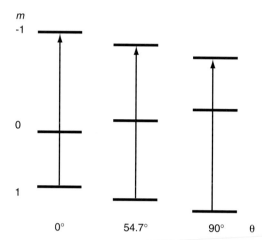

FIGURE 7.11 Variation of energy levels for $I = 1$ with angle θ and illustration of the constant energy separation for the $\Delta m = 2$ transition.

transition, $\Delta m = \pm 2$, would be a sharp line. Although such transitions are formally forbidden by the selection rules we discussed in Section 2.3, it is possible to observe them under some conditions. Also, there are pulse sequences that can effect such transitions in an indirect manner, as we shall see in later chapters.

Quadrupolar nuclei with half-integer spins provide other challenges and opportunities. For example, the first-order quadrupole interaction does not affect the frequency of the central transition, while others have the usual θ dependence. However, for appreciable quadrupole interactions the first-order treatment is inadequate, and second-order effects, which depend on higher order Legendre polynomials, come into play. MAS would be useful in some line-narrowing studies, but independent spinning about two axes is required and has been realized in *double rotation* (DOR) and *dynamic angle spinning* (DAS).

7.12 OTHER ASPECTS OF LINE SHAPES

We have outlined the basic effects by focusing almost entirely on the interactions involving a single spin (CSA and quadrupole effects) and on the dipolar and indirect spin coupling interactions between a pair of spins. In practice, more spins must usually be considered. Pairwise dipolar interactions may be summed (as in the case of J couplings), but some complications ensue as the spin functions of more than two homonuclear spins generally no longer commute. The result is that the *inhomogeneous* line broadening seen in all the examples we have examined now becomes a *homogeneous* line broadening. Inhomogeneous broadening is analogous to the broadening we observed for the narrow lines in the liquid state when the magnetic field homogeneity is inadequate to reproduce their true widths. In that case, the apparent width arises from the summation of individual lines from nuclei in different parts of the sample tube. Likewise, in the examples discussed in this chapter the spins are in molecules that have different orientations in the laboratory frame. Each is independent, as may be shown experimentally by irradiating strongly at a specific frequency to saturate only those nuclei and leave the others unchanged, thus "burning a hole" in the line. For a homogeneously broadened line, on the other hand, magnetization transfers rapidly among the interacting nuclei (a process called *spin diffusion*), so that the entire line saturates when any part is strongly irradiated. Also, homogeneously and inhomogeneously broadened lines respond differently to MAS. Spinning at even slow speeds causes an inhomogeneously broadened line to generate spinning sidebands (as in Fig. 3.2 for a line broadened by magnet inhomogeneity), whereas a homogeneously broadened line displays no change until the spinning speed is comparable with the line width.

Study of line shapes in solids often provides valuable information on molecular motion — gross phase changes, overall tumbling of molecules, or internal rotations and other motions. For a limited number of spins the dipolar, CSA, or quadrupolar interactions may be simulated and compared with experiment, whereas for multispin systems the line shape is often rather featureless and only the overall shape can be characterized.

The faithful representation of the shape of lines broadened greatly by dipolar and, especially, quadrupolar interactions often requires special experimental techniques. Because the FID lasts for only a very short time, a significant portion may be distorted as the spectrometer recovers from the short, powerful rf pulse. We saw in Section 2.9 that in liquids a $90°$, τ, $180°$ pulse sequence essentially recreates the FID in a spin echo, which is removed by 2τ from the pulse. As we saw, such a pulse sequence refocuses the dephasing that results from magnetic field inhomogeneity but it does not refocus dephasing from natural relaxation processes such as dipolar interactions. However, a somewhat different pulse sequence can be used to create an echo in a solid — a *dipolar echo* or a *quadrupolar echo* — and this method is widely employed in obtaining solid state line shapes (for example, that in Fig. 7.10). The formation of these echoes cannot readily be explained in terms of the vector picture, but we use the formation of a dipolar echo as an example of the use of the product operator formalism in Section 11.6.

7.13 ORIENTATION EFFECTS IN LIQUIDS: LIQUID CRYSTALS

Molecules that possess magnetic anisotropy experience a slight tendency toward alignment with an imposed magnetic field, but in normal isotropic solvents random thermal motions dominate, and no effects of orientation are usually observed at magnetic fields commonly employed for NMR. [At very high fields (>10 T), small orientation effects are sometimes observable, as described subsequently.] However, in liquid crystalline mesophases limited translational or rotational motions lead to a cooperative effect in the magnetic field interaction that competes successfully with thermal fluctuations to produce long-range order.

Thermotropic liquid crystals are normally pure materials that exist as a mesophase within a certain temperature range, between the solid and isotropic liquid phases. Molecules that form thermotropic mesophases are normally elongated, relatively flat, and rather rigid in the long-axis direction. They often include hydrocarbon chains attached to aromatic or cyclohexane rings, with additional substituents. *Lyotropic* liquid crystals arise from mixtures of materials, normally in an aqueous medium, one of which is often amphiphilic (hydrophobic on one end and hydrophilic on the other). Typical molecules include certain soaps and biological membranes. Whereas an isotropic liquid is characterized by random motions in the three translational and three rotational degrees of freedom, a liquid crystalline phase consists of partial order and restriction in certain degrees of freedom.

Such order can be described in terms of the preferential alignment of the *director*, a unit vector that describes the orientation of molecules in a nematic phase. Because the molecules are still subject to random fluctuations, only an average orientation can be described, usually by an ordering matrix **S**, which can be expressed in terms of any Cartesian coordinate system fixed in the molecule. **S** is symmetric and traceless and hence has five independent elements, but a suitable choice of the molecular axes may reduce the number. In principle, it is always possible to diagonalize **S**, and in such a principal axis coordinate system there are only two nonzero elements (as there would be, for example, in a quadrupole coupling tensor). In the absence of symmetry in the molecule, there is no way of specifying the orientation of the principal axes of **S**, but considerable simplification is obtained for symmetric molecules. If a molecule has a threefold or higher axis of symmetry, its selection as one of the axes of the Cartesian coordinate system leaves only one independent order parameter, with the now familiar form:

$$S = \tfrac{1}{2}(3\cos^2\theta - 1) \qquad (7.19)$$

Equation 7.16 implies that S can vary from $+1$ to $-\tfrac{1}{2}$.

When the sample is allowed to equilibrate in a strong static magnetic field, the direction of B_0 defines the z axis, and all the anisotropic interactions discussed earlier in this chapter are displayed. Many commercially important materials are liquid crystals, as are biological membranes, and NMR can be used to investigate their structures. In addition, small molecules that would tumble randomly in an isotropic solvent are impeded in a liquid crystal solvent to the extent that they, too, display anisotropic interactions. However, the order parameter for such a small molecule is usually much less than the maximum values of $+1$ or $-\tfrac{1}{2}$. Typically, values of $|S| \approx 0.1$ are observed for molecules dissolved in thermotropic liquid crystals, and in general, the order parameters for molecules dissolved in lyotropic liquid crystals are much smaller. As a result, the effective dipolar coupling are scaled down, and it is possible to analyze the spectra of rather large spin systems and to determine precise information on internuclear distances. Extensive data have been obtained for small, usually symmetric, molecules. It is also possible to orient proteins in liquid crystalline media to provide valuable structural information, as we discuss in Chapter 13.

In an isotropic solvent, small effects from the orientation of magnetically aniosotropic molecules have been observed, but only at high magnetic fields, because the effect goes as B_0^2. For example, even at $B_0 = 14\,T$ (600 MHz), benzene at room temperature has an order parameter of only -3.2×10^{-6} and gives an ^2H quadrupole splitting of about 0.5 Hz (as compared with a deuterium quadrupole coupling in the solid of $\sim$275 kHz). Dipolar splittings for protons have been observed only in molecules with large magnetic anisotropy, such as porphyrins (where a splitting of about 2 Hz has been seen) and paramagnetic species (where somewhat larger interactions are observed).

7.14 ADDITIONAL READING AND RESOURCES

As indicated at the beginning of this chapter, we have addressed only a very small portion of the rich field of NMR in solids. Of many books devoted partly or entirely to this field, two that are particularly oriented toward the aspects discussed here are *High Resolution NMR in the Solid State* by E. O. Stejskal and J. D. Memory[86] and *Solid State NMR for Chemists* by Colin Fyfe.[87] Also of interest is *NMR Spectroscopy Using Liquid Crystal Solvents* by J. W. Emsley and J. C. Lindon.[88]

The *Encyclopedia of NMR* is an excellent source, with more than 10% of the articles covering the broad area of solid state NMR. Included are a number of articles (too many to list here) that deal with high resolution NMR in solids and provide information on various sample spinning techniques that we have been unable to cover. Also in the *Encyclopedia of NMR* are several articles on NMR of liquid crystals.

7.15 PROBLEMS

7.1 If the magnetic dipolar interaction between two nuclei has a magnitude of 55 kHz when the dipoles are at an angle $\theta = 90°$, what is the magnitude of the interaction at $0°$?

7.2 Use the dipolar Hamiltonian in Eq. 7.8 to compute the energy levels and spectra for the A part of an AX spin system. Ignore chemical shifts and indirect coupling. Use wave functions and other pertinent results from Chapter 6.

7.3 Use the dipolar Hamiltonian for a homonuclear spin system, Eq. 7.6, together with symmetrized wave functions from Chapter 6, to compute the energy levels and spectrum for a single H_2O molecule in a solid.

7.4 Use data from Table 4.1 to sketch the powder pattern to be expected for solid ethylene (C_2H_4) and solid acetylene (C_2H_2). Ignore dipole and indirect spin coupling effects.

7.5 Each of the terms $A-F$ in Eq. 7.5 converts one of the basis functions— $\alpha\alpha$, $\alpha\beta$, $\beta\alpha$, $\beta\beta$—to the same or another function and is often said to "link" such pairs of functions. For example, A converts $\alpha\alpha$ to $\alpha\alpha$ because α is an eigenfunction of I_z. (a) Use the information in Section 2.3 to show the linkages for all four basis functions. (b) Convert the spin operators in term B to raising and lowering operators to demonstrate why this term is often called the "flip-flop" term.

Relaxation

In Chapter 2 we found that a perturbed nuclear spin system relaxes to its equilibrium state or steady state by first-order processes characterized by two relaxation times: T_1, the spin–lattice, or longitudinal, relaxation time; and T_2, the spin–spin, or transverse, relaxation time. Thus far in our treatment of NMR we have not made explicit use of relaxation phenomena, but an understanding of the limitations of many NMR methods requires some knowledge of the processes by which nuclei relax. In addition, as we shall see, there is a great deal of information of chemical value, both structural and dynamic, that can be obtained from relaxation phenomena.

In Chapter 7 we identified four types of interactions between pairs of nuclear spins or between a spin and its molecular environment that can influence the energy state of a nuclear spin. We shall find in this chapter that each of these interactions can provide a pathway for exchange of energy between nuclei or between a nucleus and its surroundings, that is, a pathway for relaxation. In addition, a fifth process, spin–rotation interaction, can furnish a means for relaxation for molecules in gases or in highly mobile liquids. Also, a sixth process, paramagnetic relaxation, which involves magnetic dipolar interaction between a nuclear spin and an unpaired electron spin, is an efficient path for relaxation when even a small amount of paramagnetic material is present.

Before examining these processes in turn, we must ask a fundamental question: Because energy is absorbed from an electromagnetic field only at a sharply defined resonance frequency, how can the nuclear spins give up the energy to the surroundings without the intervention of a coherent oscillating magnetic field? The answer was found very early in the development of NMR by Bloembergen, Purcell, and Pound,[89] whose theory (commonly called "BPP") we shall explore in general terms. More complex theoretical approaches of broader applicability are required to understand many NMR pheneomena, especially in solids, but we restrict our treatment to the BPP theory.

8.1 MOLECULAR MOTIONS AND PROCESSES FOR RELAXATION IN LIQUIDS

As we have noted, effective interaction between a precessing nuclear magnetization and its surroundings requires an oscillating magnetic field (or, for quadrupole interactions, an oscillating electric field) at the Larmor frequency of the nucleus. BPP recognized that in liquids or gases the oscillating field can come from molecules that are in rapid, random motion, a process well known as Brownian motion. Although such motions are incoherent, they can be Fourier analyzed (just as we analyzed the FID in Section 3.6) to obtain the "spectrum" of frequencies present in the random motions. Unlike an FID, which normally has a small number of well-separated frequencies, the range of frequencies here, described by the *spectral density function*, is continuous and broad. Nevertheless, among the range of frequencies is the Larmor frequency, and its relative contribution governs the strength of the oscillating field that can cause energy exchange and relaxation.

To make this argument more quantitative, consider the average length of time that a molecule remains in any given position before a collision with another molecule causes it to change its state of motion. For a small molecule in a nonviscous liquid, this period of time is of the order of 10^{-12} seconds; for a polymer, it is usually several orders of magnitude longer. In the BPP theory this time is called the *correlation time*, τ_c, and arises from the following treatment: The orientation of a molecule at time t can be described by $Y(t)$, a function related to second-order spherical harmonics, and its position at a later time by $Y(t + \tau)$. The extent of motion during the period τ can be expressed by the correlation function $k(\tau)$, defined by the following equation:

$$k(\tau) = \overline{Y(t)Y(t + \tau)} \tag{8.1}$$

where the bar indicates an average over the entire ensemble of molecules. Usually the correlation function is assumed to have an exponential form:

$$k(\tau) = k(0)e^{-\tau/\tau_c} \tag{8.2}$$

This equation says that the new position $k(\tau)$ is related to the initial position in an exponential manner; that is, the two positions are much more likely to be different (uncorrelated) after a long time τ. The precise mathematical form of the correlation function is the subject of considerable debate and research, but an exponential is a reasonable assumption for our purposes, and the time constant τ_c sets the scale of time. The spectral density $J(\omega)$ is the Fourier transform of $k(\tau)$:

$$J(\omega) = \int_{-\infty}^{\infty} k(\tau)e^{-i\omega t}\,dt$$

$$= \int_{-\infty}^{\infty} k(0)e^{-\tau/\tau_c}e^{-i\omega t}\,dt$$

$$= A\,\frac{\tau_c}{1 + \omega^2\tau_c^2} \tag{8.3}$$

A is a constant, which we will evaluate separately for each relaxation mechanism.

Plots of $J(\omega)$ versus ω for different values of τ_c are shown in Fig. 8.1. The values of τ_c should be compared with the reciprocal of the nuclear Larmor frequency ω_0. For either a very short or very long τ_c the value of $J(\omega)$ at ω_0 is relatively small. $J(\omega)$ reaches its maximum when $\tau_c = 1/\omega_0$, that is, when the average molecular tumbling frequency is equal to the nuclear precession

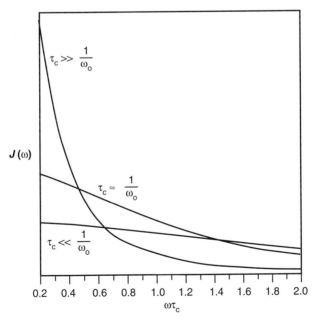

FIGURE 8.1 Spectral density function $J(\omega)$ as a function of $\omega\tau_c$ for various values of τ_c.

frequency. Under these circumstances energy transfer between precessing nuclei and randomly tumbling molecules is most efficient, the *relaxation rate* R_1 is a maximum, and T_1 is a minimum.

Our discussion here is entirely in terms of a single correlation time. Only for a small rigid molecule undergoing completely isotropic motion is this description adequate. In molecules in which motion about one axis is preferred, two or three different correlation times are applicable; in molecules in which internal rotation can occur, several correlation terms must be considered; and in more complex systems (e.g., water in biological cells) a number of translational and rotational correlation times may be pertinent. In such cases a plot of $J(\omega)$ versus ω consists of a superposition of several curves of the type given in Fig. 8.1, and a study of the frequency dependence of T_1 can be quite informative.

Returning to the simple situation of a single τ_c, we can recast the results of Fig. 8.1 into a plot of T_1 versus τ_c, as shown in Fig. 8.2. No coordinate scale is given, because it depends on the specific types of interactions to be discussed. T_1 goes through a minimum at $\tau_c = 1/\omega_0 = 1/2\pi\nu_0$, as we discussed earlier. With NMR frequencies in the range of about $1-1000$ MHz for various nuclei and magnetic field strengths, the minimum in the T_1 curve can come from about 2×10^{-10} to 2×10^{-7} seconds. Small molecules (molecular weight <200) almost always lie to the left of the minimum unless the solvent is very viscous or the observation frequency is extremely low. Small polymers, such as proteins of $10-25$ kilodaltons, may be in the vicinity of the minimum and could fall to either side depending on the Larmor frequency.

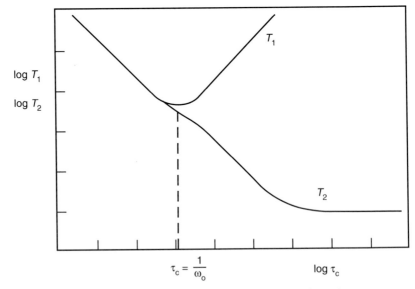

FIGURE 8.2 A log–log plot of T_1 and T_2 versus the correlation time τ_c.

In order to probe lower frequency motions, some relaxation measurements are made in instruments designed to allow relaxation to occur at very low magnetic field, where the Larmor frequency is a fraction of a MHz. Alternatively, it is possible to define and measure a *spin–lattice relaxation time in the rotating frame*, given the symbol $T_{1\rho}$, which is sensitive to motions in the kHz range. We shall return to $T_{1\rho}$ in Chapter 9.

Figure 8.2 also shows T_2 as a function of τ_c. The calculation of the functional form of T_2 proceeds in much the same way as that of T_1. However, because T_2 involves dephasing of precessing nuclear magnetizations, rather than exchange of energy between the nuclei and the environment (lattice), the dependence on molecular motion is somewhat different from that of T_1. The processes giving rise to spin–spin relaxation (T_2 processes) depend on high frequency (short τ_c) motions in the same way as T_1 processes, but low frequency motions and other low frequency processes, such as chemical exchange, significantly shorten T_2, as we discuss in Section 8.2. Hence T_2 decreases monotonically with increasing τ_c and ultimately approaches a limiting value that is characteristic of a completely rigid solid lattice.

Regardless of the types of molecular motion, relaxation occurs only if there is some specific *interaction* between the nucleus and its environment that can result in energy exchange. We now describe the six interactions that have been mentioned, beginning with the one that is always present—the interaction between nuclear magnetic dipoles. In the following treatment of each of these relaxation mechanisms, we shall focus on the relaxation rate $R_1 = 1/T_1$, as the overall relaxation rate is the sum of the rates produced by each mechanism.

8.2 NUCLEAR MAGNETIC DIPOLE INTERACTIONS

As we discussed in Chapter 7, any magnetic nucleus in a molecule supplies an instantaneous magnetic dipole field, which is proportional to the magnetic moment of the nucleus. With rapid molecular motion, the magnetic dipole interactions collectively average to zero, as we have seen, but as each individual molecule tumbles in solution under the influence of Brownian motion, its field fluctuates in magnitude and direction. Thus there is produced an oscillating magnetic field, with the frequencies of oscillation dependent on τ_c. Just as a precessing nuclear moment can interact with a coherent applied rf magnetic field, so it can interact with the molecular magnetic field component at the Larmor precession frequency, which also provides a time-dependent perturbation. As we have seen, the effectiveness of the fluctuating field in bringing about relaxation depends on the Fourier component of motion at the Larmor frequency, but it also depends (see Eq. 2.12) on the square of the matrix element describing the interaction energy.

Additional insight into the quantum processes that are involved may be obtained from Fig. 8.3, which shows the energy levels for a two-spin system, each of spin $\frac{1}{2}$. The spin operators are denoted I and S. We shall assume no spin–spin

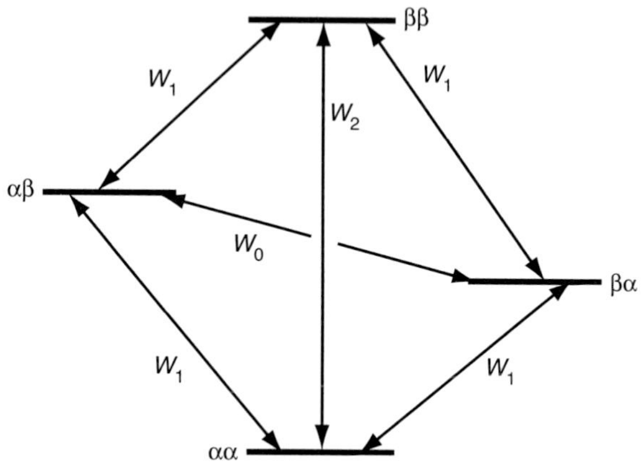

FIGURE 8.3 Energy levels for a two-spin system, I, S, showing the six possible transition pathways.

coupling, but the presence of first-order coupling does not affect the results. Unlike the similar Fig. 7.1, which emphasized the transitions allowed for coherent absorption of rf energy, this plot shows six transitions, all of which are possible pathways for the nonradiative transfer of energy in relaxation processes. The four transitions labeled W_1 are one-quantum transitions that are induced by molecular motions at the Larmor frequency, either ω_I or ω_S. W_2 is a double quantum transition that requires a component of motion at $(\omega_I + \omega_S)$, and W_0 is a zero quantum transition that results from a simultaneous flip of both spins, hence requires motion at $|\omega_I - \omega_S|$. It is easy to set up a differential equation (the analog to Eq. 2.29) to give the rate of change of I magnetization, which depends on the value of I_z at time t and its equilibrium value I_0, but because transitions W_2 and W_0 involve spin flips of S, the values of S_z and S_0 must also be taken into account:

$$\frac{dI_z}{dt} = (2W_1 + W_0 + W_2)(I_0 - I_z) + (W_2 - W_0)(S_0 - S_z) \qquad (8.4)$$

The negative sign for W_0 in the second term reflects the fact that this "flip-flop" transition increases I_z while it decreases S_z.

For a heteronuclear two-spin system, we focus on terms in the spectral density at ω_I, which do not induce transitions of the S spins, so the second term can be ignored to give an overall rate constant of $(2W_1 + W_0 + W_2)$. On the other hand, for a homonuclear system, $I = S$, so the two terms can be combined and simplified to give a rate constant $(2W_1 + 2W_2)$. Note that W_0 does not appear, a consequence of the fact that for a homonuclear system $\omega_I - \omega_S \approx 0$.

We can combine these expressions for transition rates and their associated frequencies with the expressions for the spectral density function to obtain

$$R_1{}^I = A[J(\omega_I - \omega_S) + 2J(\omega_I) + J(\omega_I + \omega_S)] \tag{8.5}$$

$$R_1{}^I = \frac{\gamma_I^2 \gamma_S^2 \hbar^2}{10 r_{IS}^6} \left[\frac{\tau_c}{1 + (\omega_I - \omega_S)^2 \tau_c^2} + \frac{3\tau_c}{1 + \omega_I^2 \tau_c^2} \right.$$
$$\left. + \frac{6\tau_c}{1 + (\omega_I + \omega_S)^2 \tau_c^2} \right] \tag{8.6}$$

for unlike spins (heteronuclear) and

$$R_1{}^I = \frac{3\gamma_I^4 \hbar^2}{10 r_{II}^6} \left[\frac{\tau_c}{1 + \omega_I^2 \tau_c^2} + \frac{4\tau_c}{1 + 4\omega_I^2 \tau_c^2} \right] \tag{8.7}$$

for a homonuclear two-spin system. Both equations can be generalized for situations involving more than two spins and/or spins $>\frac{1}{2}$. The initial constant factor is recognizable as the square of the interaction energy of two magnetic dipoles (Eq. 7.7).

A similar treatment can be carried out for R_2, but as indicated previously, processes occurring at low frequency and zero frequency contribute, along with those at the Larmor frequency. This result is reasonable, as indicated by a simple qualitative classical picture. Consider a component of a fluctuating magnetic field **b** that is moving at the Larmor frequency. In a frame rotating at this frequency, all three Cartesian components $b_{x'}$, $b_{y'}$, and $b_{z'}$ are static, but in the laboratory frame of reference b_z remains static (zero frequency), while b_x and b_y rotate at the Larmor frequency. Both $b_{x'}$ and $b_{y'}$ are orthogonal to M_z, the component of magnetization **M** along the z' axis, hence they can cause precession of **M** and change in its z component (i.e., longitudinal relaxation). The component $b_{z'}$, on the other hand, is collinear with the z component of **M** and contributes nothing to longitudinal relaxation. Thus, only the high frequency components, not the component at zero frequency, are effective in determining T_1. However, a component of **M** in the xy plane is orthogonal to $b_{z'}$ and is in general not collinear with $b_{x'}$ and $b_{y'}$, hence it can be influenced by both high frequency and zero frequency components to cause transverse relaxation.

For transverse relaxation the relations analogous to Eqs. 8.5–8.7 are

$$R_2{}^I = A[4J(0) + J(\omega_I - \omega_S) + 3J(\omega_I) + 6J(\omega_S) + 6J(\omega_I + \omega_S)] \tag{8.8}$$

$$R_2{}^I = \frac{\gamma_I^2 \gamma_S^2 \hbar^2}{20 r_{IS}^6} \left[4\tau_c + \frac{\tau_c}{1 + (\omega_I - \omega_S)^2 \tau_c^2} + \frac{3\tau_c}{1 + \omega_I^2 \tau_c^2} \right.$$
$$\left. + \frac{6\tau_c}{1 + \omega_S^2 \tau_c^2} + \frac{6\tau_c}{1 + (\omega_I + \omega_S)^2 \tau_c^2} \right] \tag{8.9}$$

$$R_2{}^I = \frac{3\gamma_I^4 \hbar^2}{20 r_{II}^6} \left[3\tau_c + \frac{5\tau_c}{1 + \omega_I^2 \tau_c^2} + \frac{2\tau_c}{1 + 4\omega_I^2 \tau_c^2} \right] \tag{8.10}$$

Unlike the equations for R_1, Eqs. 8.8–8.10 show no maximum relaxation rate at the Larmor frequency. Instead, when $\tau_c \gg 1/\omega_0$, the last four terms approach zero, and the expression is dominated by the term in τ_c alone. Thus T_2 becomes shorter with increasing τ_c, but ultimately approaches a limit (not demonstrated in this treatment). The long correlation time permits dipole–dipole interactions to become effective in leading to broad lines, that is, short T_2's, as we saw in Section 7.2.

At the other limit of correlation times, Eqs. 8.6 and 8.9 show that for small τ_c, the denominators approach unity, and $T_2 = T_1$. The region of $\tau_c < 1/\omega_0$ is often called the *extreme narrowing* condition. Note that we are considering here only dipolar interactions. Other relaxation mechanisms discussed subsequently may cause T_2 to be smaller than T_1, even under the extreme narrowing condition.

As we defined τ_c, it represents a *rotational* correlation time, which is appropriate for treating intramolecular relaxation, as it is molecular orientation, not internuclear distance, that fluctuates. Relaxation may also be mitigated by changes in internuclear distances, as molecules move relative to each other. The mathematical treatment is similar, but τ_c then represents a translational correlation time.

Equations 8.6–8.10 all contain the same initial factor, marked by a quadratic dependence on each of the magnetogyric ratios and an inverse sixth power dependence on internuclear distance. For this reason, nuclides with a small magnetic moment and/or low isotopic abundance are relaxed primarily by interaction with nearby nuclei of large γ, such as ^1H. Protons, on the other hand, are usually relaxed by other protons, both intra- and intermolecularly, depending on the average distances involved. For example, the protons in pure liquid benzene have $T_1 \approx 19$ seconds, while a dilute solution of benzene in CS_2 (which has no magnetic nuclei) shows $T_1 \approx 90$ s. (Viscosity variation, hence a change in τ_{rot}, may account for a small part of this difference, but the fivefold increase in R_1 in the neat liquid is largely due to intermolecular relaxation processes.) Relaxation of natural abundance ^{13}C in benzene, on the other hand, is almost entirely intramolecular, with $T_1(^{13}C) \approx 29$ s, nearly independent of concentration, because the distance to protons in adjacent molecules is large.

8.3 NUCLEAR OVERHAUSER EFFECT

The nuclear Overhauser effect (NOE), which is manifested in certain changes in the intensities of NMR lines, is a consequence of magnetic dipolar relaxation. The name comes from a phenomenon predicted by Albert Overhauser in 1953, when he showed theoretically that saturating the electron magnetic resonance in a metal would cause the nuclear resonance intensity to be enhanced by a factor of the order of 10^3 (the ratio of $\gamma_{electron}/\gamma_{nucleus}$). Ionel Solomon later found that a similar effect occurs between two nuclei, but with a much smaller intensity enhancement—the *nuclear* Overhauser effect.[90] Because the NOE is of great practical importance in NMR, we discuss here not only the theoretical underpinnings but also the methods for making NOE measurements.

The qualitative basis of the NOE can be readily appreciated from Fig. 8.3. If rf energy is applied at frequency ω_S, the S transitions thus induced alter the populations of the energy levels, and because of *cross relaxation* (i.e., transitions W_0 and W_2, which affect both I and S simultaneously), that change affects the intensities of the I transitions. Quantitatively, Eq. 8.4 shows what happens when the two S transitions are subjected to a continuous wave rf that is intense enough to cause saturation. At the steady state both dI_z/dt and S_z are zero, so that Eq. 8.4 (after some rearrangement) gives

$$\eta = \frac{I_z - I_0}{I_0} = \frac{\gamma_S}{\gamma_I} \frac{W_2 - W_0}{W_0 + 2W_1 + W_2} \tag{8.11}$$

We have substituted γ_S/γ_I for S_0/I_0. Eq. 8.11 gives the fractional change in intensity at steady state, which is often called the nuclear Overhauser *enhancement* and given the symbol η. (Often the value given in publications for the NOE is actually $1 + \eta$, so care must be used in interpreting the results.)

By substituting the expressions for spectral densities in Eq. 8.11, we obtain an equation that is algebraically cumbersome in general but that can be simplified in either of two regimes: (1) homonuclear spins ($I = S$) or (2) rapid tumbling (extreme narrowing limit).

For $I = S$, substitution for the transition probabilities W in Eq. 8.11 gives, after some algebraic manipulation,

$$\eta = \frac{5 + \omega^2 \tau_c^2 - 4\omega^4 \tau_c^4}{10 + 23\omega^2 \tau_c^2 + 4\omega^4 \tau_c^4} \tag{8.12}$$

Equation 8.12 is plotted as a function of τ_c in Fig. 8.4. At very small values of τ_c, η reaches its maximum of $+50\%$ enhancement, whereas at very large τ_c, η drops to -1, which means that $I_z = 0$, and there is no signal.

Although the general expression for a heteronuclear spin pair is more complex than that in Eq. 8.12, the limit for small τ_c is the same, $+\frac{1}{2}$, but the factor γ_S/γ_I remains to give

$$\eta = \frac{\gamma_S}{2\gamma_I} \tag{8.13}$$

For the homonuclear case the NOE is $+\frac{1}{2}$, as we have seen, while for the important case where I is ^{13}C and S is ^1H, $\eta = 1.988$. (Application of the complete relation analogous to Eq. 8.12 shows that η for $^{13}C/^1H$ drops as low as 0.15 for large τ_c.) Where one γ is negative, the NOE leads to a negative signal; for example, for $I = {}^{15}N$ and $S = {}^1H$, $\eta = -4.93$.

Our discussion of the NOE was based on a two-spin system with solely mutual dipolar relaxation. When there are more than two spins in magnetically

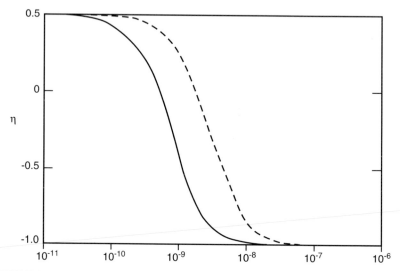

FIGURE 8.4 Dependence of the homonuclear nuclear Overhauser enhancement η as a function of correlation time τ_c at two NMR frequencies: ———, 360 MHz; – – – –, 100 MHz.

equivalent sets (e.g., A_2X_3), the results are unchanged. However, where there are three or more different kinds of spins (A observed, X irradiated, Y not irradiated) the results are more complicated but much more informative. Because the NOE arises from dipolar relaxation, the contribution of each spin pair to the magnitude of the NOE depends on the relative internuclear distances, with an inverse sixth power dependence. Thus, the NOE is dominated by very nearby nuclei with large magnetic moments, principally protons. A virtually full enhancement of 49% has been found for H_A in **I**

I

when the methoxyl protons are irradiated[91] because relaxation of H_A is due almost entirely to these very nearby nuclei.

When other relaxation mechanisms compete with dipolar relaxation (often the case, as we see subsequently), the magnitude of η is reduced according the fraction of the total relaxation rate represented by the dipolar contribution: $R_1^{\text{Dipolar}}/R_1^{\text{Total}}$, where

$$R_1^{\text{Total}} = R_1^{\text{Dipolar}} + R_1^{\text{Other}} \tag{8.14}$$

with this relation, NOE data can clearly be used to assess the relative importance of dipolar relaxation, particularly for nuclei such as ^{13}C and ^{15}N. Also, the NOE is used routinely to enhance the signals from less sensitive nuclei according to Eq. 8.13.

The homonuclear NOE between specific protons is used very frequently to estimate relative internuclear distances and thus obtain information on molecular conformation. For organic molecules that tumble rapidly, such measurements are usually made by difference between a spectrum in which a specific resonance is continuously irradiated and one in which a "dummy" irradiation is applied well off resonance. For larger molecules, including biopolymers, this method fails, because *spin diffusion* causes the nonequilibrium polarization to be rapidly transferred through the entire proton spin system of the molecule. We used the term "spin diffusion" in Section 7.12 to describe the spread of polarization via static dipole–dipole couplings in solids and the creation of a homogeneously broadened line. In liquids the process of spin diffusion is somewhat different. It is mediated by cross relaxation, specifically by the "flip-flop" motions described in the term W_0, which dominates at long τ_c. Hence, for polymers, pulse sequences are used, usually with two-dimensional NMR methods, to create a local nonequilibrium polarization that can be measured in a short, well-defined period (usually a few milliseconds). Also, it is possible to use such methods to measure an NOE in the rotating frame (analogous to $T_{1\rho}$ described in Section 8.1) and overcome the negative value of η found for large molecules. We shall return to these pulse techniques in Chapters 10 and 12.

8.4 RELAXATION VIA CHEMICAL SHIELDING ANISOTROPY

In Chapter 7 we examined the aniosotropy in chemical shielding, which leads to a dispersion of resonance frequencies for molecules fixed in a solid. As the molecules tumble in a fluid state, the field at the nucleus is continually changing in magnitude. The components of random tumbling motion at the Larmor frequency can then lead to spin–lattice relaxation, just as we have seen in the preceding sections. The general results of Section 8.1 apply here and in the following sections describing other relaxation mechanisms. Relations similar to Eqs. 8.5–8.10 can be derived for CSA relaxation, but we limit the presentation here to only a few major points.

As with dipolar relaxation, the interaction energy for CSA (in this case from Eqs. 6.3 and 7.17) enters quadratically, leading to the following expression for an axially symmetric shielding tensor in the extreme narrowing limit:

$$R_1^{CSA} = R_2^{CSA} = \frac{2}{15}\, \gamma^2 B_0^2 (\sigma_\parallel - \sigma_\perp)^2 \tau_c \tag{8.15}$$

The interesting point in Eq. 8.15 is that R_1^{CSA} increases quadratically with B_0, so CSA relaxation has become of more significance as higher magnetic fields have become available. CSA relaxation is the dominant mechanism for some nuclei with large chemical shift ranges and small magnetic moments, such as ^{57}Fe and ^{113}Cd, and can be quite significant for others, including ^{31}P and ^{15}N. The shorter value of T_1 at high field is generally advantageous in permitting more rapid repetition of exciting pulses and resultant enhancement of signal, but the concomitant shortening of T_2 leads to line broadening. For example, for ^{31}P in biologically interesting molecules with only a moderate correlation time, spectral resolution often has an optimum around $8-10$ T, because at higher field the separation of chemically shifted lines, which increases linearly with B_0, is overtaken by the quadratically increasing line widths. Since the range of ^1H chemical shifts is small, $\sigma_\parallel - \sigma_\perp$ is too small to be significant. For ^{13}C, the CSA is significant for nontetrahedral carbon atoms, but relaxation by CSA is often masked by more rapid dipolar relaxation if there is a directly bonded proton.

8.5 ELECTRIC QUADRUPOLE RELAXATION

We saw in Chapter 7 that the resonance frequency of a quadrupolar nucleus is dependent on the orientation of the molecule in which it resides. Molecular tumbling now causes fluctuating *electric* fields, which induce transitions among the nuclear quadrupole energy levels. The resulting nuclear relaxation is observed in the NMR just as though the relaxation had occurred by a magnetic mechanism.

The general theory of quadrupole relaxation is somewhat complex but simplifies for the case of rapid molecular tumbling and axial symmetry of the molecular electric field. The interaction energy between the nucleus and a surrounding electric field gradient (Eq. 7.18) appears quadratically:

$$R_1^Q = R_2^Q = \frac{3\pi^2}{10}\, \frac{2I + 3}{I^2(2I - 1)}\, \chi^2 \tau_c \tag{8.16}$$

As we saw in Section 7.11, the quadrupole coupling constant $\chi = (e^2 Qq/h)$ is zero for a highly symmetric situation (e.g., Cl^- ion, tetrahedral $^{14}NH_4^+$) but can be very large in other cases. For asymmetric ^{14}N bonds, χ is typically a few MHz, leading to T_1 and T_2 in the range of $10-20$ ms; for ^2H, χ is generally only $100-200$ kHz, with relaxation times of several hundred milliseconds; and for chlorine and bromine in asymmetric covalent bonds, values over 100 MHz may

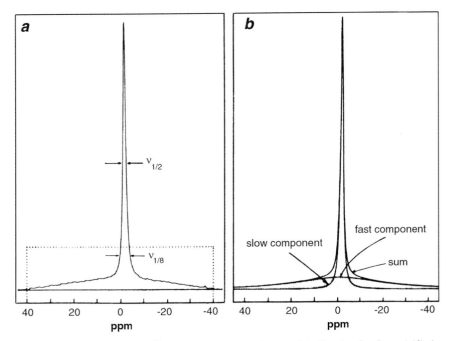

FIGURE 8.5 Non-Lorentzian ^{23}Na line shape from a solution of NaCl and malonyl gramicidin in lysolecithin micelles. (*a*) Experimental spectrum. (*b*) Computer-generated Lorentizian components. From Venkatachalam and Urry.[92]

be found, giving T_1 in the range of microseconds. It is apparent that where quadrupole relaxation exists it is normally the dominant relaxation process.

Because quadrupolar nuclides have $I > \frac{1}{2}$, there are more energy levels to consider, and the probability of a relaxation transition between one pair of levels in a single nucleus may not be equal to that between another pair of levels. For example, nuclides with $I = \frac{3}{2}$ (such as ^{23}Na) have distinctly different relaxation rates for the $m - \frac{1}{2} \rightarrow \frac{1}{2}$ transition and the $\pm\frac{1}{2} \rightarrow \pm\frac{3}{2}$ transitions. In an even slightly anisotropic environment, such as a liquid crystal solvent or a biological cell, the spectrum of a "free" ^{23}Na ion has two components, as indicated in Fig. 8.5, with quite different values of both T_1 and T_2.

8.6 SCALAR RELAXATION

The indirect spin–spin coupling can provide a relaxation mechanism by two separate processes. As we saw in Chapter 7, J is a tensor, and its aniosotropy can cause relaxation as molecules tumble, just as with CSA. However, for most nuclei, values of J (and the resultant anisotropy) are generally much smaller than chemical shift aniosotropy (expressed in Hz); hence J anisotropy is only very rarely of any significance as a relaxation mechanism.

The second mechanism results from the scalar coupling itself, which transmits information on the spin state of one nuclear spin to its coupled partner. As we observed in Section 8.1, any process that gives rise to a fluctuating magnetic field at a nucleus might cause relaxation. For example, when two nuclei I and S are spin coupled, the value of J measures the magnitude of the magnetic field at I arising from the spin orientation of S. As S relaxes, I experiences a magnetic field fluctuation. Likewise, if the magnitude of J changes because of bond breaking in chemical exchange processes, I experiences a similar fluctuation. We shall discuss three situations in which relaxation might occur by these processes:

1. *Relaxation rate $R_1^S \gg 2\pi J$ and $R_1^S \gg R_1^I$.* This situation occurs, for example, where $I = \frac{1}{2}$ and is coupled with a moderately large value of J to a quadrupolar nucleus S that relaxes rapidly. (Because of the rapid relaxation of S, the expected splitting of the resonance lines of I does not occur, as we have seen in previous chapters.) Equations for this type of scalar relaxation have been derived:

$$R_1^I = \frac{8\pi^2 J^2}{3} S(S+1) \frac{T_1^S}{1 + (\omega_I - \omega_S)^2 (T_1^S)^2} \tag{8.17}$$

$$R_2^I = \frac{4\pi^2 J^2}{3} S(S+1) \left[T_1^S + \frac{T_1^S}{1 + (\omega_I - \omega_S)^2 (T_1^S)^2} \right] \tag{8.18}$$

Here S is the spin of the quadrupolar nucleus, ω_I, and ω_S are the Larmor frequencies of the two nuclei, and T_1^S is the longitudinal relaxation time of S. Generally the denominator is very large because the frequency term corresponds to many megahertz. Hence, the scalar coupling contribution to R_1 is usually negligible and can be significant only if $\omega_I \approx \omega_S$—a situation that occurs only for specific combinations of nuclides, such as ^{13}C and ^{79}Br, where the resonance frequencies at 1.4 T differ by only 0.054 MHz.

Scalar coupling can contribute substantially to R_2 as a result of the linear term in T_1^S in Eq. 8.18, provided T_1^S is short enough to meet the conditions given earlier for the validity of this equation but is also long enough to provide a significant contribution. For example, the relaxation of ^{14}N (in the millisecond range) can lead to a shortening of T_2 of a spin-coupled proton by this process, thus accounting for the broad lines often found for protons attached to a nitrogen. However, for nuclides such as ^{35}Cl, with T_1 in the microsecond range, there is little effect on T_2 of the coupled nucleus (accounting for the sharp ^1H and ^{13}C lines in $CHCl_3$).

2. *Relaxation rate $R_1^S \ll 2\pi J$ and $R_1^S > R_1^I$.* This situation occurs, for example, where $I = {}^{13}C$ and $S = {}^1H$. The proton relaxation is not nearly rapid enough to affect the spin–spin splitting, nor does it influence $T_1(^{13}C)$, but it does shorten $T_2(^{13}C)$ substantially. For example, in $^{13}CH_3COOCD_3$, where $J = 130$ Hz and

$T_1(^1\text{H}) = 12.5$ s, $T_1(^{13}\text{C}) = 19.2$ s but $T_2(^{13}\text{C})$ is only 6.1 s. In the absence of scalar relaxation, $T_1(^{13}\text{C})$ and $T_2(^{13}\text{C})$ might have been expected to be equal from dipolar relaxation in such a small molecule in the extreme narrowing regime (Eqs. 8.6 and 8.9).

3. *J itself is a function of time.* In the case of a chemical exchange that causes bond breaking, the magnetic field at I fluctuates because J is modulated between its normal value and zero. The resulting effects are clearly analogous to case 1 or 2, depending on the rate of exchange relative to J. As viewed from nucleus I, it makes little difference whether nucleus S relaxes or exchanges, so the same equations apply.

8.7 SPIN–ROTATION RELAXATION

The spin–rotation interaction arises from magnetic fields generated at a nucleus by the motion of a *molecular* magnetic moment that arises from the electron distribution in the molecule. The Hamiltonian for this interaction is given by

$$\mathscr{H} = \boldsymbol{I} \cdot \mathbb{C} \cdot \boldsymbol{J}_R \tag{8.19}$$

where $\boldsymbol{I}$ is the spin operator, $\boldsymbol{J}_R$ is the angular momentum operator for the molecule, and $\mathbb{C}$ is the spin–rotation tensor. The elements of $\mathbb{C}$ can often be estimated from molecular beam resonance results or from nuclear shielding. Because the spin–rotation tensor components depend on the electron distribution in the molecule, as do the components of the chemical shift tensor, nuclei with large chemical shift ranges are generally found to have significant spin–rotation interactions. Small molecules that rotate rapidly in the gas or liquid are most affected, and this is the dominant relaxation mechanism for H_2 and other small molecules in the gas phase.

The theory developed initially by Hubbard[93] for spherical top molecules in liquids shows that collisions cause the molecule to experience changes in the magnitude and direction of its angular momentum, which then causes the fluctuations required to create magnetic fields oscillating at the Larmor frequency. For a small molecule in a liquid it can be shown that

$$R_1{}^{SR} = R_2{}^{SR} = \frac{2I_M kT}{\hbar^2} [C_0{}^2 + 2(\Delta C)^2] \tau_J \tag{8.20}$$

where I_M is the moment of inertia of the molecule, C_0 and ΔC are related to the isotropic and aniosotropic parts of $\mathbb{C}$, and τ_J is the angular momentum correlation time, which is a measure of the time the molecule remains in a given rotational state. It can be shown that τ_J and τ_c, the correlation time that we have discussed previously, are related. For a spherical top molecule,

$$\tau_J \tau_c = I_M / 6kT \tag{8.21}$$

τ_c has an exponential temperature dependence and becomes shorter as molecules rotate more rapidly. From Eq. 8.21 we see that this dependence must be approximately balanced by an exponential dependence in τ_J such that τ_J becomes *longer* with increasing sample temperature. The result is that for spin–rotation relaxation, T_1 becomes shorter with increasing temperature, which is opposite to the temperature dependence found for all other relaxation mechanisms discussed here.

Spin–rotation interaction is known to be the dominant relaxation mechanism for ^{13}C in CS_2 (except at very high fields and low temperatures) and is important for ^{13}C in methyl groups, which reorient rapidly by *internal* rotation, even in large molecules.

8.8 RELAXATION BY PARAMAGNETIC SUBSTANCES

Paramagnetic atoms or molecules (i.e., those with one or more unpaired electrons) can contribute to relaxation of nuclei in two ways: (1) dipolar relaxation by the *electron* magnetic moment; (2) transfer of unpaired electron density to the relaxing atom itself. The basic equations for relaxation by a paramagnetic agent stem from work by Solomon, Bloembergen, and others. These equations, which are not reproduced here, consist of terms analogous to those in Eqs. 8.6 and 8.9 for dipolar relaxation and terms similar to those in Eqs. 8.17 and 8.18 for transfer of electron density because a scalar coupling with the unpaired electrons is involved. A complication arises regarding the correlation times, as several processes can determine the values—the usual rotational correlation time τ_c, the spin–lattice relaxation time of the electron, and the rate of any ion–ligand exchange processes. As in nuclear scalar relaxation, the scalar portion of the interaction contributes little to T_1 but often dominates T_2.

As we pointed out in Section 4.11, lanthanide shift reagents owe their utility partly to the fact that the electron spin–lattice relaxation time for the lanthanides is very short, so that NMR lines are not exceptionally broad. On the other hand, there are "shiftless" paramagnetic reagents that shorten both T_1 and T_2 to a moderate degree without causing contact or pseudocontact shifts.

Because relaxation by paramagnetic agents is highly effective, even a small concentration of a paramagnetic substance can influence the appearance of an NMR spectrum. For example, the most common paramagnetic relaxation agent is molecular oxygen. At atmospheric pressure enough oxygen dissolves in most solvents to provide a relaxation rate for protons of about $0.5\ s^{-1}$ and to make it the dominant mode of relaxation for most protons in small diamagnetic molecules. For measurement of NOEs or to obtain the narrowest possible lines in small molecules, it is essential to degas the sample.

Likewise, small concentrations of paramagnetic ions that complex to suitable ligands and exchange rapidly among ligand molecules can dramatically alter the

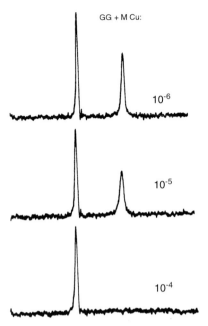

GG + M Cu:

10^{-6}

10^{-5}

10^{-4}

FIGURE 8.6 ^{1}H NMR spectrum of glycylglycinate ion, $NH_2CH_2C(O)NHCH_2COO^-$, in D_2O (0.5 M) with addition of Cu^{++} ion (10^{-6} to 10^{-4} M).

appearance of a spectrum, as illustrated in Fig. 8.6, which shows proton NMR spectra of the glycylglycinate anion in the presence of Cu^{2+} ion. The paramagnetic cupric ion is known to bind to the amino terminal end of the glycylglycinate, so that the CH_2 protons adjacent to the nitrogen are principally relaxed by the copper. The ligand is present in large excess ($\sim10^5$:1), and the copper ion exchanges rapidly among various ligand molecules. The effect on T_2 is observable because the line width for a 1:1 ratio of ligand to Cu^{2+} would be several kHz; hence the average that is observed is still quite large. Spin–spin relaxation by many paramagnetic ions is dominated by the scalar coupling interaction, whereas spin–lattice relaxation is normally dipolar. Measurements of T_1 then can sometimes be used to obtain relative distances from a paramagnetic ion to various protons in a ligand.

8.9 OTHER FACTORS AFFECTING RELAXATION

In this chapter we have provided only a simple introduction to a complex subject. From our discussion of the several relaxation mechanisms, it might be supposed that the relaxation rates simply add to give the total relaxation rate for the molecular system, that is, that

$$R_1^{\text{Total}} = R_1^D + R_1^{CSA} + R_1^Q + R_1^{SC} + R_1^{SR} + R_1^P \qquad (8.22)$$

For most applications covered in this book, Eq. 8.22 is an adequate approximation, but a closer look at relaxation mechanisms discloses many instances in which there are substantial departures from additivity, as there are *cross correlation* effects. In Eq. 8.1 we defined an *auto*correlation function, which we applied separately to any pair of spins relaxing by dipolar interactions, and we treated other relaxation mechanisms independently on the assumption that a similar correlation function could be applied independently. In fact, however, these motions are often highly correlated. For example, in a rigid molecule the fluctuations of the vector joining one pair of nuclei are related to those of a vector joining a second pair of nuclei or to the axes defining a shielding anisotropy. Depending on the relative orientations of the axes defining these interactions, the effects of relaxation on a given transition may add or subtract. The result is that individual lines in a spin multiplet may relax at very different rates, sometimes so different that the observed line widths are quite disparate, as illustrated in Fig. 8.7. In this example, dipolar and CSA relaxations add for one transition to give a very broad line and subtract for the other component of the spin doublet.

The effects of cross correlation can be exploited as a rich source of information on molecular dynamics by measuring relaxation behavior as a single line of a spin multiplet is perturbed. In addition, experiments can be designed to take advantage of the partial cancellation of relaxation effects and thus to obtain narrower lines than might otherwise occur. For example, in two-dimensional NMR it has been possible to utilize the quadratic dependence of CSA on B_0 to

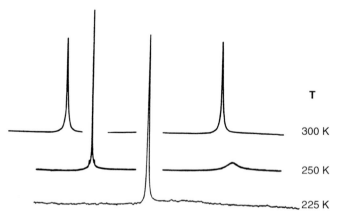

FIGURE 8.7 ^{31}P NMR spectra of Na_2HPO_3 (202 MHz, 11.7 T) at the temperatures indicated. At 225 K, where $\omega_0\tau_c \approx 1$, the low frequency line is so broad that it is unobservable. From Farrar and Stringfellow,[94] *Encyclopedia of Nuclear Magnetic Resonance*, D. M. Grant and R. K. Harris, Eds. Copyright 1996 John Wiley & Sons Limited. Reproduced with permission.

demonstrate that an optimum magnetic field exists where certain line widths are minimized for large proteins. The narrower lines translate directly to an improvement in sensitivity, with a 4- to 10-fold improvement demonstrated for certain aromatic $^{13}C-^1H$ correlation peaks.[95]

Even without the effects of cross correlation, the assumption of a single correlation time for a given relaxation mechanism is usually oversimplified, as we indicated in Section 8.1. Figure 8.8a shows *partially relaxed* spectra resulting from an inversion-recovery pulse sequence (Section 2.9) for the ^{13}C spins in a small organic molecule. It is apparent that the values of T_1 are somewhat different, because the null times vary from one carbon atom to another. From the analysis of similar partially relaxed spectra, the values of $T_1(^{13}C)$ shown in Fig. 8.8b and c have been obtained for phenol and *n*-decanol. Both molecules tumble rapidly in

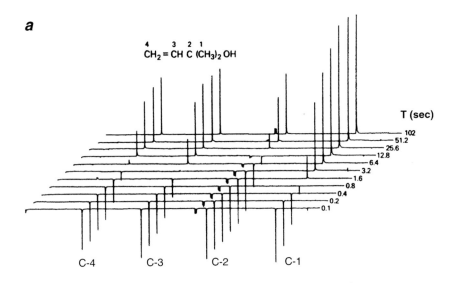

a

$$CH_2 = CH\ C\ (CH_3)_2\ OH$$
      4   3   2   1

T (sec)

102
51.2
25.6
12.8
6.4
3.2
1.6
0.8
0.4
0.2
0.1

C-4   C-3   C-2   C-1

b $CH_3 -CH_2 -CH_2 -CH_2 -CH_2 -CH_2 -CH_2 -CH_2 -CH_2 -CH_2\ OH$

3.1   2.2   1.6   1.1   0.84   0.84   0.84   0.77   0.77   0.65

c

OH
17

2.4

2.4

1.4

FIGURE 8.8 (*a*) Partially relaxed ^{13}C NMR spectra from an inversion-recovery pulse sequence. (*b*) Values of T_1 (seconds) for ^{13}C nuclei in *n*-decanol. (*c*) Values of T_1 for ^{13}C nuclei in phenol.

the extreme narrowing limit, but the motions are characterized by more than one correlation time. In phenol the molecule is rigid, but molecular motion is anisotropic. The more rapid motion about the Cl–C4 axis has no effect on C4 but reduces τ_c for C2 and C3 below that resulting from tumbling of the molecule about the other two axes. In n-decanol the increase in T_1 along the chain results from substantial segmental motion, so that each carbon has its own effective τ_c. Near the polar end of the molecule τ_c represents nearly the value for overall tumbling of the molecule, but farther down the chain the effective τ_c becomes shorter, leading to a longer T_1.

In some polymers such segmental motions can be important, whereas in others (e.g., proteins) the overall skeleton is rigid, but there are rapid internal motions of moieties relative to the skeleton. In this case, relaxation and NOE data are often analyzed by the Lipari–Szabo formalism,[96] which yields values for an overall correlation time τ_M, a correlation time for fast motions τ_e, and a generalized order parameter S (see Eq. 7.16), which describes the amplitudes of the internal motions.

8.10 ADDITIONAL READING AND RESOURCES

In addition to the specific references given in the chapter, much of the "classic" treatment of relaxation comes from the book by Abragam,[33] but there are many other discussions of this subject in almost every book on NMR. Additional details along the lines presented here are given in *Pulse and Fourier Transform NMR* by Thomas C. Farrar and Edwin D. Becker,[97] *Nuclear Magnetic Resonance Spectroscopy* by Robin K. Harris,[32] and *The Nuclear Overhauser Effect in Structural and Conformational Analysis* by D. Neuhaus and M. P. Williamson.[98]

The serial publications listed in Section 1.3 contain a number of articles on various aspects of relaxation. The *Encyclopedia of NMR* includes 13 articles under titles beginning with the word *Relaxation* and several other related articles, including *Nuclear Overhauser Effect*,[99] *Carbon-13 Relaxation Measurements*,[100] *Paramagnetic Relaxation in Solution*,[101] *Brownian Motion*,[102] and *Spin–Rotation Relaxation Theory*.[103] Many of these articles are based on density matrix formulations, which can readily be understood with the background provided in Chapter 11.

8.11 PROBLEMS

8.1 Acetonitrile, CH_3CN, has a known geometry, with distances from the nitrile carbon to each hydrogen of 2.14 Å and to the nitrogen of 1.14 Å. Find the relative values of R_1 for (a) dipolar relaxation of the nitrile carbon by the protons, (b) dipolar relaxation of this carbon by the nitrogen (^{14}N), and (c) chemical shift anisotropy relaxation at 68 MHz.

Take $\sigma_\parallel - \sigma_\perp = 300$ ppm. Assume the extreme narrowing limit, with τ_c to be the same for the three processes.

8.2 Calculate the scalar coupling contribution to R_1 and R_2 of the nitrile carbon in acetonitrile, $CH_3{}^{13}C^{14}N$. Take T_1 $(^{14}N) = 1.8$ ms. The C–N coupling constant was measured in the isotopomer $CH_3{}^{13}C^{15}N$, and $^1J(^{13}C^{15}N) = -17$ Hz. What is the resulting width of the ^{13}C resonance for this carbon?

8.3 Find the ratio T_1/T_2 for homonuclear dipolar relaxation at frequencies $\nu = 0.2\nu_0; \nu_0; 1.2\nu_0; 1.8\nu_0$.

8.4 For the methyl carbon in toluene $T_1(^{13}C)$ has been measured as 17.4 s, with an NOE $\eta_{CH} = 0.63$. Find $R_1{}^{Dipolar}$ and $R_1{}^{Other}$.

8.5 Find η for ^{29}Si when ^1H is saturated in a molecule tumbling fast enough to be in the extreme narrowing range.

8.6 Beginning with Eq. 8.11, derive an expression for η_{CH} for ^{13}C that is analogous to Eq. 8.12, and show that $\eta_{CH} = 0.15$ in the limit of very large τ_c. Note that the three terms inside the brackets in Eq. 8.6 represent W_0, $2W_1$, and W_2, respectively.

8.7 For ^{13}C in CH_3OH, T_1 is found to increase with temperature from $-75°C$, pass through a maximum at $35°C$, and decrease at higher temperatures. What mechanisms of relaxation are most likely to account for these results?

8.8 For deuterium in perdeuterated toluene $(C_6D_5CD_3)$ $T_1(ring) = 0.94$ s and $T_1(methyl) = 4.9$ s at $25°C$. (a) Assuming isotropic rotation for the ring, find $\tau_c(ring)$ and $\tau_c(methyl)$ if the quadrupole coupling constants χ are 193 kHz for the ring deuterons and 165 kHz for the methyl deuterons. (b) Predict the approximate value of $T_1(^{13}C)$ for the methyl carbon due to dipolar relaxation by the bonded protons. Assume a C—H bond distance of 1.09 Å.

8.9 To measure T_1 of a sample a $90°$, τ, $90°$ pulse sequence is applied for various values of τ. The signal amplitudes obtained from Fourier transformation of the FIDs following the pulse sequence are as follows:

τ (s)	0.5	1.0	1.5	2.0	2.5	3.0	100
Signal	4.6	6.8	8.4	9.8	10.9	11.8	14.9

From these data find the value of T_1. (Recall the equation derived in Problem 3.5.)

Pulse Sequences

In previous chapters we have referred to sequences of pulses that can be used to make particular measurements, such as T_1 and T_2 (Section 2.9), and more complex sequences that narrow lines in solids (Section 7.8). In this chapter we explore the use of pulse sequences in more detail and investigate the behavior of magnetizations when subjected to arrays of pulses. We follow these "spin gymnastics" about as far as possible with the classical picture of magnetization vectors and set the stage for invoking the more powerful formalisms described in Chapter 11.

As the illustrations in Section 2.9 showed, application of a sequence of pulses permits us to manipulate the spin system in a variety of ways. The magnetization is rotated by the first pulse, allowed to precess for some period, and rotated further by subsequent pulses. Most of the remainder of this book is devoted to the application of a variety of pulse sequences that are carefully chosen to provide spectral information of interest.

As we saw in Section 2.9, most rf pulses are short (usually a few microseconds), with high power and a square wave profile. They are designed to rotate magnetizations over a large range of resonance frequencies. However, in some instances we wish to obtain pulse excitation over only a narrow region. In this chapter we also explore such *selective pulses.*

9.1 THE SPIN ECHO

One of the simplest, yet most important, pulse sequences used in NMR experiments is the 90°, τ, 180° sequence, which produces a spin echo, as we showed in Section 2.9. The basic concept, as illustrated in Fig. 2.13, is that magnetizations that precess with slightly different frequencies dephase in the $x'y'$ plane, but application of a 180° (or π) pulse along the y' axis causes the magnetizations to rephase. Thus the resultant macroscopic M_{xy}, which diminishes in magnitude during the dephasing process with concomitant decay of free induction signal, is restored during the rephasing, and the signal builds up to a maximum (the echo) before decaying as the magnetizations again dephase. We saw that the spin echo can thus reverse the dephasing resulting from inhomogeneity in $\mathbf{B}_0$.

Effect of Chemical Shifts and Spin Coupling

The rephasing effect of a 180° pulse not only removes the effects of magnetic field inhomogeneities, it also rephases signals from nuclei that differ in precession frequency because of differing chemical shifts, as illustrated in Fig. 9.1. However, if the two magnetizations differ in frequency because of scalar coupling between them, the situation is more complex.

We consider two spins I and S, which may be homonuclear or heteronuclear, and focus attention on spin I in a rotating frame, where it precesses with a chemical shift ω and coupling $2\pi J$, both in radians/second. Figure 9.2 shows the two components of $\mathbf{M}_I$ precessing as a result of spin coupling with S. We now apply a 180° pulse at time τ and must distinguish three cases: The pulse may affect only nucleus I (the nucleus that is being observed), only nucleus S, or both I and S. Figure 9.2 (top) illustrates the situation in which nucleus S is unaffected by the π pulse. This situation arises when I and S are different nuclear species with different resonance frequencies or when a selective pulse is applied to I. Rephasing is complete, just as shown in Fig. 9.1 for two chemically shifted nuclei, and inhomogeneity effects are eliminated, as we illustrated in Fig. 2.13.

However, if nucleus S is also affected by the 180° pulse, we must consider the quantum mechanical consequences. The pulse inverts the S spin states, α and β, thus causing the slower moving and faster moving magnetizations to interchange their frequencies and achieve incomplete rephasing at 2τ, as shown in Fig. 9.2 (center). In fact, the magnitude of the echo can be seen to be multiplied by $\cos 2\pi J\tau$, and a series of echoes has a modulation at frequency πJ superimposed on its otherwise exponential decay. Such modulation is important in both one- and two-dimensional experiments that we describe later.

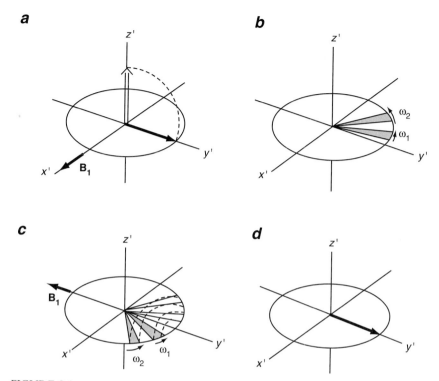

FIGURE 9.1 Use of a spin echo pulse sequence to refocus two magnetizations precessing at different frequencies because of difference in chemical shift.

The third possibility is that the π pulse affects only S, as illustrated in Fig. 9.2 (bottom). In this case the interchange of S spin states causes the dephasing from the coupling between I and S to be reversed at time 2τ, and Fourier transformation would show that the nuclei are "decoupled" just as though a continuous rf had been applied (Section 5.5). However, there is no echo because the I spins are not affected, and chemical shifts continue to appear in the spectrum.

It is important to note that in this description of spin-coupled nuclei, we begin to see the limitations of the vector approach to understanding the NMR processes involved. In this case it is fairly simple to graft on the quantum aspect of spin orientation to provide an intuitively satisfying picture of decoupling and of the J modulation of spin echoes. Other cases will not be so clear and will force us to adopt a more powerful approach in Chapter 11.

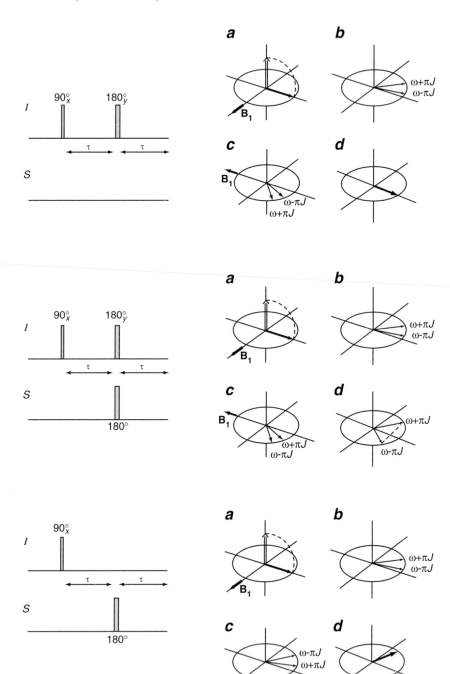

Enhancement of Signal/Noise Ratio

As we saw in Section 3.12, the signal/noise ratio is often improved by time averaging, but the rate at which pulses can be repeated is governed by T_1. In instances in which $T_1 \approx T_2 \gg T_2^*$, the formation of a spin echo can be used to enhance sensitivity. Consider the 90°, τ, 180° sequence. The initial 90° pulse leads to an FID, and the 180° pulse causes an echo at 2τ, the leading portion of which is a reversed FID. With suitable data manipulation, the two FIDs can be coherently added to gain $\sqrt{2}$ in S/N. Moreover, if a 90° pulse is applied along the negative y' axis precisely at the peak of the echo, the refocused magnetization **M** is restored to the z axis. If T_2 is sufficiently long, the magnitude of **M** is close to the original equilibrium value, so that the 90°, τ, 180°, τ, 90° sequence can be repeated. This technique, originally used to drive the magnetization back to equilibrium along the z axis for studies of relaxation in ^{129}Xe and later developed for signal enhancement in multiline spectra under the name driven equilibrium Fourier transform (DEFT) NMR, is useful, but only under the conditions given. In the presence of spin coupling, echo modulation effects interfere and even in decoupled spectra T_2 is often much less than T_1.

90° Echoes

Although not as readily visualized as the echo from a 90°, τ, 180° pulse sequence, a 90°, τ, 90° sequence also produces a spin echo. Thus, if T_2 is long enough, echo artifacts may appear in a simple succession on 90° pulses intended only to excite and time average FIDs. Also, somewhat analogous to the DEFT sequence just described, a sequence of three equally spaced 90° pulses produces a *stimulated echo*, in which the refocused magnetization is stored along the z axis for an intermediate τ period. Several techniques for 2D NMR in solids and for NMR imaging make use of the ability of a sequence of pulses to store the magnetization along z where it decays only by T_1 processes, which are often relatively slow.

An alternative method for storing magnetization along the z axis consists of using the same sequence of three 90° pulses but with different timing. After the first pulse has rotated magnetization into the xy plane, the second pulse rotates it to the $-z$ axis for some period τ, after which the third pulse restores **M** to the

FIGURE 9.2 The effect of various 90°, τ, 180° pulse sequences on a spin I, which is coupled to spin S. *Top:* Complete rephasing and echo formation from the 180° pulse applied only to I. *Center:* Rephasing of inhomogeneity effects but modulation of echo magnitude from 180° pulses applied to both I and S. The magnitude of the echo along the y' axis is proportional to the projection of the magnetizations, as indicated. *Bottom:* Refocusing of coupling at 2τ but lack of echo formation from a 180° pulse applied only to S. As indicated, the magnetizations continue to precess at frequency ω and do not refocus along y'.

xy plane. By repeating the second pair of pulses with variation of τ and carrying out suitable cycling of phases (Section 3.5), unwanted magnetizations of different phases initially in the *xy* plane can be eliminated. The sequence 90°, τ, 90°, called a *z-filter*, is used in some 2D NMR pulse sequences to discriminate against unwanted magnetizations.

Gradient-Recalled Echo

As we shall see, it is often desirable to apply a large magnetic field gradient for a short period (a *pulsed field gradient*) in order to make the magnetic field very in-homogeneous for a limited period in a well-controlled manner, which dominates any modest uncontrolled gradients in the static field. It is clear that such a pulsed gradient causes a rapid dephasing of magnetization in the *xy* plane. The magnetization could be refocused, as we have seen, by a 180° rf pulse, but it can just as easily be refocused by merely applying an identical pulsed magnetic field gradient in the opposite direction without any rf pulse. The echo produced in this way is usually termed a "gradient-recalled echo," and the term "spin echo" is usually reserved for an echo produced by rf pulses. A gradient-recalled echo, which is particularly useful in NMR imaging (Chapter 14), has no effect in refocusing chemical shifts or spin couplings.

Effect of Motion on an Echo

An implicit assumption underlies our discussion of echoes thus far, namely that dephasing and rephasing occur under opposite but otherwise precisely identical conditions. If a molecule moves in such a way that it experiences different values of the static field during the two phases, refocusing of the signal is incomplete. This potential problem can be turned to advantage by controlling the variables in such a way that molecular motion, either bulk flow or molecular diffusion, can be measured by NMR.

For example, bulk flow of liquid in a pipe or blood vessel can be followed by applying in the direction of flow a *bipolar gradient*, that is, a pair of oppositely polarized pulsed field gradients, each of duration δ, with no time between them. All nuclei in the moving liquid thus experience different resonance frequencies during the two halves of the bipolar gradient, hence at the end have accumulated a phase different from that of nuclei that do not move. To first order, the accumulated phase is proportional to the velocity of flow. If the static surroundings also give an NMR signal (for example, in a biological system), a simple phase cycling with the bipolar gradient reversed in sign permits cancellation of the signal from the nonflowing material.

Molecular diffusion can also be determined by NMR, as we describe in the next section.

9.2 THE CARR–PURCELL PULSE SEQUENCE

Figures 2.13, 9.1, and 9.2 demonstrate the formation of an echo following a π pulse. Application of additional π pulses can be used to form a *train* of echoes. It is clear that the dephasing of magnetizations following an echo is of the same form as the initial dephasing during the FID and that application of a second π pulse at 3τ causes a second echo at 4τ, etc. The envelope formed by the echo peaks decays according to the real T_2, rather than T_2^*, and Fourier transform of each echo provides a set of "partially relaxed" spectra, from which T_2 of each line may be determined. (Carr and Purcell first recognized the value of such a long sequence of π pulses,[104] and their names are usually used to depict the method, but the technique that we described for the spin echo in Chapter 2 and that discussed here include a refinement by Meiboom and Gill,[105] as discussed later.)

Effect of Diffusion and Exchange

As already noted, diffusion of molecules from one part of an inhomogeneous magnetic field to another inhibits complete refocusing of magnetizations and can significantly reduce the intensity of the echo according to Eq. 9.1:

$$S = Ae^{-t/T_2}e^{-\gamma^2 G^2 D \tau^2 t/3} \tag{9.1}$$

Fortunately, the variable τ can be adjusted, independent of the desired total length of the echo train, so that during the period τ little diffusion occurs and the second exponential factor can be made arbitrarily small, so that valid measurements of T_2 can be made.

Just as motion of molecules within a magnetic field gradient changes the precession frequencies of their nuclear spins, so can chemical exchange alter their precession frequencies, as described in Section 2.10. Again, the relevant period for exchange is τ, so the effect of exchange can be minimized by choosing a small value of τ. On the other hand, with measurements at larger values of τ, exchange rates can be measured, particularly for instances in which exchange is too rapid to affect spectra, as shown in Fig. 2.14.

Measurement of Molecular Diffusion

The impact of diffusion on echo intensity can be turned to advantage as a sensitive and versatile method for measuring diffusion coefficients. Although Eq. 9.1 could be applied with a Carr–Purcell sequence, the measurement is best done by using a relatively homogeneous static field B_0 on which a pair of pulsed field gradients G is imposed for short periods in the sequence shown in Fig. 9.3. During the first period δ the magnetizations rapidly dephase and are rapidly rephased during the second period δ after the 180° pulse. Diffusion that occurs

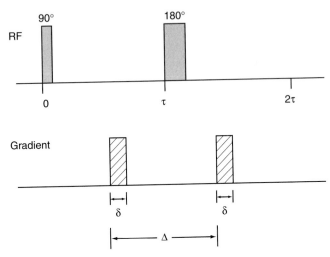

FIGURE 9.3 Pulse sequence for measurement of diffusion coefficient D by the pulsed field gradient spin echo technique. The gradient G is applied for a period δ both before and after the 180° pulse, with separation Δ.

during the period Δ reduces the echo according to the relation

$$S(2\tau) = S(0)e^{-\gamma^2 G^2 D \delta^2 \Delta} \tag{9.2}$$

Δ can be measured accurately, and the value of $G\delta$ need not be known precisely so long as it is the same for the two magnetic field pulses. This technique permits accurate measurement of diffusion coefficients for one substance dissolved in another or for self-diffusion of a single substance. By Fourier transforming the echo, diffusion coefficients for several substances can be determined. In conjunction with spatial localization methods provided by NMR imaging, studies of diffusion can be quite valuable, as we point out in Chapter 14.

9.3 CORRECTING FOR PULSE IMPERFECTIONS

Inhomogeneity of the $\mathbf{B}_1$ rf field causes nuclei in different parts of the sample to experience values of $\mathbf{B}_1$ that are not what is desired. In addition, as we have seen, nuclei that are off resonance experience a $\mathbf{B}_{\text{eff}}$ that has a magnitude and direction not coincident with the applied $\mathbf{B}_1$. Thus the effective pulses may be different from the nominal 90° and 180° pulses that are desired. The result of such *pulse imperfections* can be particularly troublesome if there is a cumulative effect in successive applications of pulses in a sequence. We describe two ways in which imperfections can be overcome in different situations.

The Meiboom–Gill Method

Both the original Hahn spin echo experiment and the Carr–Purcell sequence were carried out with *all* pulses applied along a single axis (say, positive x'), so that successive π pulses rotate **M** continually in one direction in the $y'z'$ plane. There are two disadvantages. First, odd-numbered echoes result from rephasing along the negative y' axis, hence appear as negative signals. Second, inhomogeneity in the B_1 field causes some nuclei to experience a pulse that is smaller than 180°. With repetition of the π pulses, the errors are cumulative and can quickly render measurements meaningless. The modification introduced by Meiboom and Gill, which we have incorporated in our explanations, applies the 180° pulses along the y' axis, a $\pi/2$ phase shift from the initial exciting 90° pulse. It is easy to see in Fig. 9.1 that a pulse of $(180° - \theta)$ leaves the magnetizations above the $x'y'$ plane, so that in (*d*) the projection in the $x'y'$ plane is less than the full magnetization. However, the next pulse of the same length puts the magnetizations into the plane. Thus the error introduced with one 180° pulse is automatically corrected in the following 180° pulse. Odd-numbered echoes are slightly in erro, but there is no cumulative effect. In practice, this modification of the Carr–Purcell technique (often abbreviated "CPMG") is now used almost exclusively.

Composite Pulses

We consider here the concept of a *composite pulse*, which might be considered as a kind of special pulse sequence in which only a negligible time elapses between the pulses. Thus **M** is moved through a particular path where it does not precess freely and in a time period during which virtually no relaxation occurs. A simple example, illustrated in Fig. 9.4, is designed to substitute for a single 180° pulse in inverting magnetization and to compensate for off-resonance effects. As we have seen, B_{eff} varies in magnitude and direction as a function of the resonance offset Ω, and a 180° pulse on resonance would result in off-resonance magnetizations undergoing trajectories of the sort shown in Fig. 9.4*a*, with most components being far from the desired position along the $-z$ axis. However, this particular composite pulse is seen in Fig. 9.4*b* to provide much better inversion of the entire magnetization. Composite pulses may also be designed to overcome the effects of B_1 inhomogeneity.

Such pulses can be used in place of the simple 90° and 180° pulses in most of the pulse sequences that we describe in this and later chapters. The trade-off of better effective rf homogeneity must be weighed in each instance against the additional complexity (which can usually be handled by modern spectrometers) and the extra time taken (which may not be acceptable in certain experiments).

a   **b**

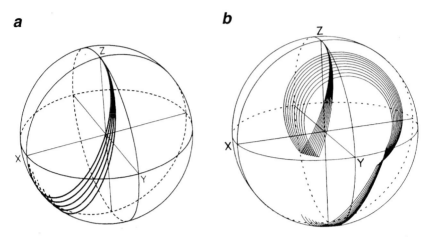

FIGURE 9.4 (a) Rotation of magnetization vectors in response to a 180_x pulse. The vectors shown are off resonance, with $(\omega - \omega_{rf})/\gamma B_1$ ranging from 0.4 to 0.6. (b) Rotation of the same off-resonance magnetization vectors by a composite pulse, $90_x 240_y 90_x$, produces much more effective inversion of the magnetization than the simple 180° pulse. From Freeman.[19]

9.4 SPIN LOCKING

We have seen that the period τ in a CPMG experiment may be made arbitrarily small. As τ decreases, the magnetizations move less from the y' axis, and in the limit as $\tau \rightarrow 0$, **M** remains precisely along y'. Of course, this hypothetical series of π pulses with no spacing is simply a single long, high power pulse **B**$_1$ along y'. Viewed in this way, it is also clear from Eq. 2.52 that near resonance **B**$_{eff} \approx$ **B**$_1$, and **M** should remain "locked" along y', as it cannot precess about a collinear magnetic field. When **B**$_1$ is turned off, **M** precesses, a normal FID results, and Fourier transformation provides a spectrum. However, during the period when the spin-locking field is on, relaxation occurs, so these are "partially relaxed" spectra.

Relaxation in the Rotating Frame

Because the spin-lock sequence is just the limiting case of a CPMG sequence, it might be expected that the relaxation time measured in this way would be T_2. For liquids, in which molecules move rapidly and T_2 can be properly defined, this expectation is realized, and spin locking furnishes an alternative method for measuring T_2, and it can be shown that by varying the magnitude of **B**$_1$, exchange rates can also be measured. In solids, however, the spin interactions that we discussed in Chapter 7 lead to a more complex situation, in which relaxation depends on the value of **B**$_1$. Because **M** relaxes along a magnetic field **B**$_1$ that is static in the rotating frame, the process is called "relaxation in the rotating frame" and the relaxation time is given the symbol $T_{1\rho}$. As we saw in Chapters 7 and 8, values of $T_{1\rho}$ are important in application of cross polarization techniques and can provide information on molecular motions in the kHz range.

9.5 SELECTIVE EXCITATION

In most NMR studies we wish to subject magnetizations to a uniform excitation, even though they might differ somewhat in resonance frequency; hence the need for such devices as composite pulses. However, in certain experiments it is important to affect magnetization over only a narrow range, while leaving other magnetizations undisturbed. A selective pulse is often called a *soft* pulse, in contrast to the normal *hard* pulse designed for excitation over a wide frequency range. In the simplest instance, this calls for the use of a small $\mathbf{B}_1$ at the center of the range to be excited, so that magnetizations very near resonance receive nearly the desired pulse, while those away from resonance generate only a small component in the xy plane, as illustrated in Fig. 9.5.

With the hard pulses that we have encountered, the time variation of excitation is simply a "square" pulse, in which the rf power is turned on and off as fast as possible, and the amplitude is constant during the pulse. However, the use of a simple square pulse leads to a far from ideal frequency excitation profile for a soft pulse, because the Fourier transform of the square pulse is a sinc function with sidelobes, as we saw in Fig. 3.8. Over the millisecond range used for a selective (soft) pulse, it

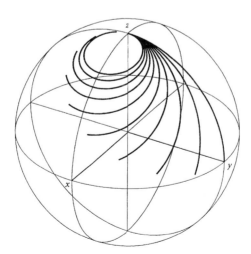

FIGURE 9.5 Trajectories of magnetizations from 0 to 900 Hz off resonance, excited by a rectangular pulse of width 1 ms with $\gamma B_1/2\pi = 250$ Hz, which is a 90° pulse on resonance. A magnetization at 1000 Hz from resonance would complete almost a full circle and give no signal. From Freeman.[106]

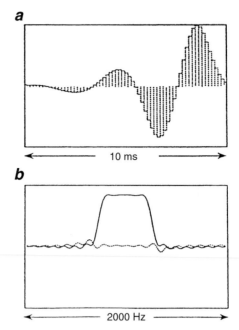

FIGURE 9.6 Illustration of the E-BURP selective excitation pulse. (*a*) Depiction of the time excitation function. (*b*) Frequency domain excitation profiles for absorption (solid line) and dispersion (dotted line). From Freeman.[106]

is feasible to tailor the excitation as a function of time. A large number of pulse functions that are much more complex than a square wave have been devised. For example, with on optimized computer simulation, it is possible to generate quite good selective pulses, the properties of which can be readily specified. In particular, the soft pulse can be designed for pure phase absorption excitation, with little or no dispersion contribution. These band-selective, uniform response, pure-phase (*BURP*) pulses can be optimized for excitation (E-BURP), inversion (I-BURP), and other manipulations of magnetizations. An E-BURP pulse is illustrated in Fig. 9.6.

DANTE

In some spectrometers it is not feasible to generate BURP or related tailored soft pulses. A useful alternative is the DANTE pulse sequence (delays alternating with nutation for tailored excitation). Normal high excitation power is used, but rather than apply a single 90° pulse, the DANTE method uses a sequence of small angle pulses that sum to 90° but with a short time between pulses during which nuclei

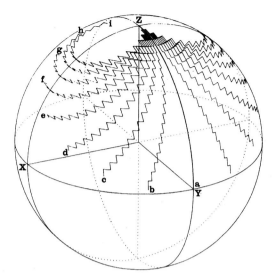

FIGURE 9.7 Trajectories for the on-resonance magnetization and several off-resonance magnetizations from a DANTE pulse sequence consisting of 18 pulses along the x' axis, each of 5°. From Freeman.[19]

precess freely, as indicated in Fig. 9.7. Thus, in the rotating frame a nucleus that is precisely on resonance experiences a 90° rotation in the $y'z'$ plane in series of steps and ends on the y' axis, just as it would with an ordinary 90° pulse. However, nuclei that are off resonance precess a fraction of a revolution after each small pulse step, hence experience a series of rotations about different axes, their magnetizations become randomized, and they do not contribute significantly to the FID. Typically, 10–200 pulses are used, with precession periods of perhaps 2 ms. The DANTE sequence generates sidebands that can also be used for excitation but must be spaced (by choice of the precession period) to prevent undesired excitations. A number of variants of on- and off-resonance DANTE have been devised.

BIRD Pulse and X-Filters

Many NMR studies of nuclides of low sensitivity (nucleus X) that are spin coupled to protons are now carried out by *indirect detection* of the proton resonance, primarily by multidimensional NMR methods that we discuss later. For such detection to be effective for an X that is in low natural abundance (such as ^{13}C) or is selectively enriched, it is essential to discriminate against the much larger signal arising from proton resonances of molecules with a nonmagnetic form of the nuclide (e.g., ^{12}C). We describe here two simple ways to achieve such discrimination.

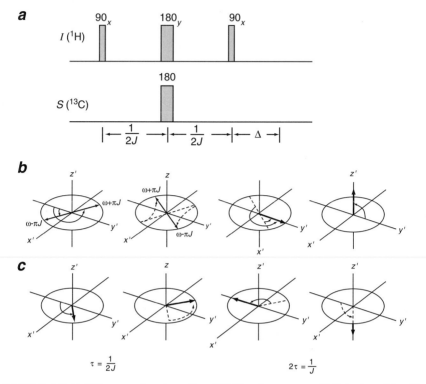

FIGURE 9.8 (*a*) Bilinear rotation decoupling (BIRD) pulse sequence. (*b*) Behavior of the ^{1}H magnetization from a proton coupled to ^{13}C. (*c*) Behavior of the ^{1}H magnetization from a proton not coupled to ^{13}C. See text for details.

The first, known as *bilinear rotation decoupling* (BIRD), uses a pulse sequence (shown in Fig. 9.8) of $90_{x'}$, τ, $180_{y'}$, τ, $90_{x'}$, Δ, all applied to ^{1}H, and $180°$ applied to ^{13}C, with τ chosen to be $1/2J$. With the $90°$ pulses applied along x' and $180°$ pulse applied along y' as in the Meiboom–Gill method, the α and β components of the magnetization for the ^{13}C–H system continue to precess after the $180°$ pulse (as in the central part of Fig. 9.2) but come into phase along $+y'$ after the total evolution period of $2\tau = 1/J$, as indicated in Fig. 9.8*b*. The next $90°$ pulse then rotates the magnetization to $+z$, where it started. For a ^{12}C–H system, however, the $180°$ pulse causes an echo along $-y'$, and the next $90°$ pulse then rotates this magnetization to $-z$. After a delay period of $\Delta = T_1 \ln 2$, the ^{12}C–H magnetization is reduced to zero by longitudinal relaxation (see Fig. 2.12), so that a subsequent $90°$ pulse at this time excites only the ^{13}C–H system.

Other pulse sequences, known collectively as *X-filters*, are also used to discriminate in favor of or against resonances from ^{1}H coupled to X. For example, in one such sequence protons are again subjected to a sequence $90_{x'}$, τ, $180_{y'}$, with $\tau = 1/2J$, while the X (^{13}C) spins are alternately subjected to a 180° or 0° pulse sequence concurrent with the 180° proton pulse. The data for successive repetitions are then alternately added or subtracted. For a ^{12}C$-$H fragment, the X pulses are irrelevant, so the resultant signal after two pulse sequences is zero. But for a ^{13}C$-$H spin-coupled system, the magnetizations at 2τ lie along $+y'$ and $-y'$ in successive scans (as in Fig. 9.8), so that subtraction of the signals that are 180° out of phase gives a net positive result. Here, magnetization for the ^{13}C$-$H system alone is obtained, this time directly, without the delay time for relaxation. Other X-filter sequences are also available. X-filter or BIRD sequences can be inserted at appropriate points in complex sequences for 1D or 2D NMR studies.

Solvent Suppression

In many instances we wish to excite the entire spectrum uniformly except for a narrow frequency range, usually to avoid exciting the resonance of water as a solvent. DANTE can be configured to achieve that objective, but a number of other approaches to solvent suppression are also available. One simple but effective approach is the "jump and return" (JR) technique, which uses the same rationale as DANTE. JR consists of a 90° pulse followed by a short precession period τ and then a second 90° pulse of opposite phase. If the solvent line is on resonance, it is restored to the z axis, whereas all off-resonance magnetizations retain substantial components in the xy plane, hence give rise to the desired FID without the solvent resonance. (There are some advantages to using flip angles of 45°, but the principle is the same as for 90° pulses.)

Some methods of solvent suppression use a selective pulse that is applied long enough to saturate the water signal prior to excitation of the entire spectrum with a 90° pulse. This method is often effective but has a significant disadvantage in that protons in the sample that exchange with water or cross-relax water protons are also partially or completely saturated, hence do not appear in the spectrum with full intensity. Alternatively, a preliminary selective 90° pulse, followed by a pulsed field gradient to dephase the water magnetization, is also widely used. Such a selective pulse can be restricted to a few milliseconds and thus avoid the problems associated with actual saturation of the water signal.

Some methods take advantage of a difference in a particular property between water and the molecule to be studied. In particular, a macromolecule usually has a shorter value for proton T_1 than water and a much lower diffusion coefficient. One of the oldest methods for water signal suppression is WEFT (water elimination Fourier transform), in which an inversion recovery sequence is applied (see Fig. 2.12) with τ chosen to be the time that the water signal goes through zero ($T_1 \ln 2$), just as in the BIRD pulse sequence. Another method makes use of the technique described in Section 9.3 to measure diffusion coefficients.

With careful selection of gradient strength and timing, it is feasible to minimize the signal at the echo from small, rapidly diffusing molecules such as water, while obtaining almost full refocusing of the magnetization from slow-moving macromolecules.

Another method, with the acronym WATERGATE (*water* suppression by *g*radient-*t*ailored *e*xcitation), uses a pulse sequence similar to that in Fig. 9.3, but with selective pulses applied to water just before and after the 180° pulse. As we saw in Section 9.2, the sequence of Fig. 9.3 can be used to measure molecular diffusion, but here we make τ small enough so that diffusion can be ignored. Thus, without the selective pulses, the signal following the echo simply reproduces the entire spectrum. However a selective 180° pulse applied only to water, essentially concurrently with the nonselective 180° pulse, inverts the water magnetization and results in its further dephasing, rather than rephasing, in the latter half of the sequence. WATERGATE can reduce the water signal by a factor of 10^4–10^5 in a single scan, hence is a useful component in 2D NMR pulse sequences.

9.6 DECOUPLING

As we mentioned in Chapters 2 and 5, single frequency decoupling can usually be carried out most readily by simple continuous wave irradiation at the decoupling frequency. Likewise, for solids, dipolar decoupling is obtained by high power irradiation at a single frequency, because irradiation at any frequency within a homogeneously broadened line affects the entire line. However, for multiline high resolution spectra of liquids, it is often necessary to provide decoupling that affects all lines in the spectrum. For example, in the study of ^{13}C, ^{15}N, and many other nuclei we usually wish to decouple all protons. For many years such decoupling was carried out by modulating a single decoupling frequency, usually with pseudorandom noise. However, suitable pulse methods are now available to carry out decoupling far more effectively and with the use of less rf power and its concomitant heating of the sample.

We saw in Section 9.1 that a 180° pulse applied at time τ to a nucleus S that is coupled to the observed nucleus I causes a refocusing of magnetization that eliminates the coupling at time 2τ (Fig. 9.2, bottom). If the two nuclei are of different species, an effective heteronuclear decoupling procedure can be based on this concept. To be effective, a string of 180° pulses must be spaced at a time $\tau \ll 1/J$, the value of B_2 must be adequate to decouple over the spectral range, and the pulses must be perfect 180° pulses. For example, for ^1H decoupling with values of J less than a few hundred hertz, a set of 100 μs pulses ($\gamma B_2/2\pi \approx 5$ kHz) with no spacing between them easily meets the first two requirements. However, "perfection" over a wide frequency range requires the use of composite pulses with careful phase cycling. Several very effective multiple pulse cycles have been developed, of which the best known are *MLEV-16, WALTZ-16, GARP*, and *DIPSI*. The prototype decoupling sequence, MLEV, uses a composite 180° pulse of the type that we described in Section 9.3. MLEV-16 uses a composite pulse designated R, where $R = 90^\circ_{x'}180^\circ_{y'}90^\circ_{x'}$ and its phase inverted form $\overline{R} = 90^\circ_{-x'}180^\circ_{-y'}90^\circ_{-x'}$, permuted in a four-pulse cycle further permuted into a 16-pulse supercycle:

$$\text{MLEV-16: } R R \overline{R}\, \overline{R} \quad \overline{R} R R \overline{R} \quad \overline{R}\, \overline{R} R R \quad R \overline{R}\, \overline{R} R$$

WALTZ-16 is based on an element $R = 90^\circ_{x'}180^\circ_{-x'}270^\circ_{x'}$. WALTZ-16 is more effective than MLEV-16, primarily because it includes only 180° phase shifts, not the 90° phase shift inherent in MLEV. To simplify the notation, these pulse cycles are usually abbreviated in terms of multiples of 90° pulses, with a phase inversion denoted by a bar, as in the $R,\overline{R}$ notation. In these terms the basic WALTZ element R becomes $1\overline{2}3$ (hence the acronym WALTZ). Permutations of R and $\overline{R}$ lead to WALTZ-16. Further computer-optimized improvements have been devised and are not restricted to integral multiples of 90° pulses.

9.7 POLARIZATION TRANSFER METHODS

We now describe a somewhat more complex pulse sequence, *INEPT* (an acronym for *i*nsensitive *n*uclei *e*nhanced by *p*olarization *t*ransfer). INEPT was devised to enhance the signals from a nucleus with a relatively low magnetic moment (such as ^{13}C) that is scalar coupled to a nucleus with a large magnetic moment (usually ^1H) by transferring the larger nuclear polarization. As a one-dimensional NMR technique, INEPT and its several variants have been widely used, both for sensitivity enhancement and for discriminating among primary, secondary, tertiary, and quaternary carbon atoms (*spectral editing*). Moreover, as we shall see, pulse sequences closely related to that used in INEPT often provide extremely valuable components of more complex multidimensional NMR experiments that transfer polarization between different nuclear species.

Selective Population Transfer

The rationale for INEPT can best be understood by looking first at a somewhat simpler continuous wave experiment for transferring polarization, *selective population transfer* (SPT). SPT can be understood simply in terms of the populations of energy levels, whereas INEPT requires consideration of coherent precessing magnetization. Figure 9.9 shows the energy levels and populations of an AX spin system, which we take to be ^1H and ^{13}C, respectively, in this example. At equilibrium the populations conform to a Boltzmann distribution. Because the ^1H

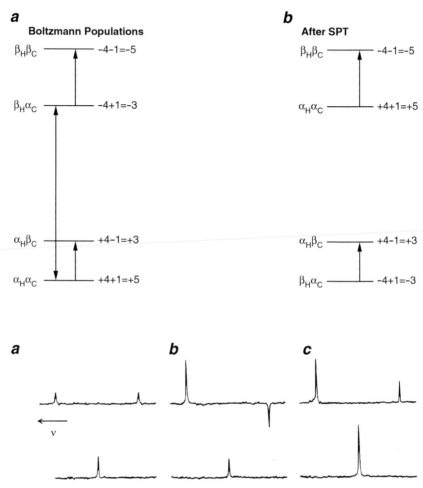

FIGURE 9.9 Energy levels, populations, and ^{13}C spectra in the SPT experiment. *Top:* Energy levels and "excess" populations in (*a*) the normal Boltzmann distribution and (*b*) the distribution after a selective pulse on line ν_1 has inverted the populations of the $\alpha\alpha$ and $\beta\alpha$ levels. *Bottom:* ^{13}C spectra (coupled and decoupled) obtained with (*a*) Boltzmann distribution; (*b*) altered population distribution; (*c*) altered distribution; with phase inversion of line ν_3.

energy levels are separated by approximately four times the separation of the ^{13}C energy levels, Fig. 9.9 gives the "excess" populations in the four levels as $(4 + 1)$, $(4 - 1)$, $(-4 + 1)$ and $(-4 - 1)$. If a selective 180° pulse is applied at ν_1 (the frequency of the $\alpha\alpha \leftrightarrow \beta\alpha$ transition), the populations of these levels are interchanged with no effect on the populations of the other two states. The intensities of the two ^{13}C transitions, ν_3 and ν_4, which are proportional to the differences in populations between the $\alpha\alpha$ and $\alpha\beta$ levels and the $\beta\alpha$ and $\beta\beta$ levels, respectively, are easily seen to be altered from their 1:1 values to -3:5. Thus, while the algebraic sum of the two ^{13}C intensities remains at $+2$, the two individual lines are considerably enhanced in intensity. This experiment demonstrates an important technique for transferring polarization by selectively inverting the populations of half of the protons, those coupled to ^{13}C's with spin state α, while leaving the protons that are coupled to ^{13}C's in state β with their normal population difference. However, in this form SPT cannot easily be applied to real molecules containing several nuclei with different chemical shifts.

INEPT

INEPT achieves the same result as SPT but is much more widely applicable because it uses nonselective pulses to manipulate spins over a wide range of frequencies. Again consider a coupled AX spin system, $^{1}H-^{13}C$, as an example. Figure 9.10 shows the pulse sequence and depicts the ^{1}H and ^{13}C magnetizations, each of which has two components, depending on the spin state of the coupled nucleus. (a) The initial 90° pulse rotates the ^{1}H magnetization to the xy plane but does not affect the ^{13}C magnetization. (b) The ^{1}H components precess at the frequency of the ^{1}H chemical shift $\pm J/2$ Hz. (c) At time $\tau = 1/4J$, when the two vectors are at 90° to each other, simultaneous 180° pulses are applied along the x' axis to the ^{13}C and ^{1}H spins. (d) The effect of the two 180° pulses is to interchange the ^{1}H vectors but also to change their frequencies (as in Fig. 9.2) so that they continue to diverge. (e) At 2τ (i.e., $1/2J$) an ^{1}H spin echo forms in which all chemical shifts are refocused, but the α and β components are now directed opposite to each other along x' and $-x'$. (f) At 2τ, a 90° ^{1}H pulse is applied along y' to rotate the two components, as shown, to $+z$ and $-z$. Thus, this experiment achieves what was obtained directly by the selective ^{1}H pulse in the SPT experiment—inversion of the magnetization of those protons coupled to α ^{13}C spins, while magnetization of protons coupled to β ^{13}C spins is unchanged. If we now think about this system in terms of its energy levels, it is clear that the ^{13}C populations are altered just an in the SPT experiment. Finally, a ^{13}C 90° pulse applied at 2τ then samples the ^{13}C magnetization to produce the expected doublet of intensities $-3:+5$. Alternatively, if the second 90° pulse is phase cycled along $+y'$ and $-y'$, the natural magnetization component is canceled in successive scans, and a doublet of intensity $-4:+4$ is obtained.

The sequence can easily be modified to allow the two separate components to combine—*refocused INEPT*. After the final 90° pulses but before data acquisition, a period 2Δ is added, with $\Delta = 1/4J$, during which the two vectors combine. ^{1}H and ^{13}C 180° pulses are placed in the middle of this precession period to refocus chemical shifts. If decoupling is now applied during data acquisition, a single line of intensity 8 is obtained, that is, four times as large as would be obtained

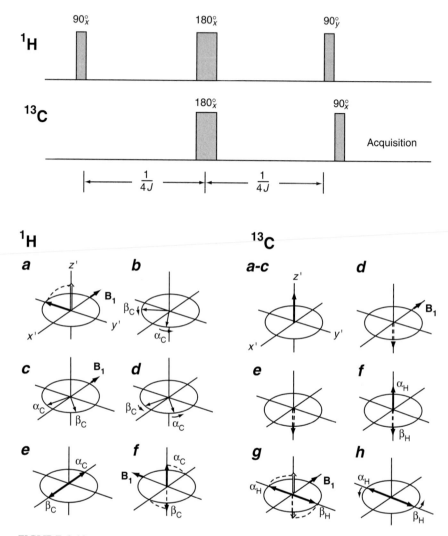

FIGURE 9.10 Summary of the INEPT experiment as applied to a coupled $^1H-^{13}C$ system. *Top:* Pulse sequence. *Bottom:* Depiction of the behavior of ^1H and ^{13}C magnetizations, as described in the text. In practice, the 90° ^{13}C pulse may be applied coincident with the final ^1H pulse but is shown slightly displaced to illustrate in (*g*) and (*h*) the behavior of the ^{13}C magnetization after the population inversion has been established in (*f*).

directly. Moreover, the repetition time can be based on T_1 (H), which is usually much shorter than T_1 (^{13}C), just as in the cross polarization technique in solids (Section 7.6).

As in the case of a spin echo with coupled nuclei, our explanation of INEPT grafted the quantum concept of energy levels onto the classical picture of precessing magnetization in a manner that is barely satisfactory. We shall revisit INEPT after developing more powerful mathematical techniques in Chapter 11.

Antiphase Magnetization

INEPT transfers polarization successfully because the final 90° ^{1}H pulse is applied when the α and β magnetization vectors are exactly 180° out of phase or in *antiphase*. In general, precession of vectors corresponding to magnetizations from nuclei that are spin coupled causes them repeatedly to go into and out of phase with each other. Let's look at this process more closely and see how to express the result mathematically.

Figure 9.11 shows the precession of two such vectors for the A nuclei in an AX system precessing in a frame rotating at the chemical shift of A. If a pulse has placed the magnetizations (of magnitude $|\mathbf{M}|$) along the y' axis at time 0, the vectors precess in the xy plane as shown, according to the usual relations:

$$\mathbf{M}_\alpha = |\mathbf{M}|[-x' \sin 2\pi(J/2)t + y' \cos 2\pi(J/2)t]$$
$$\mathbf{M}_\beta = |\mathbf{M}|[+x' \sin 2\pi(J/2)t + y' \cos 2\pi(J/2)t] \tag{9.3}$$

It is clear from Eq. 9.3 and Fig. 9.11 that the *net* magnetization, that is, the vector sum $\mathbf{M}_\alpha + \mathbf{M}_\beta$, never has a component in the x' direction at any time. Because the component along y' oscillates according to the cosine relation, at time $t = 1/2J$ (and at subsequent times $3/2J$, $5/2J$, etc.) the total net magnetization is zero. Although mathematically correct, this result obscures the physical reality that each component retains its absolute value $|\mathbf{M}|$ at all times.

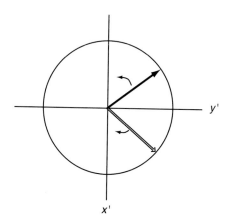

FIGURE 9.11 Precession of the two spin components of magnetization, $\mathbf{M}_\alpha$ and $\mathbf{M}_\beta$, in a frame rotating at the chemical shift frequency, as described in the text.

We can resolve this paradox by defining an *in-phase* magnetization vector **P** and an *antiphase* magnetization **A**:

$$\mathbf{P} = 2|\mathbf{M}|\mathbf{y}'\cos 2\pi(J/2)t$$
$$\mathbf{A} = 2|\mathbf{M}|\mathbf{x}'\sin 2\pi(J/2)t$$

$$(9.4)$$

Equations 9.4 represent the actual situation correctly. At all times the magnitude of the vector sum $|\mathbf{P} + \mathbf{A}| = 2|\mathbf{M}|$ as it should, and each component has the required oscillatory behavior. In Chapter 11 we discover how to express **P** and **A** in terms of the familiar spin operators I_x, I_y, and I_z, and in Chapter 12 we see that the presence of both in-phase and antiphase components is fundamental to the operation of many two-dimensional NMR experiments.

Spectral Editing by INEPT

Our description of INEPT dealt with an AX system, for example, ^{13}C coupled to a single ^{1}H. The ^{13}C spectrum of an organic molecule arises from ^{13}C nuclei coupled through one bond to zero, one, two, or three protons. Analysis of the behavior of the ^{13}C magnetizations for the four cases in a manner analogous to that in Figure 9.10 shows that they refocus at different times, so that in practical application of refocused INEPT to enhance ^{13}C signals a compromise value of Δ must be utilized to account for this behavior and for the fact that the value of $^1J_{CH}$ varies (Section 5.4). However, the first disadvantage can be turned to advantage in using the differing behavior to discriminate among ^{13}C signals from CH_3, CH_2, CH, and quaternary carbons, hence to "edit" a spectrum into subspectra. We discuss this use of INEPT, along with other editing sequences, in Chapter 12.

9.8 ADDITIONAL READING AND RESOURCES

Virtually all NMR books describe many of the pulse sequences covered in this chapter. *A Handbook of NMR*[19] and *A Dictionary of Concepts in NMR*,[18] both mentioned previously, provide particularly clear and succinct discussions of most of the topics in this chapter.

9.9 PROBLEMS

9.1 Consider the formation of a spin echo following a 180° pulse applied to a spin system consisting of two protons. (a) Assume that the protons have the same chemical shift (as in H_2O, for example). Explain why the echo forms and eliminates the effect of magnetic field inhomogeneity. (b) Assume that the two protons differ in chemical shift but are not spin coupled. Will the echo amplitude be changed relative to its value in case (a)? Explain. (c) Assume that the two protons differ in chemical shift but are spin coupled. Will the echo amplitude be affected? Explain.

9.2 Explain, in words and with the aid of vector diagrams, why a gradient echo does not refocus chemical shifts.

9.3 Assume that because of B_1 inhomogeneity a nominal 180° pulse causes a weighted average 176° pulse applied to nuclei throughout the sample. Compare the reduction in intensity of the 5th and 20th echoes in the original Carr–Purcell method and in the Meiboom–Gill modification.

9.4 Consider a soft pulse, as illustrated in Fig. 9.5, applied to a nucleus with magnetogyric ratio γ. Take the value of B_1 to be such that a magnetization **M** on resonance executes a 90° precession in a pulse time T_p. Using Eq. 2.52 and Fig. 2.8, determine how far off resonance a magnetization must be to execute a 360° precession about $\mathbf{B}_{\text{eff}}$ in time T_p.

9.5 The center portion of Fig. 9.2 shows the modulation effect on the spin echo of spin coupling between nuclei I and S. How would this vector picture differ for an IS_2 spin system? What would be the modulation frequency?

9.6 (a) Draw an energy level diagram, analogous to Fig. 9.9, for a $^{13}CH_2$ system, and illustrate the allowed transitions. Take symmetry into account in forming the basis functions (as for an A_2B spin system, Table 6.4) and in limiting the allowed transitions. (b) Determine the "excess" populations, as in Fig. 9.9, for a Boltzmann distribution. (c) Determine the populations after applying a 180° pulse to one of the ^1H transitions, and find the intensities of the three ^{13}C lines, as compared with their normal intensities.

Two-Dimensional NMR

For 20 years center stage has been occupied by two-dimensional (and now three- and four-dimensional) NMR techniques. 2D NMR and its offshoots offer two distinct advantages: (1) relief from overcrowding of resonance lines, as the spectral information is spread out in a plane or a cube rather than along a single frequency dimension, and (2) opportunity to correlate pairs of resonances. In the latter respect 2D NMR has features in common with various double resonance methods, but as we shall see, 2D NMR is far more efficient and versatile. Hundreds of different 2D NMR techniques have been proposed in the literature, but most of these experiments can be considered as variations on a rather small number of basic approaches. Once we develop familiarity with the basic principles, it will be relatively easy to understand most variations of the standard 2D experiments.

10.1 GENERAL ASPECTS OF 2D SPECTRA

In Chapter 3 we saw how a one-dimensional spectrum, $S(\omega)$, is obtained from Fourier transformation of a time domain signal, $s(t)$. By analogy, a 2D spectrum, $S(\omega_1,\omega_2)$, is obtained from a two-dimensional Fourier transformation of a time

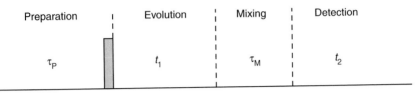

FIGURE 10.1 Schematic representation of a basic two-dimensional NMR experiment in terms of four periods: preparation, τ_P; evolution, t_1; mixing, τ_M; detection, t_2. For a given experiment, τ_P and τ_M are usually fixed periods, while t_1 and t_2 are variable time periods, as described in the text.

domain signal, $s(t_1,t_2)$, that is function of two independent time variables, t_1 and t_2. Figure 1.6 showed a simple 2D spectrum presented as a *stacked line plot*, which is useful for some purpose but makes exact measurement of frequency coordinates difficult, especially in the case of overlapping resonances. Hence 2D spectra are most commonly presented as *contour plots*, where contours are drawn through points of equal intensity, analogous to elevation lines on a geographic map. We shall see a number of examples.

The Basic 2D Experimental Procedure

For a typical 2D experiment four time intervals can be distinguished, as shown in Fig. 10.1: A preparation period (τ_P), an evolution period (t_1), a mixing period (τ_M), and a detection period (t_2). During the preparation period the nuclear spin system is simply allowed to come to equilibrium in many experiments, while in others elaborate pulse sequences may be used to "prepare" the system for the remainder of the 2D experiment. The final step in the preparation period causes the magnetization **M** to be rotated to the xy plane, often by a simple 90° pulse but occasionally by a more complex pulse sequence. During t_1 the magnetization evolves freely, so that M_{xy} precesses at its Larmor frequency, and each nuclear magnetization is "tagged" according to its Larmor frequency, which depends on the particular Hamiltonian that is operative in the nuclear spin system. Sometimes additional pulses may be applied or other conditions may be changed during t_1 to alter the overall Hamiltonian during this period. The period τ_M gives nuclear magnetizations in the xy plane an opportunity to interact or to "mix" their wave functions. τ_M may be short and involve only the application of, for example, a single 90° pulse, or it may involve a sequence of pulses to rotate **M** out of the xy plane and time for the spin system to relax or otherwise alter its state. Finally, t_2 is the usual data acquisition period in which an FID is acquired, just as in a one-dimensional experiment.

    This procedure is then repeated a large number of times, with different durations of the evolution period, keeping all other settings constant. For each value of t_1, the signal that is acquired during t_2 is digitized and stored in computer memory. Upon completion of the entire 2D experiment, a data matrix representing the two-dimensional time domain signal $s(t_1,t_2)$ is available in the memory (or on the magnetic disk) of the computer. Two Fourier transforms, with respect to t_1 and t_2, plus other data processing similar to that in one-dimensional NMR then provide the desired 2D spectrum, as illustrated in the following section.

It is easy to see from Fig. 10.1 how a 2D experiment can be generalized to three or more dimensions by adding additional independent time periods for further evolution (and usually additional mixing periods). In this way, as we see in later chapters, it is possible to correlate additional physical parameters and to display 3D data in a cube or as a set of planes, and 4D NMR spectra can be displayed by spreading the data in each plane of the 3D plot into a separate set of planes. In principle, there is no limit to the number of dimensions that may be obtained, but there are serious practical consequences of adding dimensions. Each independent time period requires that a set of FIDs be obtained while all other times are held constant. Even carefully optimized 4D experiments require of the order of $32 \times 32 \times 128$ separate repetitions, each taking about 3 seconds, or a total of about $4\frac{1}{2}$ days of instrument time. Such a set of experiments produces about 64×10^6 points before zero-filling and data processing. Although this amount of data is well within the capability of modern computer workstations, processing time is, of course, increased.

Behavior of Magnetization and NMR Signals

Some examples may help to clarify the steps involved in 2D NMR. Consider first the pulse sequence shown in Fig. 10.2 applied to a molecule with a single NMR frequency (e.g., the ^1H resonance of $CHCl_3$). The preparation period is simply a

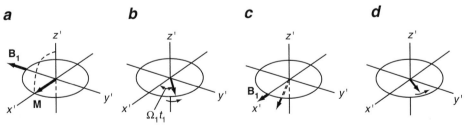

FIGURE 10.2 Prototype 2D NMR pulse sequence and behavior of the magnetization from a single NMR frequency Ω_1. (a) **M** is rotated by a 90° pulse to x' (preparation period). (b) **M** precesses at frequency Ω_1 (radians/second) during the evolution period, reaching an angle $\Omega_1 t_1$ with respect to x' at time t_1. (c) **M** is rotated by a second 90° pulse (mixing period) to the $x'z'$ plane, where it continues to make an angle $\Omega_1 t_1$ with x'. (d) The portion of **M** remaining in the $x'y'$ plane, **M** cos $\Omega_1 t_1$, precesses at a rate Ω_2 radians/second during the detection period.

90° pulse applied to the system at equilibrium, where it has magnetization **M**. For example, as indicated in Fig. 10.2, we can apply $\mathbf{B}_1$ along the $-y'$ axis, so that **M** rotates 90° about y' to the x' axis and then precesses in the $x'y'$ plane of the rotating frame during the period t_1 at angular frequency Ω_1. (We use ω as the angular frequency variable and denote a specific value of frequency by Ω.) At the end of the evolution period, **M** makes an angle $\Omega_1 t_1$ with the positive x' axis. For this example, τ_M is only a few microseconds and consists entirely of a $90°_{-x'}$ pulse (i.e., $\mathbf{B}_1$ applied along the positive x' axis) that rotates **M** into the $x'z'$ plane, thus producing a longitudinal component with an amplitude $M_0 \sin \Omega_1 t_1$ and leaving a transverse component with amplitude $M_0 \cos \Omega_1 t_1$ parallel to x'. (Note that this $\pi/2$ phase change in the transmitter has no fundamental significance in this example; it simply keeps **M** in Fig. 10.2 in a region that is easily visualized).

During the detection period only transverse magnetization M_{xy} is detected, so the initial magnitude of the signal is proportional to $M_0 \cos \Omega_1 t_1$. During the detection period t_2, M_{xy} rotates with angular frequency Ω_2, which in this simple example is equal to Ω_1 because nothing has caused it to change precession frequency. The quadrature detected signal is then described by

$$s(t_1, t_2) = M_0 \cos \Omega_1 t_1 e^{i\Omega_2 t_2} e^{-t_1/T_2(1)} e^{-t_2/T_2(2)} \tag{10.1}$$

where $T_2(1)$ and $T_2(2)$ are the transverse decay constants during evolution and detection times, respectively. In this example (and in most other homonuclear 2D experiments) these two time constants are equal.

For a given value of t_1, we can Fourier transform the FID with respect to the variable t_2 to obtain a spectrum

$$S(t_1, \omega_2) = M_0 \cos \Omega_1 t_1 e^{-t_1/T_2(1)} [A(\omega_2) + iD(\omega_2)] \tag{10.2}$$

where $A_2(\omega_2)$ and $D_2(\omega_2)$ are the absorptive (real) and dispersive (imaginary) parts of the spectrum, given by

$$A(\omega_2) = \frac{T_2}{1 + [T_2(\omega_2 - \Omega_2)]^2} \tag{10.3}$$

$$D(\omega_2) = \frac{[T_2]^2(\omega_2 - \Omega_2)}{1 + [T_2(\omega_2 - \Omega_2)]^2} \tag{10.4}$$

After the system has relaxed back to equilibrium, the process is repeated, but now **M** is allowed to precess for a longer value of t_1, and the signal is detected and Fourier transformed. The absorptive portion of a set of spectra obtained for a set of different t_1 values is sketched in Fig. 10.3a. The amplitude of the resonance at $F_2 = \Omega_2/2\pi$ is modulated by $\cos \Omega_1 t_1$, as shown more clearly in Fig. 10.3b, where the data matrix has been transposed in the computer. The value of this modulation frequency can be obtained from a second Fourier transformation, this time with respect to t_1. If $2N$ data points are acquired for each value of t_1, the

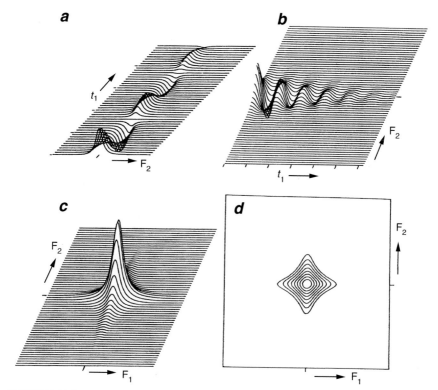

FIGURE 10.3 Data from the 2D pulse sequence in Fig. 10.2. (*a*) Spectra resulting from Fourier transform of $S(t_2)$ at various values of t_1. (*b*) Rotation of axes in (*a*) to show modulation in signal amplitude as a function of t_1. (*c*) Signal after second Fourier transform with respect to t_1, showing a single peak at coordinates $(\Omega_1/2\pi, \Omega_2/2\pi)$. (*d*) Contour plot, showing the peak from (*c*). Courtesy of Ad Bax (National Institutes of Health).

absorptive part of each spectrum obtained from the t_2 Fourier transformation consists of N points. Therefore, N cross sections parallel to the t_1 axis can be taken to produce the plots (usually called interferograms) in Fig. 10.3*b*. If $2M$ values of t_1 are used (where M and N are not necessarily equal), each interferogram contains $2M$ data points. It is clear from Fig. 10.3*b* that only interferograms near frequency $F_2 = \Omega_2/2\pi$ show significant modulation, by $\cos \Omega_1 t_1$. Thus, Fourier transformation with respect to t_1 yields a resonance in the F_1 dimension at $F_1 = \Omega_1/2\pi$ for the interferograms at or near $F_2 = \Omega_2/2\pi$ but zero-intensity spectra for all interferograms that do not show modulation. Hence, the final 2D spectrum, displayed as a stacked plot in Fig. 10.3*c* and a contour plot in Fig. 10.3*d*, shows a single resonance at coordinates $(F_1, F_2) = (\Omega_1/2\pi, \Omega_2/2\pi)$. Overall, this absorption mode plot contains MN points from the $(2M) \times (2N)$ data points acquired in the time domains.

This simple example demonstrates the general manner in which 2D time data are acquired and the procedure by which 2D Fourier transform spectra are obtained, but it also illustrates several other important points. First, it is apparent that a nuclear spin whose precession frequency is the same during the evolution and acquisition periods yields a single resonance on the diagonal of the (F_1, F_2) plot. Had there been several *noninteracting* nuclei in the sample, each would provide a separate resonance on the diagonal. Clearly, no more information is available in such a 2D experiment than could be obtained from a simple 1D spectrum. In general, 2D experiments are designed to provide useful information on the *interaction* of nuclear spins during the mixing time (τ_M), usually by scalar coupling but sometimes by relaxation or exchange processes. When such interactions occur, the magnetization may be transferred from one nucleus to another, thus introducing additional modulation frequencies into the interferograms of Fig. 10.3*b* and generating *off-diagonal* peaks in the 2D spectrum. We now look at an illustration of one process that causes such off-diagonal peaks to appear.

A Real Example: Chemical Exchange and Cross Relaxation

The proton NMR spectrum of *N,N*-dimethylformamide (DMF) was shown in Fig. 2.15 as an illustration of the exchange of nuclei that are chemically non-equivalent. As the rate of internal rotation about the C—N bond increases, the spectrum changes drastically. However, even near room temperature, where two distinctly separate resonances occur, there is exchange—too slow to alter the ordinary one-dimensional spectrum but fast enough to show effects in a suitably designed 2D study.

Consider now the application to DMF of the pulse sequence $90°, t_1, 90°, G, \tau,$ $90°$ (where G represents a pulsed magnetic field gradient), followed by data acquisition, as illustrated in Fig. 10.4. The behavior of a magnetization **M** during the $90°, t_1, 90°$ portion was described previously. However, after the second $90°$ pulse, we now focus attention on the component along the z axis, $M_0 \sin \Omega_1 t_1$, rather than that remaining in the xy plane. In fact, the pulsed field gradient G is inserted to dephase magnetization and ensure that M_{xy} decays rapidly to zero so that it can be ignored. The final $90°$ pulse then rotates this z component back to the xy plane, where it precesses and gives rise to a FID.

In our DMF example, there are three separate magnetizations, $\mathbf{M}_1$ and $\mathbf{M}_2$ from the two chemically nonequivalent methyl groups and $\mathbf{M}_F$ from the formyl proton. If there were no interactions among these protons, the 2D spectrum would be as uninteresting as that in Fig. 10.3 and would consist of three separate lines along the principal diagonal. However, if the "mixing" period τ is long enough, exchange of methyl groups occurs. Hence, $\mathbf{M}_1$ spends the evolution period t_1 precessing at an angular frequency Ω_1, but a fraction of the nuclei

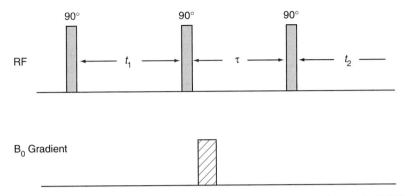

FIGURE 10.4 Pulse sequence for measuring chemical exchange and cross relaxation (NOESY).

contributing to $\mathbf{M}_1$ exchange during τ so that they precess during the detection period at frequency Ω_2. After the double Fourier transformation, the spectrum, as shown in Fig. 10.5, displays not only the two diagonal peaks from nuclei that did not exchange but also off-diagonal peaks whose intensity indicates the amount of exchange occurring during the period τ.

The example illustrates clearly the basic objective of 2D NMR—the correlation of two NMR frequencies because magnetization precesses at two different frequencies during periods t_1 and t_2. The challenge in designing 2D experiments is not only to find a process that will cause a magnetization to alter its frequency but also to be able to relate that process to some intrinsically interesting molecular features. In the DMF example the magnetization changes precession frequency because of an actual physical exchange of nuclei in space, and such studies can be useful in elucidating exchange pathways in multisite systems, such as fluxional molecules. However, in the vast majority of 2D (and multidimensional) NMR studies the process leading to change in precession frequency is more subtle, as well shall see.

In DMF, the magnetization $\mathbf{M}_F$ certainly does not undergo any physical exchange with $\mathbf{M}_1$ or $\mathbf{M}_2$, so it should give rise to only a diagonal peak. However, close examination of Fig. 10.5 shows a weak off-diagonal peak that indicates a correlation between $\mathbf{M}_F$ and $\mathbf{M}_2$. How does this peak arise? If we consider further what can happen during the mixing period τ, we can recognize that all magnetizations are aligned along $\mathbf{B}_0$, so longitudinal relaxation can occur during τ. As we saw in Section 8.2, two protons (with large magnetic moments) that are sufficiently close relax each other by magnetic dipolar interactions. That is exactly what happens here. The formyl proton is relaxed by the nearby protons from the *cis* methyl group. (The inverse process, relaxation of methyl protons by the formyl proton, is of little importance because the nearby protons within a methyl group predominantly relax each other.) In Section 8.3 we saw that dipolar relaxation

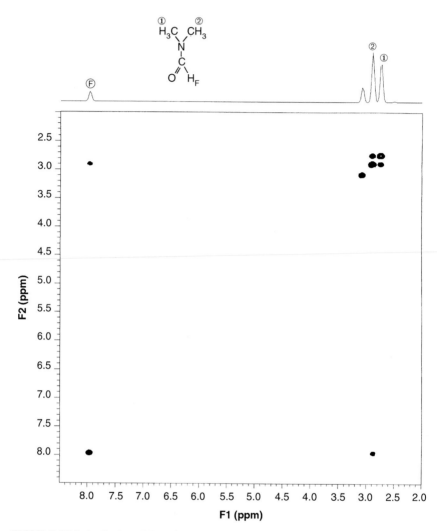

FIGURE 10.5 Application of the pulse sequence of Fig. 10.4 to dimethylformamide in DMSO-d_6. The diagonal peaks from $\mathbf{M}_1$, $\mathbf{M}_2$, and $\mathbf{M}_F$ are indicated; the other diagonal peaks arise from H_2O impurity and DMSO-d_5 (very weak). The cross peaks close to the diagonal result from methyl exchange; the weaker peaks near (2.9, 8.0) are from the NOE. Shown at the top is a projection of intensity in the F_2 direction. Courtesy of Herman J. C. Yeh (National Institutes of Health).

causes intensity changes—the nuclear Overhauser effect, NOE—and it is the NOE that is responsible for the off-diagonal peaks.

The pulse sequence shown in Fig. 10.4 is sometimes used to study exchange, but more than 99% of the use of this sequence is to study the effect of dipolar cross relaxation via the NOE. As a result, this type of study is given the name *nuclear Overhauser effect spectroscopy*, NOESY, and the pulse sequence of Fig. 10.4

is called the NOESY sequence. As we see in Chapter 13, NOESY is the principal 2D NMR technique for providing data on proximity of protons in macromolecules and determining their precise three-dimensional structure.

10.2 A SURVEY OF BASIC 2D EXPERIMENTS

A very large number of 2D and multidimensional NMR techniques are available, and new variations are continually being developed and refined. Some, such as NOESY, can be readily understood and interpreted on the basis of our magnetization vector picture. In particular, a number of methods used in liquids and in solids depend on changing the Hamiltonian between the evolution and data acquisition periods by altering some factor that is external to the spin system—for example, turning on a decoupler or spinning the sample. Generally these are amenable to a simple treatment. However, many others depend on interactions between the spins via spin coupling. As we have seen in Chapters 6 and 7, such interactions require a quantum mechanical approach, but to understand the dynamic mixing of spin wave functions as the 2D experiment evolves, we require the additional power of the density matrix and product operator formalisms, which we discuss in Chapter 11.

In this section we provide a summary of the commonly used 2D methods, principally in liquids, but with some reference also to solids. Our objectives are (1) to describe the purpose of the particular type of study (known in 2D NMR jargon as an *experiment*), (2) to indicate qualitatively the sort of data that can be obtained, and (3) to provide an explanation of the spin physics within the framework of the theory that we have at hand so far. In some instances we shall find that we need to apply density matrix or product operator treatments and that we must revisit these experiments in Chapter 12.

J-Resolved Spectra

Among the simplest 2D experiments to understand and to implement is one that uses a spin echo pulse sequence to produce a 2D spectrum that separates chemical shifts from splitting related to spin couplings. The pulse sequence is that shown in Fig. 9.2 (center), but with the evolution time t_1 replacing the period 2τ. At the end of the evolution period, the X chemical shifts are refocused, but the α and β components precess during t_1 so that after Fourier transformation the frequencies J_{AX} appear on the F_1 axis. During t_2 the spins precess fully, but if AX is a heteronuclear system, such as $H^{13}C$, broadband proton decoupling can be applied during t_2 to produce a decoupled ^{13}C spectrum in the F_2 dimension. An example of a heteronuclear *J*-resolved spectrum is shown in Fig. 10.6. The splittings along F_1 immediately identify each ^{13}C resonance along F_2 as CH_3, CH_2, or CH.

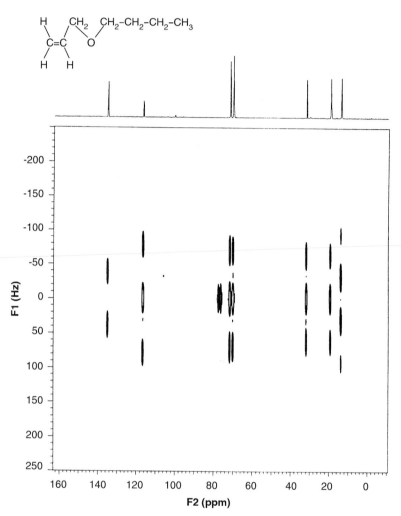

FIGURE 10.6 Heteronuclear *J*-resolved $^{13}C-H$ spectrum (75 MHz) of allylbutyl ether in CDCl$_3$. The ordinate F_1 shows the splitting from $^1J_{CH}$ in Hz, and the abscissa F_2 displays the ^{13}C chemical shifts in ppm. At the top is the completely decoupled one-dimensional ^{13}C spectrum. Courtesy of Herman J. C. Yeh (National Institutes of Health).

The spin-echo pulse sequence can also be applied to produce homonuclear (e.g., ^1H) *J*-resolved spectra. For such a system it is not possible to apply broadband decoupling, so the F_2 dimension might be expected to display the ordinary coupled ^1H spectrum. However, because the coupling information is independently available, it is not difficult to process the data in such a way that only the chemical shifts are displayed in the F_2 dimension, as illustrated in Fig. 10.7.

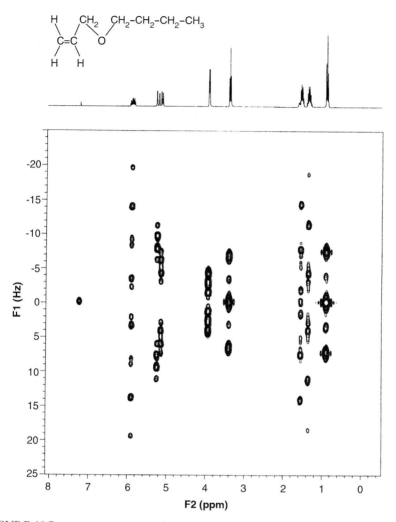

FIGURE 10.7 Homonuclear *J*-resolved ^{1}H spectrum (300 MHz) of allylbutyl ether in CDCl$_3$. The ordinate F_1 shows splittings from H–H couplings in Hz, and the abscissa F_2 displays the ^{1}H chemical shifts in ppm. At the top is the one-dimensional ^{1}H spectrum. Courtesy of Herman J. C. Yeh (National Institutes of Health).

The projection on the F_2 axis would show the equivalent of a completely decoupled proton spectrum.

J-resolved spectra can be useful in disentangling or editing spectra with overlapping peaks, but other 2D experiments are generally used more frequently in the structure elucidation of organic molecules.

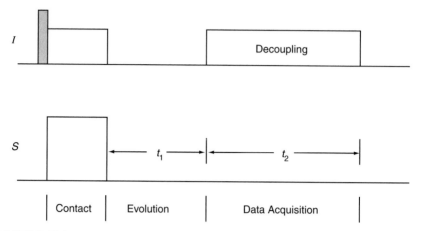

FIGURE 10.8 Pulse sequence for the separated local field experiment. After cross polarization, spin S precesses under the influence of dipolar interactions during t_1 but is decoupled from I during t_2.

2D Experiments in Solids

Many two-dimensional experiments in solids are also conceptually very simple. In Chapter 7 we examined several methods for producing narrow lines in solids by suppressing interactions from dipolar coupling and chemical shielding anisotropy. However, these interactions provide valuable information if they can be viewed separately and correlated with each other. The separated local field (SLF) experiment serves as the prototype for a large number of 2D techniques that have been developed for solids. Figure 10.8 shows that the SLF experiment for a ^{13}CH system begins with ordinary cross polarization, as described in Section 7.6, but the nuclei are allowed to precess during t_1 under the influence of dipolar coupling, and during t_2 decoupling is applied to remove the effect of dipolar coupling but leave chemical shift anisotropy active. Thus the 2D plot shows a dipolar broadened spectrum in the F_1 dimension, with CSA displayed on the F_2 axis. In practice, refinements to SLF are employed by spinning the sample slowly at the magic angle and observing rotational echoes that define the CSA envelope and by using a multiple pulse sequence, such as MREV-16 or BLEW-24, to eliminate the effect of H−H interactions.

Exchange in solids can be studied by NOESY, except that the preparation period consists of a cross polarization sequence, rather than just a 90° pulse. Similar studies can also be carried out with ^2H NMR by combining the 2D pulse sequence with the solid echo described in Chapter 7.

Several 2D methods in solids make use of discrete jumps in sample orientation, rather than MAS, whereas others change the rate of spinning between evolution and detection periods. One key component in such experiments is the inclusion of pulses to rotate the magnetization to the z axis and store it there while the mechanical changes are made. 2D experiments in solids are very useful, but in keeping with our emphasis on liquid state NMR, we shall not further explore the applications in solids.

Correlation via Spin Coupling

Prior to the advent of 2D methods, selective spin decoupling was used extensively in both proton NMR and in heteronuclear (especially ^{13}C) NMR to ascertain which sets of nuclei contribute to observed spin coupling. Such information is critical to assignment of resonances and to the elucidation of the structure of an unknown molecule. 2D methods now largely supply this information much more efficiently, by correlations that depend on the existence of spin coupling. The homonuclear version of one such experiment is called COSY (*correlation spectroscopy*), and the heteronuclear version is known by several acronyms, most commonly HETCOR (*heteronuclear correlation*).

The pulse sequence for COSY is very simple: 90°, t_1, 90°, t_2. In fact, we used this sequence in Fig. 10.2 to illustrate a basic 2D experiment. As we saw, a single resonance line has the same frequency in both evolution and detection periods, hence provides a line on the principal diagonal of the 2D spectral plot. However, if magnetization can be transferred from one spin to another spin with a different resonance frequency during the mixing period, then we expect to see off-diagonal peaks, as we did in NOESY. As we see in Chapter 12, the second 90° pulse serves as a mixing pulse and generates off-diagonal elements, but only if the two spins interact by dipolar coupling in solids or by "indirect" spin–spin coupling observed in liquids. We concentrate here on spin–spin coupling in liquids but recognize that similar effects can be obtained from dipolar coupling in partially ordered molecules dissolved in liquid crystals.

It is difficult to deduce from the picture of precessing vectors exactly why spin coupling is necessary for magnetization transfer, but qualitatively the process involves the antiphase magnetization generated as coupled spins precess in the xy plane (see Section 9.6). We shall look at this process in Chapter 12 after we have developed additional theoretical tools. Meanwhile, we present here some of the general features of COSY spectra.

COSY is very widely used. A simple COSY spectrum is illustrated in Fig. 10.9. Note that the off-diagonal peaks provide a direct indication of the coupling patterns among the protons. The usual COSY pulse sequence is best for values of J of about 3–15 Hz, which is typical of most geminal and vicinal $^1H-^1H$ couplings. However, the technique can be extended to respond to small, often longer range, couplings of 1 Hz or less by inserting short fixed time intervals of about 100–400 ms before and after the second pulse.

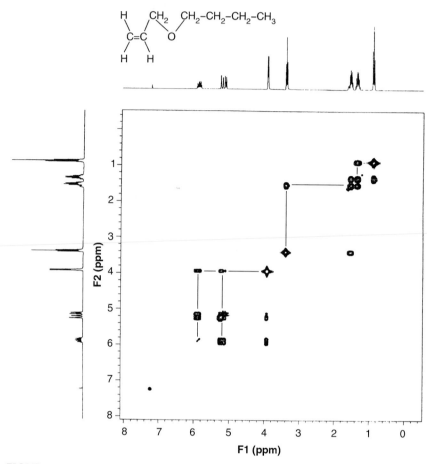

FIGURE 10.9 COSY spectrum (300 MHz) of allylbutyl ether in $CDCl_3$. The one-dimensional spectrum is displayed at the top and left side of the 2D spectral plot. Note that the peaks on the diagonal fall at the chemical shifts of each of the protons, while the cross peaks indicate spin coupling between the protons whose chemical shifts are at the corresponding diagonal peaks. For example, the protons at 3.4 ppm are coupled only to those at 1.5 ppm, but the latter are also coupled to protons at 1.3 ppm. Other correlations are indicated by connecting lines. Courtesy of Herman J. C. Yeh (National Institutes of Health).

As we see in Section 10.3, 2D NMR lines, including those in COSY, have a complex shape. A number of variants of COSY have been developed to improve the spectral presentation and to avoid unwanted signals. We shall return to COSY in Chapter 12.

HETCOR, the heteronuclear analog of COSY, is widely used, particularly in ^{13}C NMR to correlate ^1H and ^{13}C resonance frequencies via their one bond couplings $^1J_{CH}$. The experiment can be carried out with several pulse sequences, one of which is shown in Fig. 10.10a. The initial ^1H 90° pulse begins the evolution period, and as in COSY, mixing occurs as a result of the concurrent application of the ^1H and ^{13}C 90° pulses. The delay time Δ immediately before the mixing pulses is set to $\Delta = 1/2J$ to allow precession of antiphase magnetization in a manner similar to INEPT and for proton polarization to be transferred to ^{13}C during the mixing period. Details of the magnetization transfer are similar to those in INEPT and COSY, which are discussed in Chapters 11 and 12. The delay time Δ' is to obtain optimum signals from CH, CH_2, and CH_3 groups, as in INEPT.

The 180° ^{13}C pulse at the middle of the evolution period interchanges the precession frequencies of the α and β spins (see Fig. 9.2, bottom) and effectively decouples the spins during t_1, and a broadband decoupling sequence, such as WALTZ or GARP, is applied during t_2. Thus, the ^{13}C spectrum in the F_2 dimension is decoupled, and the ^1H spectrum in the F_1 dimension retains homonuclear couplings but is also decoupled from ^{13}C, as illustrated in Fig. 10.10b.

As in COSY, modifications in delay times can make HETCOR sensitive to smaller two-bond and three-bond $^{13}C-H$ couplings, but indirect detection methods described later are usually more effective.

Isotropic Mixing

The value of COSY stems from its dependence on the presence of spin coupling between the nuclei involved in the correlation. As we have seen, such coupling for protons is usually limited to three or four chemical bonds, hence provides some specificity that is helpful for structure elucidation. On the other hand, useful complementary information can be obtained from longer range interactions among a set of coupled nuclei. The standard method for obtaining such information is described by two acronyms—TOCSY (for *total correlated spectroscopy*, which best describes the aim of the experiment) and HOHAHA (for *homonuclear Hartmann–Hahn*, which better describes the mechanisms employed).

In Section 7.6 we saw that spin locking ^1H and ^{13}C with values of B_1 that conform to the Hartmann–Hahn condition permits internuclear polarization transfer via the heteronuclear dipolar coupling in a solid. In HOHAHA, the object is to use the isotropic spin coupling $J_{ij}I_i \cdot I_j$ in the liquid state, rather than the dipolar coupling, to effect a mixing of spin states analogous to that in the solid state cross polarization technique. Under suitable spin-lock conditions, the Zeeman interaction disappears, and the effective Hamiltonian is just the sum of the isotropic coupling terms. The result is that energy and polarization can be interchanged among *all* nuclei within a coupled set—a process called *isotropic mixing*. Thus, if nucleus r is coupled to s and s to t, an $r–t$ cross peak appears in the TOCSY-HOHAHA spectrum even though r is not coupled to t. By reducing the mixing time, the

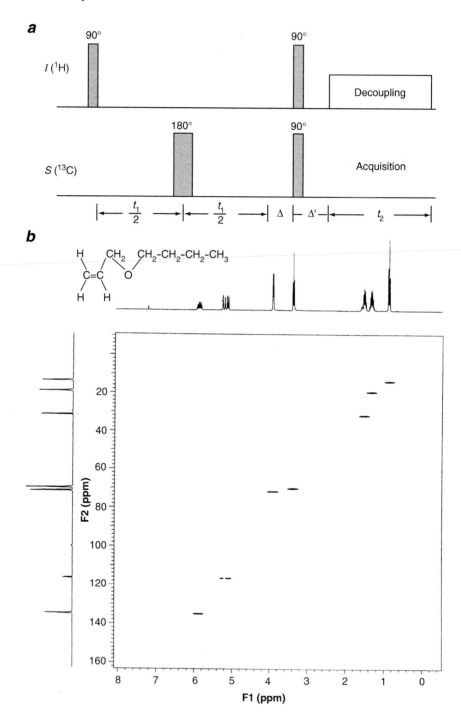

experiment may be made as restrictive as COSY, but with longer times (~100 ms) a larger number of spins are involved. As we see in Chapter 13, this method is particularly useful in assigning spectral lines to specific residues within an oligomer or polymer.

In solid state cross polarization the spin lock is obtained with a long, high power pulse, but for HOHAHA such a single, unmodulated pulse is not effective. Instead, the pulse cycle MLEV-16, as described in Section 9.6, or a variant with an additional pulse, MLEV-17, covers a sufficiently broad frequency range, much as in the broadband decoupling methods we discussed in Section 9.6.

NOESY and ROESY

We described the basic aspects of NOESY in Section 10.1 as an introductory example of a 2D experiment. NOESY is very widely used in measuring macromolecular conformation, as we see in Chapter 13. However, as shown in Fig. 8.4, the $^1H-^1H$ nuclear Overhauser enhancement η varies from its value of $+0.5$ in small molecules to a limiting value of -1 in large polymers with very long τ_c, and at intermediate values of τ_c the NOE may vanish. An alternative is to use the NOE measured in the rotating frame, as this quantity is always positive. By analogy to NOESY, this technique has the acronym ROESY (*r*otating frame *O*verhauser *e*nhancement *s*pectroscop*y*),

The experimental method for obtaining ROESY is essentially the same as that for HOHAHA, application of a spin-lock pulse sequence for mixing at the end of the evolution period. HOHAHA effects can interfere with ROESY measurements but are minimized by using lower rf power and offsetting the pulse frequency to interfere with the Hartmann–Hahn condition.

INADEQUATE

Combination of HETCOR with COSY and TOCSY provides very useful but indirect information on the proximity of particular carbon atoms. For structure elucidation, a direct method based on $^1J_{CC}$ is desirable, but with a natural abundance of only 1.1%, the probability of having two ^{13}C nuclei directly bonded to each other is only about 1 in 10,000. Nevertheless, with sufficient sample, it is possible to obtain the information via INADEQUATE (*i*ncredible *n*atural

---

FIGURE 10.10 (*a*) Pulse sequence for HETCOR, as described in text. (*b*)$^1H-^{13}C$ HETCOR spectrum (300 MHz) of allylbutyl ether in $CDCl_3$. The one-dimensional ^1H spectrum is shown at the top and the proton-decoupled ^{13}C spectrum at the left. The 2D plot clearly indicates the H–C correlations. Spectrum courtesy of Herman J. C. Yeh (National Institutes of Health).

abundance *double quantum transfer experiment*). This name describes the method very well: It is incredible (and a tribute to the sensitivity of modern NMR spectrometers) that natural abundance ^{13}C can produce the necessary signals, which come from a transfer of polarization by a double-quantum process. We shall defer discussion of the technique to Chapter 12, after we have developed the background to understand the double-quantum transfer.

Indirect Detection

The ability to manipulate spins in two-dimensional experiments and to transfer magnetization between spins has made it possible to use a sensitive nucleus (primarily ^1H) to measure the spectral features of less sensitive nuclei, such as ^{13}C and ^{15}N. Several methods are commonly used, but each begins with a ^1H pulse sequence, often resembling the one in INEPT (Section 9.7). As in INEPT, a combination of ^1H and X pulses transfers polarization to the X spin system. In some instances further transfers are made to another spin system (Y), then back through X to ^1H, where the signal is detected. Thus, the large polarization of the proton is used as the basis for the experiment, and the high sensitivity of ^1H NMR is exploited for detection. Such indirect detection methods use two-, three-, and sometimes four-dimensional NMR.

In Chapter 12 we take up in more detail three indirect detection methods, HSQC, HMQC, and HMBC. As we shall see, *h*eteronuclear *s*ingle *q*uantum *c*oherence transfers polarization (or coherence, as we discuss in Chapter 11) by a route that deals only with the single quantum transitions that we have encountered thus far. Heteronuclear *m*ultiple *q*uantum *c*oherence is quite similar to HSQC superficially but depends on the generation and evolution of zero quantum and double quantum coherence (also discussed in Chapter 11). The name *h*eteronuclear *m*ultiple *b*ond *c*orrelation is somewhat misleading, because it has nothing to do with multiple bonds (i.e., C=C, C≡C) but rather uses parameters that provide correlations of nuclei that are coupled through two or more (single or multiple) bonds. HMBC is complementary to HETCOR and HMQC and provides often indispensable structural information.

10.3 DATA ACQUISITION AND PROCESSING

In this section, we consider some aspects of recording and processing 2D NMR spectra. Almost all of the points discussed in Chapter 3 apply here as well, but there are some important features that we have not previously encountered. Most modern NMR spectrometers are capable of carrying out a wide range of 2D experiments. A variety of software for processing and analyzing the data is provided with each instrument, and much more software is available from other commercial sources and from published research reports and computer programs. As the details of hardware and software (and even some aspects of the terminology) vary from one source to another, we can provide here only a general overview.

Effects of Modulation during the Evolution Period

We observed in the examples given in previous sections that the magnetization responsible for the FID obtained during the acquisition period t_2 depends on the magnitude of **M** and its location in the $x'y'$ plane (i.e., its phase) at the end of the evolution period. Thus, for a given value of t_1 the signal depends in some way on the value of Ω_1, the precession frequency during the period t_1, and it is therefore modulated at this frequency. *Amplitude modulation* is shown as a sine or cosine term in $\Omega_1 t_1$ that multiplies the amplitude of **M**, as in the examples we saw in Section 10.1. *Phase modulation* usually appears as an imaginary exponential term in $\Omega_1 t_1$, as will be seen in some later examples. Although these two types of modulation affect the NMR signals in somewhat different ways, they are related through the mathematical identities

$$\cos \Omega_1 t_1 = \tfrac{1}{2}[e^{i\Omega_1 t_1} + e^{-i\Omega_1 t_1}]$$

$$\sin \Omega_1 t_1 = -\tfrac{i}{2}[e^{i\Omega_1 t_1} + e^{-i\Omega_1 t_1}]$$

(10.5)

Thus, an amplitude-modulated signal can always be considered as the sum of two signals modulated in phase by opposite frequencies. Equation 10.5 permits one type of modulation to be converted to the other provided both phase components or both amplitude components are known.

In the example given in Fig. 10.2, the amplitude modulation of M_{xy} gave a signal proportional to $\cos \Omega_1 t$, as shown in Eq. 10.1, and Fourier transform in the t_2 dimension gave the expression for $S(t_1,\omega_2)$ given in Eq. 10.2. If only the absorption mode $A(\omega_2)$ is retained, a real (cosine) Fourier transform carried out with respect to t_1 yields a real portion with absorption mode character as given in Eq. 10.3. The overall two-dimensional spectrum is then given by

$$S(\omega_1,\omega_2) = A(\omega_1) \cdot A(\omega_2)$$

(10.6)

which is a simple absorption spectrum in both dimensions, exactly as we desire. However, there is a drawback to dealing only with the cosine modulation in that positive and negative frequencies cannot be distinguished—a situation analogous to that found for a single phase sensitive detector in a one-dimensional experiment (Section 3.7). The spectrum described in Eq. 10.6 is obtained by placing the transmitter at the edge of the spectrum, rather than at the center, with the same disadvantages noted in Section 3.4. However, with additional sine-modulated data, we can obtain the advantages of quadrature phase-sensitive detection. We return to a consideration of the processing required after looking at an example of a phase-modulated signal.

We have already observed phase modulation in the acquisition of data in a 1D spectrum, where finite rf power results in off-resonance magnetizations giving rise to a phase error proportional to the off-resonance frequency (see Eq. 2.55), and similar effects occur in the t_2 dimension of 2D spectra. Here our interest is in modulation that arises in the t_1 dimension. A simple example (but with no utility as a 2D pulse sequence) is 90°, t_1, t_2, which resembles a 1D experiment in which the data acquisition period is delayed by the time t_1. Thus, there is a phase modulation in the acquired data that depends on $\Omega_1 t_1$, as shown by the expression for the interferogram signal

$$S(t_1,\omega_2) = Ce^{i\Omega_1 t_1}e^{-t_1/T_2}[A(\omega_2) + iD(\omega_2)] \tag{10.7}$$

where A and D were defined in Eqs. 10.3 and 10.4. By expanding the imaginary exponential in Eq. 10.7 and gathering real and imaginary terms, we can see that

$$S(t_1,\omega_2) = \frac{C}{2}e^{-t_1 T_2}[(\cos \Omega_1 t_1)A(\omega_2) - (\sin \Omega_1 t_1)D(\omega_2)]$$
$$+ i\frac{C}{2}e^{-t_1/T_2}[(\sin \Omega_1 t_1)A(\omega_2) - (\cos \Omega_1 t_1)D(\omega_2)] \tag{10.8}$$

Because both $\sin \Omega_1 t_1$ and $\cos \Omega_1 t_1$ modulations appear, Fourier transformation with respect to t_1 gives

$$S(\omega_1,\omega_2) = C[A(\omega_2)A(\omega_1) - D(\omega_2)D(\omega_1)]$$
$$+ iC[A(\omega_2)D(\omega_1) + D(\omega_2)A(\omega_1)] \tag{10.9}$$

Equation 10.9 represents a complicated line shape, which is a mixture of absorptive and dispersive contributions. Figure 10.11 gives an example of such a *phase-twisted* line shape. The broad base of the line, caused by the dispersive contribution, and the difficulty in correctly phasing such a resonance make it unattractive for practical use. The phase twist problem can be alleviated by displaying only the *absolute value* mode

$$(A(\omega_1)^2 + D(\omega_1)^2)^{1/2}(A(\omega_2)^2 + D(\omega_2)^2)^{1/2} \tag{10.10}$$

but the broad base of the resonance from the dispersive part remains in the absolute value mode.

Processing of Time Response Signals

Equation 10.5 indicates how amplitude modulation may be converted to phase modulation, which provides considerable flexibility in data processing. In order to obtain both sine and cosine components of amplitude modulation, we usually

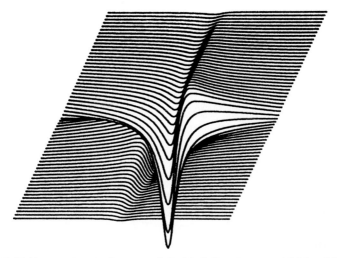

FIGURE 10.11 Example of a phase-twisted 2D NMR line. Courtesy of Ad Bax (National Institutes of Health).

repeat the experiment with some sort of phase cycling. If we want to simulate the effect of having a second phase detector *for the evolution period*, we must sample the magnetization at t_1 with a pulse that is out of phase by $\pi/2$ with that used in the initial experiment, and we must also increment the phase of the actual phase-sensitive detection by $\pi/2$.

For example, the COSY experiment illustrated in Fig. 10.2 can be repeated with the first pulse applied in the same direction but with $\mathbf{B}_1$ for the second pulse applied along y', rather than x'. $\mathbf{M}$ is then rotated into the $y'z'$ plane, where its projection along y', $M_0 \sin \Omega_1 t_1$, is detected. The signals for the two experiments are, then,

$$s(t_1, t_2) = M_0 \cos \Omega_1 t_1 e^{i\Omega_2 t_2} \tag{10.11a}$$

$$s'(t_1, t_2) = iM_0 \sin \Omega_1 t_1 e^{i\Omega_2 t_2} \tag{10.11b}$$

We have dropped the relaxation terms to simplify the expression. In Eq. 10.11a, the factor i results from the change in detector phase by $\pi/2$, because

$$e^{\pi/2} = \sqrt{-1} \tag{10.12}$$

(In principle, this $\sqrt{-1}$ should be represented by a different symbol, such as j, to emphasize that this refers to the complex plane ω_1, whereas i in the exponential refers to the complex plane ω_2. For our purpose we do not require this distinction.)

The sum and difference of the expressions in Eq. 10.11 then lead to the following expressions that demonstrate phase modulation in the t_1 dimension:

$$s_P(t_1,t_2) = M_0 e^{i\Omega_1 t_1} e^{i\Omega_2 t_2}$$
$$s_N(t_1,t_2) = M_0 e^{-i\Omega_1 t_1} e^{i\Omega_2 t_2}$$

(10.13)

Either S_P (called *positive* detection) or S_N (*negative* detection) may be Fourier transformed to give a 2D peak, with a phase twist, as we saw previously, or displayed as the magnitude peak (Eq. 10.10). Because S_N arises from magnetization that precesses first clockwise during t_1, then counterclockwise during t_2, the effect of B_0 inhomogeneity is removed. The line arising from S_N is often called an *echo* peak and that from S_P an *antiecho* peak. In some types of experiments where it is not possible to phase all peaks, a magnitude spectrum is required, and the echo (N-type) mode is almost always preferred to reduce the width somewhat.

States and TPPI Methods

Two methods are commonly used, separately or together, to process amplitude-modulated data in a manner that provides the benefits of quadrature detection in the t_1 dimension but also leads to absorption line shapes in both dimensions. Again, the key is to repeat the experiment with suitable phase cycling. Let us return to the example of the COSY sequence illustrated in Fig. 10.2. The experiment can be repeated with the *first* pulse cycled by $\pi/2$ radians, while the second pulse is not cycled, and the receiver phase is kept the same. For the particular choice of pulse phases in Fig. 10.2, the first 90° pulse can be applied along the x' axis, rather than along the y' axis as shown in the figure, thus rotating **M** to the $-y'$ axis. After precessing at Ω_1 during the t_1 period, **M** is rotated by the $90°_{-x'}$ pulse into the $x'z'$ plane, but this time its projection along x' is given by $\sin \Omega_1 t_1$, rather than $\cos \Omega_1 t_1$. The 2D time responses (ignoring the relaxation terms) from these two amplitude-modulated experiments are

$$s(t_1,t_2) = M_0 \cos \Omega_1 t_1 e^{i\Omega_2 t_2}$$
$$s'(t_1,t_2) = M_0 \sin \Omega_1 t_1 e^{i\Omega_2 t_2}$$

(10.14)

Note that these expressions are the same as those in Eq. 10.11, without the $\sqrt{-1}$ factor from the change in detector phase. The signals may be processed by a very useful method, introduced as the *hypercomplex* Fourier transform method and popularized as the *States* method (or SHR method, as described by States, Haberkorn, and Ruben[107]). The two signals in Eq. 10.14 (with relaxation terms restored) are *separately* Fourier transformed with respect to t_2 to give

$$S(t_1,\omega_2) = \cos \Omega_1 t_1 [A(\omega_2) + iD(\omega_2)]$$
$$S'(t_1,\omega_2) = \sin \Omega_1 t_1 [A(\omega_2) + iD(\omega_2)]$$

(10.15)

The imaginary parts are discarded to obtain

$$S_R(t_1,\omega_2) = \cos\Omega_1 t_1 A(\omega_2)$$
$$S_R'(t_1,\omega_2) = \sin\Omega_1 t_1 A(\omega_2)$$

(10.16)

and these expressions are then combined to give a complex signal:

$$S''(t_1,\omega_2) = e^{i\Omega_1 t_1} A(\omega_2) \qquad (10.17)$$

Fourier transform with respect to t_1 gives

$$S''(\omega_1,\omega_2) = A(\omega_1)A(\omega_2) + iD(\omega_1)A(\omega_2) \qquad (10.18)$$

The real part $A(\omega_1)A(\omega_2)$ is an absorption spectrum in both dimensions.

Some 2D experiments provide phase modulation directly as N-type and/or P-type signals, as in Eq. 10.13. These signals can also be processed along the lines that we have described. The signals in Eq. 10.13 are separately Fourier transformed with respect to t_2 and the imaginary parts discarded. Then s_P is added to s_N^*, the complex conjugate of s_N, to give

$$s_P + s_N^* = 2e^{i\Omega_1 t_1} A(\omega_2) \qquad (10.19)$$

which is analogous to Eq. 10.17 and on Fourier transformation gives a real spectrum in absorption mode in both dimensions.

Recording and processing the two signals of Eq. 10.14, which are in quadrature, is the analog in the t_1 dimension of the use of two phase-sensitive detectors in the t_2 dimension. As we showed in Section 3.7, there is an alternative to use of two detectors. The "Redfield" method uses one detector but increments the receiver phase by 90°, as illustrated in Fig. 3.8. An analogous technique is available to treat t_1 data — *time-proportional phase incrementation*, TPPI.[38,108]

In TPPI the time increment in the t_1 dimension is made half as large as normally required to satisfy the sampling theorem (Section 3.7), so that $\Delta t = 1/4f_{max}$, and the phase of the pulse preceding each t_1 modulation is incremented by $\pi/2$. As in the Redfield method, this combination of sampling at twice the normal rate and changing phase causes the spectrum to shift to lower frequency by $f_{max}/2$ and makes it appear that the rf is at the edge of the spectral width, rather than in the center of the range where it is actually applied. Aliasing of frequencies above f_{max} involves the same type of folding in the TPPI method as with the Redfield technique, and the States method leads to a cyclic aliasing, as in the use of two phase detectors (Section 3.7). Contrary to the situation in conventional one-dimensional FT experiments, such folded resonances are not attenuated by audio frequency filters.

Axial Peaks

The fundamental premise of 2D NMR is that magnetization precesses during the t_1 period and, as we have seen, thus introduces a modulation into the magnetization that is measured during the t_2 period. However, while **M** precesses during t_1, spin–lattice relaxation occurs at rate $1/T_1$, so that at the end of the t_1 period a small component of M_z has developed. The subsequent pulse or pulses that create the measurable magnetization usually cause some or all of this M_z component to rotate into the xy plane, hence provide a signal. Because this magnetization remained along the z axis during t_1, it is unmodulated and Fourier transforms to zero frequency in the ω_1 dimension, where a signal appears along the axis. Such *axial peaks* can be reduced by phase cycling and can be moved to the extreme edge of the spectrum by application of TPPI.

In general, the States method is particularly valuable in providing a flat baseline but leaves axial peaks in the center of the spectrum, where they can interfere with true cross peaks. However, a combination States–TPPI method, which provides the advantages of both techniques, is now widely used. It is essentially the States method, with sampling at $\Delta t = 1/2f_{max}$, but the phases of the preparation pulse and receiver are changed by π for successive values of t_1. The transmitter phase change has no effect on the axial peaks, which are not modulated during t_1, but the alternation of the receiver phase introduces a modulation at the Nyquist frequency, thus moves axial peaks from the center to the edges of the spectrum.

Zero-Filling and Digital Filtering

The various procedures that we described in Chapter 3 for treating data in the time and frequency domains in order to improve signal/noise ratio, resolution, and/or spectral appearance apply equally to 2D spectra. For small molecules, 2D absorption mode lines can be quite narrow, so digital resolution (number of data points per Hz) must be large enough to define the peak; otherwise the highest point of a resonance in a 2D spectrum can be reduced by as much as 70%, thereby decreasing the apparent peak intensity of the line in a contour plot dramatically. Problems are encountered particularly when the ranges of t_1 and/or t_2 are limited in order to speed up the experiment. Zero-filling provides the simplest method to ensure an adequate number of data points, but digital filtering to produce Gaussian broadening (Section 3.9) is also desirable to avoid a strong discontinuity in the time domain signal that gives rise to "sinc wiggles" as we saw in Chapter 3.

10.4 SENSITIVITY CONSIDERATIONS

Two-dimensional NMR experiments are sometimes thought to be much less sensitive than their one-dimensional analogs. However, the sensitivity of 2D experiments can be quite high with well-executed experimental procedures. In this section we discuss briefly several of the general factors that determine sensitivity, but each type of 2D experimental procedure has its own unique sensitivity characteristics.

Signal / Noise in 1D and 2D Spectra

In a regular 1D FT NMR experiment, the signal-to-noise ratio for a single time domain data point can be very poor, but the Fourier transformation takes the signal energy of all data points and usually puts it into only a few narrow resonance lines. Similarly, in a 2D experiment a spectrum with poor S/N may be recorded for each t_1 value, but the Fourier transformation with respect to t_1 combines the signal energy of a particular resonance from all spectra obtained for different t_1 values and concentrates it into a small number of narrow lines in the 2D spectrum.

Usually our interest is in the signal/noise that can be achieved in a particular total experimental time. Equation 3.1 shows that the signal amplitude $S(\omega)$ depends on the integral over the envelope of the FID $s(t)$. As we saw in Section 3.12, repetition of the experiment n times with coherent addition of the data gives a Fourier-transformed signal that is n times S (with random noise increasing only as $\sqrt{n}$). If the time spent making n repetitions in a 1D study with all parameters fixed is instead used in a 2D study to obtain n interferograms as t_1 is incremented, the overall signal accumulated in the same total time is thus of the same magnitude (nS) as in the 1D experiment, while the random noise contributions are the same. If we look at the comparison more closely, we can identify factors that decrease the 2D signal. As shown in Section 10.3, quadrature phase detection in the t_1 dimension requires two repetitions to achieve the signal that one FID provides in the t_2 dimension. Also, our simple calculation assumed no loss of signal from relaxation processes, but as t_1 is incremented, there is more time for relaxation to occur. The result is a progressively smaller overall signal as t_1 increases. For this reason, it is usually desirable to limit the maximum value of t_1 to no longer than is absolutely necessary for obtaining good enough resolution in the ω_1 dimension after Fourier transformation. Overall, these processes may lead to a lower S/N in the 2D study by a factor of about 2 compared with the 1D study, but this reduction is much less than the factor of $\sqrt{n}$ that one might have intuitively expected.

As we have seen in the examples in Section 10.2, most 2D experiments involve transfer of polarization or coherence, sometimes in multistep processes in which the efficiency of each is far less than 100%. Thus, great care must be used in designing such experiments to craft each step carefully to optimize the final signal. Also, extensive phase cycling that is required in some multidimensional NMR experiments extends the minimum time required for the study. If signals are weak and extensive time averaging is needed anyway to obtain adequate sensitivity, then this requirement presents few problems, but once adequate S/N has been achieved extra repetitions are often objectionable.

Noise and Artifacts Peculiar to 2D Spectra

In 2D NMR a source of noise that is not evident in 1D NMR has to be considered. Instabilities in the spectrometer system can cause small unwanted fluctuations in the amplitude, phase, or frequency of a resonance line. Such random fluctuations during the evolution period t_1 (hence, called t_1 *noise*) are of significance only at frequencies Ω_2 where there is a signal. On Fourier transformation this noise appears in sections parallel to the ω_1 axis.

Probably the most serious source of t_1 noise on most modern spectrometers is the effect of sample spinning. If this sample spinning induces (very small) translations of the sample in the plane perpendicular to the static magnetic field, small changes in the Q of the receiver coil are induced, causing an unwanted change in the phase of the observed signal. This change in Q depends critically on the geometry of the coil; for a good design the Q becomes less dependent on such translations of the sample. To avoid spinner-induced t_1 noise, many 2D spectra are recorded without spinning the sample provided magnet homogeneity is adequate.

Small fluctuations in magnetic field strength that are not compensated for by the lock system and changes in magnetic field homogeneity can also be a source of t_1 noise. If the changes are slow compared with the repetition rate of the experiment, the t_1 noise in the ω_1 dimension is restricted to a narrow band around the resonance in the ω_1 dimension. Coherent addition of a large number of signals tends to reduce the effect of t_1 noise by partial cancellation.

It is difficult to disentangle t_1 noise from a baseline artifact that causes an underlying *ridge* in the ω_1 dimension. This ridge is the analog of the DC offset and baseline roll found in a 1D spectrum when the first few data points are distorted. This effect can be eliminated or at least minimized by ensuring that the FT algorithm treats the first data point correctly and that the spin system is in a steady state when the first data are recorded.

Experiment Repetition Rate

The delay time between successive scans in a 2D experiment is, of course, of major importance for the sensitivity that can be obtained per unit time. For the case of a one-dimensional experiment in which a single 90° pulse is applied to a spin system and time averaging is used, we saw in Section 3.11 that the optimum time Δ between consecutive pulses is $1.27T_1$, or alternatively that the Ernst flip angle can be used with more rapid repetition. For most 2D experiments, $\Delta = 1.27T_1$ is also close to optimum for a 90° pulse. This repetition time actually starts at the time when the spin system begins returning to thermal equilibrium, for example, for the experiment represented in Fig. 10.2, Δ begins just after the second 90° pulse.

In some studies the number of scans necessary for adequate phase cycling provides more than enough sensitivity, and the repetition time may be decreased below the optimum to save overall experimental time. However, in some experiments too short a repetition time can introduce false lines into the 2D spectrum because the state of the spin system achieved in the preparation period becomes a function of t_1. For example, residual HDO protons often have a long T_1 value, and rapid repetition rate causes an array of spurious resonances in the ω_1 dimension at the HDO ω_2 frequency.

10.5 ADDITIONAL READING AND RESOURCES

A large number of books are devoted largely to two-dimensional NMR. *Multidimensional NMR in Liquids* by Frank van de Ven[109] provides excellent coverage of many aspects. Most of the treatment requires background that we develop in Chapter 11, but the book also discusses the processing of data along the lines given here but in more detail. *A Complete introduction to Modern NMR Spectroscopy* by Roger Macomber[110] provides an introduction to 2D methods, along with practical discussions of other topics from previous chapters. *Two-Dimensional NMR Spectroscopy* edited by William Croasmun and Robert Carlson[111] includes good introductory articles on the concept and experimental aspects of 2D NMR. *Two-Dimensional NMR in Liquids* by Ad Bax[112] was the first book in the field and provides very good coverage of the basic concepts but has been somewhat superseded by later experimental developments.

We list other books in 2D NMR in later chapters, where they have particular applicability.

10.6 PROBLEMS

10.1 Distinguish clearly between a one-dimensional NMR experiment that uses a time increment, such as the inversion-recovery technique (Section 2.9 and Fig. 8.8*a*), and a two-dimensional NMR experiment, such as NOESY.

10.2 Use the data in Fig. 10.6, along with chemical shift correlations from Chapter 4, to assign the ^{13}C chemical shifts to the seven carbon atoms. Are there ambiguities?

10.3 Use the data in Fig. 10.7 (also depicted from a different perspective in Fig. 1.6), along with chemical shift correlations from Chapter 4, to assign the ¹H chemical shifts to the protons in the molecule. Are there ambiguities?

10.4 Use the data in Fig. 10.9, along with chemical shift correlations from Chapter 4, to assign the ¹H chemical shifts to the protons in the molecule. Are there ambiguities?

10.5 Use the data in Fig. 10.10 to correlate ¹H and ¹³C chemical shifts in allylbutylether and to resolve any ambiguities from correlations in Problems 10.2–10.4.

10.6 The absolute value mode of the expression in Eq. 10.9 is formed from the sum of the squares of the real part and the imaginary part of this expression. Show that this reduces to the expression given in Eq. 10.10.

10.7 Distinguish between *echo* and *antiecho* peaks.

CHAPTER 11

Density Matrix and Product Operator Formalisms

In our introduction to the physics of NMR in Chapter 2, we noted that there are several levels of theory that can be used to explain the phenomena. Thus far we have relied on (1) a quantum mechanical treatment that is restricted to transitions between stationary states, hence cannot deal with the coherent time evolution of a spin system, and (2) a picture of moving magnetization vectors that is rooted in quantum mechanics but cannot deal with many of the subtler aspects of quantum behavior. Now we take up the more powerful formalisms of the density matrix and product operators (as described very briefly in Section 2.2), which can readily account for coherent time-dependent aspects of NMR without sacrificing the quantum features.

As in previous chapters, we do not attempt to provide a rigorous treatment here, but rather develop the concept of a density matrix, which is often unfamiliar to chemists, and show how it may easily be used to understand the behavior of one-spin and two-spin systems. As in the treatment of complex spectra in Chapter 6, we shall see that the density matrix approach can be readily extended to larger spin systems, but with a great deal of algebra and often with little physical insight. However, in the course of treating the simple spin systems, we will notice that some of the results can be obtained more succinctly by certain manipulations of the spin operators I_x, I_y, and I_z, with which we are familiar.

This leads to the postulation of algebraic rules dealing with the products of such operators.

11.1 THE DENSITY MATRIX

Before defining the density matrix, let's review some background from previous chapters. In Section 2.3 we gave an introduction to the essential quantum mechanical features of nuclear magnetic moments interacting with an applied magnetic field. We saw that a spin can be described in terms of its eigenfunctions α and β (which are sometimes expressed in the "ket" notation $|{}^{1}\!/_{2}>$ and $|-{}^{1}\!/_{2}>$, respectively), and we summarized the basic properties of these eigenfunctions. In Sections 2.3, 6.3, and 6.4, we developed the relevant aspects of the quantum mechanics for a system of N spins differing in chemical shift and interacting with each other by spin coupling. In this chapter we make use of many of these results.

In Section 2.6 we saw that a single spin $\frac{1}{2}$ can be described as a linear combination, or *coherent superposition*, of α and β, and in Eq. 2.31 we showed that a time-dependent wave function Ψ can be constructed. We found that the expectation value of the spin I led to a picture of a single nuclear spin vector precessing coherently at the Larmor frequency about an applied magnetic field $\mathbf{B}_0$, taken along the z axis. We then noted that in a sample containing a large number of identical nuclei, the vectors precess at the same frequency but at random phase, because at equilibrium there is nothing to establish a preferred direction in the xy plane. In Fig. 2.3, we showed that this collection of precessing nuclei gives rise to a macroscopic magnetization $\mathbf{M}$ along the z axis, and we have been able to examine the behavior of $\mathbf{M}$ under the effect of pulses, relaxation, and other perturbations.

Now we wish to pursue the pictorial presentation of Fig. 2.3 in a mathematical manner that permits us to retain explicitly the quantum features that tend to be obscured in the graphical presentation. In more explicit terms, this corresponds to an *incoherent superposition* of the magnetizations of individual spins or of individual sets of N interacting spins (spin systems). Incoherent or random motions are commonly treated by statistical methods that deal with an ensemble of molecules, each containing N interacting spins.

Expectation Values and Ensemble Averaging

As we saw in Chapter 6, in a spin system consisting of N nuclei, each of spin $\frac{1}{2}$, there are 2^N spin product functions, labeled ϕ_1, ϕ_2, etc. (or alternatively in the ket notation, $|1>, |2>$, etc.). Each of the 2^N wave functions $\psi^{(i)}$ describing the state of all nuclear spins in the molecule can always be described as a linear combination of the ϕ's, which form a complete set of basis functions:

$$\psi^{(i)} = \sum_{k=1}^{2^N} c_k \phi_k \tag{11.1}$$

where the c_k's are complex coefficients. Equation 11.1 is a coherent superposition (as in Eq. 2.31), and ψ represents a "pure state" of the spin system. (As we saw in Chapter 6, several of the c_k's might be zero.) Now, we consider an ensemble of molecules in which each of the 2^N pure states defined in Eq. 11.1, for example, $\psi^{(i)}$, has a certain probability of occurring, $p^{(i)}$. We can describe an average over the ensemble by

$$\Xi = \sum_{i=1}^{2^N} p^{(i)} \psi^{(i)} \tag{11.2}$$

The sum of the probabilities $p^{(i)}$ is, of course, unity.

It is well known from basic quantum theory that the value of an observable quantity A, represented by the operator $\mathbf{A}$, can be found for any pure state i as the expectation value, $\langle A \rangle^{(i)}$:

$$\langle A \rangle^{(i)} = \langle \psi^{*(i)} | A | \psi^{(i)} \rangle$$

$$= \sum_{m,n=1}^{2^N} c_m^{*(i)} c_n^{(i)} \langle \phi_m^* | A | \phi_n \rangle$$

$$= \sum_{m,n=1}^{2^N} c_m^{*(i)} c_n^{(i)} A_{mn} \tag{11.3}$$

Note that the matrix element A_{mn} is a constant that is independent of the particular state of the molecule, and the products of the c_k describe the extent to which each matrix element contributes to state i.

For a macroscopic sample, consisting of a large number of chemically identical molecules, the spin product functions are the same for all individual molecules, but the expectation value of the operator must be averaged over all molecules (spin systems) in the ensemble. With the notation that has been introduced, this average may be obtained by considering the probability of occurrence of each pure state, as expressed in terms of the basis functions:

$$\langle A \rangle = \sum_i p^{(i)} \langle A \rangle^{(i)}$$

$$= \sum_i \sum_{m,n} p^{(i)} c_m^{*(i)} c_n^{(i)} A_{mn} \tag{11.4}$$

We can collect some of the terms in Eq. 11.4 to define the nmth element of a new matrix ρ by

$$\rho_{nm} = \sum_i p^{(i)} c_n^{(i)} c_m^{*(i)} \tag{11.5}$$

Equation 11.5 defines the elements of a matrix $\boldsymbol{\rho}$, called the *density matrix*, which completely specifies the state of the ensemble, because it includes all the variables of Eq. 11.4.

We shall find it very useful to express this information in one matrix. For example, to express the expectation value of operator **A** succinctly, we combine Eqs. 11.4 and 11.5 and rearrange terms to obtain:

$$\langle A \rangle = \sum_{m=1}^{2^N} \sum_{n=1}^{2^N} \rho_{nm} A_{mn}$$

$$= \sum_{n=1}^{2^N} [\rho A]_{nn}$$

$$= \mathrm{Tr}[\rho A] = \mathrm{Tr}[A\rho] \tag{11.6}$$

Thus, the ensemble average of any observable quantity A can be obtained simply by taking the trace of the matrix product of **A** and $\boldsymbol{\rho}$ in either order.

Properties of the Density Matrix

The term *density matrix* arises by analogy to classical statistical mechanics, where the state of a system consisting of N molecules moving in a real three-dimensional space is described by the *density* of points in a $6N$-dimensional *phase space*, which includes three orthogonal spatial coordinates and three conjugate momenta for each of the N particles, thus giving a complete description of the system at a particular time. In principle, the density matrix for a spin system includes all the spins, as we have seen, and all the spatial coordinates as well. However, as we discuss subsequently we limit our treatment to spins. For simplicity we deal only with application to systems of spin $1/2$ nuclei, but the formalism also applies to nuclei of higher spin.

From Eq. 11.5 it is apparent that $\rho_{mn}^{*} = \rho_{nm}$, which defines a Hermitian matrix, so the density matrix is seen to be Hermitian. For a system of N spins $1/2$, ρ is $2^N \times 2^N$ in size. From the fact that the probabilities $p^{(i)}$ in Eq. 11.5 must sum to unity and the basis functions used are orthonormal, it can be shown that the trace of the density matrix is 1.

We can simplify the mathematical expressions by recognizing that the sum in Eq. 11.5 is just an ensemble average, so this equation can be rewritten as

$$\rho_{nm} = \overline{c_n c_m^{*}} \tag{11.7}$$

where the bar indicates an average. The coefficients c_n are complex and can be written as

$$c_n = |c_n| e^{i\alpha_n} \tag{11.8}$$

As we observed previously, the phases are random; hence there is no correlation between the values of α_i and c_i, and we may average separately to give

$$\rho_{nm} = \overline{|c_n||c_m^*|}\ \overline{e^{i(\alpha_n - \alpha_m)}}$$ (11.9)

Thus, $\rho_{nm} \neq 0$ *only* if both the average value of $|c_n||c_m^*|$ and the average value of $\exp[i(\alpha_n - \alpha_m)]$ are nonzero. The latter requirement means that there must be a *coherence* between the phases α_m and α_n for the corresponding element of ρ to be nonzero.

This is a very important result. It means that *at equilibrium, where the phases are truly random, all off-diagonal elements of the density matrix must be zero.* The diagonal elements are then given by $|c_n|^2$, which we recognize as the probability of occurrence of state $|n>$, or the *population* of state $|n>$.

As we see in later sections, rf pulses can alter the density matrix in such a way that specific off-diagonal elements become nonzero. This means that a coherence has been introduced between the corresponding eigenstates, $|m>$ and $|n>$. In spectroscopy we are accustomed to thinking of transitions between such eigenstates, occurring in a very short time and resulting in population changes in the eigenstates. A coherence, on the other hand, represents a long-lived persistent constructive interference between the states. When states $|m>$ and $|n>$ differ in quantum number by one unit, we speak of a *single quantum coherence*; when the states differ by more than one unit, we have a *multiple quantum coherence*; and when they have the same quantum number, we have a *zero quantum coherence*. As we shall see, a single quantum coherence arising from one spin corresponds to the now-familiar macroscopic magnetization. Other coherences are not so easily pictured as magnetization, but they can be manipulated in NMR experiments and play a very important role in two-dimensional NMR.

Evolution of the Density Matrix

We now have a formula for constructing the density matrix for any system in terms of a set of basis functions, and from Eq. 11.6 we can determine the expectation value of any dynamical variable. However, the real value of the density matrix approach lies in its ability to describe coherent time-dependent processes, something that we could not do with steady-state quantum mechanics. We thus need an expression for the time evolution of the density matrix in terms of the Hamiltonian applicable to the spin system.

The time-dependent Schrödinger equation is

$$\mathcal{H}\Psi = i\hbar\frac{\partial\Psi}{\partial t}$$ (11.10)

Provided $\mathcal{H}$ is not an explicit function of time, Ψ can be expanded as in Eq. 11.1, except that the c_k now become functions of time. Insertion of the expansion into

Eq. 11.10 gives

$$i\hbar \sum_k \frac{dc_k(t)}{dt} \phi_k = \sum_k \mathcal{H} c_k(t) \phi_k \tag{11.11}$$

Multiplying both sides of Eq. 11.11 by ϕ_n^*, integrating, and taking into account the orthonormality of the basis functions ϕ_k, we obtain

$$i\hbar \frac{dc_n}{dt} = \sum_k c_k \mathcal{H}_{nk} \tag{11.12}$$

where $\mathcal{H}_{nk}$ is the matrix element $\langle n|\mathcal{H}|k\rangle$.

To develop an expression for the time dependence of ρ, we begin by examining the time derivative of $c_n c_m^*$:

$$\frac{d}{dt}(c_n c_m^*) = c_n \frac{dc_m^*}{dt} + \frac{dc_n}{dt} c_m^*$$

$$= \frac{i}{\hbar} c_n \sum_k c_k^* \mathcal{H}_{km} - \frac{i}{\hbar} \sum_k c_k \mathcal{H}_{nk} c_m^*$$

$$= \frac{i}{\hbar} \sum_k (c_n c_k^* \mathcal{H}_{km} - \mathcal{H}_{nk} c_k c_m^*) \tag{11.13}$$

Multiplying Eq. 11.13 by the populations p_i and summing, we obtain an expression for the time derivative of the density matrix:

$$\frac{d}{dt}\rho_{nm} = \frac{d}{dt} \sum_i p^{(i)} c_n c_m^*$$

$$= \frac{i}{\hbar} \sum_k [\rho_{nk} \mathcal{H}_{km} - \mathcal{H}_{nk} \rho_{km}]$$

$$= \frac{i}{\hbar} [(\rho\mathcal{H})_{nm} - (\mathcal{H}\rho)_{nm}] \tag{11.14}$$

As a matrix equation, this becomes

$$\frac{d\rho}{dt} = \frac{i}{\hbar} [\rho\mathcal{H} - \mathcal{H}\rho] \tag{11.15}$$

Equation 11.15, the *Liouville−von Neumann equation*, is really a restatement of the time-dependent Schrödinger equation for an ensemble of systems. It indicates that the density matrix describing the ensemble will evolve in time whenever the density operator and Hamiltonian do not commute.

With a time-independent Hamiltonian, Eq. 11.15 may be integrated to give

$$\rho(t) = e^{-i\mathcal{H}t/\hbar} \rho(0) e^{i\mathcal{H}t/\hbar} \tag{11.16a}$$

If we express $\mathcal{H}$ in the angular frequency units normally used in NMR, this relation becomes

$$\rho(t) = e^{-i\mathcal{H}t}\,\rho(0)\,e^{i\mathcal{H}t} \qquad (11.16b)$$

We have carried out the integration in a formal way, with $\mathcal{H}$ treated as though it were a variable rather than an operator. In the next section we define such an exponential operator and establish the validity of Eq. 11.16.

Equation 11.16 is a general result that can be applied to any spin system. In addition, it is helpful to have an expression for the time dependence of each element of ρ. Equation 11.12 is applicable to a set of basis functions ϕ_k that are otherwise not limited. However, we know from Chapter 6 that we can (in principle, at least) solve the Schrödinger equation to obtain the true eigenfunctions of $\mathcal{H}$. If these eigenfunctions are used as the basis functions, then Eq. 11.12 simplifies to give

$$i\hbar\frac{d}{dt}c_n(t) = \sum_k c_k(t)E_k\langle n\mid k\rangle$$

$$= E_n c_n(t) \qquad (11.17)$$

On integration, this gives an expression for the time dependence of the coefficients

$$c_n(t) = c_n(0)e^{-iE_nt/\hbar} \qquad (11.18)$$

By substituting these expressions for c_n and c_m into Eq. 11.5 and collecting terms, we obtain

$$\rho_{nm}(t) = \rho_{nm}(0)e^{i(E_m - E_n)t/\hbar} = \rho_{nm}(0)e^{i\omega_{mn}t} \qquad (11.19)$$

Equation 11.19 shows how each element of the density matrix evolves when the Hamiltonian itself is not varying with time, for example, during free precession. Clearly, the diagonal elements ($m = n$) do not change with time, but the off-diagonal elements (coherences) evolve with a frequency ω_{mn} that depends on the energies of the corresponding basis functions. We saw previously that at equilibrium ρ contains only diagonal elements. In later sections we shall see how off-diagonal elements may be created by the action of rf pulses.

Exponential Operators

In Eq. 11.16, we encountered a familiar operator $\mathcal{H}$, but in an exponential form. Many operators $\mathbf{R}$ can most conveniently be expressed in an exponential form, where the exponential is defined in terms of an infinite power series:

$$\mathbf{R} = e^A \equiv \sum_{r=0}^{\infty}(A^r/r!) = 1 + A + (A^2/2!) + (A^3/3!) + \ldots \qquad (11.20)$$

1 is the identity operator or unit matrix. $\mathbf{A}^r$ means that the operator $\mathbf{A}$ is applied successively r times or that the matrix representing $\mathbf{A}$ is multiplied r times. In Eq. 11.20 the exponential can be real or complex. From the definition it can be shown that if two operators $\mathbf{A}$ and $\mathbf{B}$ commute,

$$e^{(A+B)} = e^A e^B = e^B e^A \qquad (11.21)$$

Also,

$$e^{-iAt} \mathbf{B} e^{iAt} = \mathbf{B} \qquad (11.22)$$

These relations turn out to be quite useful. For example, when spins are weakly coupled, the Zeeman and scalar coupling portions of the Hamiltonian commute, so that these effects may be treated separately and in any order.

Notations for the Density Matrix and Its Subsets

The density matrix and density operator concept has extremely wide application in physics and chemistry. (In 1998, for example, Walter Kohn won the Nobel Prize in Chemistry for applying this concept, known in this case as density functional theory, to determining the electronic structure of complex molecules.) Most literature uses the symbol ρ for the density matrix and for the density operator (sometimes with the more precise $\hat{\rho}$ as the operator). In NMR many authors use σ as a "reduced" density matrix containing only the spin terms in order to distinguish this part from the "full" density matrix ρ, which also contains spatial terms. Thus, ρ is partitioned, with the spatial terms treated in a separate relaxation matrix. Although the density matrix approach represents a powerful means for dealing with relaxation, our treatment ignores relaxation, and this chapter is focused entirely on the reduced density matrix. However, we continue to use the symbol ρ so as not to confuse the symbol for the density matrix with σ used for chemical shielding and σ_0, σ_x, σ_y, and σ_z used by some authors to denote the Pauli spin matrices.

In Section 11.2 we note that ρ is usually expressed in a rotating frame. Unless there is a specific application in which it is important to distinguish between the laboratory and rotating frames, we continue to use ρ as the symbol, regardless of coordinate system.

In Section 11.3 we find that the density matrix for a spin system at equilibrium can be separated into the unit matrix **1** and other terms pertaining to populations of spin states. The unit matrix is unaffected by rf pulses or any other evolution of the density matrix; hence it is conventional to delete it, and we do so. Some authors introduce a new symbol for the truncated matrix, but most do not. We continue to use the symbol ρ for this truncated density matrix.

11.2 TRANSFORMATIONS OF THE DENSITY MATRIX

We know that rf pulses rotate magnetization vectors and alter the state of the spin system. To examine such effects quantitatively in terms of alterations in the density matrix, it will be helpful to express the density matrix in the rotating frame of reference.

Transformation to the Rotating Frame

So far, the equations in this chapter are based on the laboratory frame of reference. In Section 2.8 we saw that the description of magnetic resonance can often be simplified by using a frame rotating with angular frequency ω_{rf} about the z axis, where $\omega_{rf}/2\pi$ is usually chosen to be the pulse frequency (and reference frequency) used to observe the spin system. Now we want to express the density matrix in the rotating frame in order to facilitate our handling of time-dependent Hamiltonians that arise when radio frequency fields are applied.

The transformation can be made by applying Eq. 11.16, with a Hamiltonian appropriate to a "fictitious" magnetic field (see Section 2.8) that would cause procession at a frequency of $-\omega_{rf}$ (equivalent to the frame moving at $+\omega_{rf}$):

$$\rho_{rf} = e^{-i\omega_{rf}F_z t} \rho e^{i\omega_{rf}F_z t} \qquad (11.23)$$

The operator F_z was defined in Eq. 6.2 and has eigenvalues f_n that give the total spin of a wave function. The result of interest from Eq. 11.23, as might be expected, is that the frequencies in the density matrix elements are reduced in the rotating frame, so that all coherences have frequencies in the usual "audio" range that is observed after detection. As indicated in the previous section, we continue to use the symbol ρ for the density matrix in the rotating frame, and to simplify the notation we generally use ω to denote frequencies measured in the rotating frame.

The Effect of a Radio Frequency Pulse

As we pointed out in Section 2.8, a sufficiently strong radio frequency field $\mathbf{B}_1$ applied near resonance leads to an effective field $\mathbf{B}_{eff}$ in the rotating frame that is nearly equal in magnitude and direction to $\mathbf{B}_1$ (see Fig. 2.8). Under this condition, the Hamiltonian in the rotating frame, $\mathcal{H}_{rf}$ has only negligible contributions from the Zeeman, chemical shift, and scalar coupling terms. Consider a pulse of duration τ along the x' axis in the rotating frame, applied to a spin system with total spin F_z. $\mathcal{H}_{rf}$ (in angular frequency units) is, then,

$$\mathcal{H}_{rf} = \gamma B_1 F_x \tag{11.24}$$

where F_x is defined (analogous to F_z in Eq. 6.2) as

$$F_x = \sum_{n=1}^{N} I_x^{(n)} \tag{11.25}$$

During the period τ, typically a few microseconds, $\mathcal{H}_{rf}$ is essentially constant, so Eq. 11.16 is applicable. The argument in the exponential becomes

$$\mathcal{H}_{rf} t = \gamma B_1 \tau = \theta \tag{11.26}$$

[θ is the angle that (classically) the magnetization would traverse.] The density matrix thus becomes

$$\rho(\tau) = e^{-i\theta F_x} \rho(0) e^{i\theta F_x} \tag{11.27}$$

For a weakly coupled system of N nuclei of spin $\frac{1}{2}$, we can easily obtain explicit expressions for the exponential operators in Eq. 11.27 by applying Eq. 11.25 and the expansion of Eq. 11.20:

$$e^{i\theta F_x} = \prod_n e^{i\theta I_x^{(n)}} \tag{11.28}$$

$$e^{i\theta I_x^{(n)}} = 1 + i\theta I_x^{(n)} + (i\theta I_x^{(n)})^2/2! + (i\theta I_x^{(n)})^3/3! + \cdots$$

$$= [1 - (\theta/2)^2/2! + \cdots] + 2i[(\theta/2) - (\theta/2)^3/3! + \cdots] I_x^{(n)}$$

$$= \cos(\theta/2) + 2i\sin(\theta/2) I_x^{(n)} \tag{11.29}$$

where the relations $i^2 = -1$, and $I_x^2 = \frac{1}{4}$ have been used.

For manipulations of the density matrix by Eq. 11.27, it is convenient to express Eq. 11.29 in matrix form:

$$e^{i\theta I_x^{(n)}} = \cos(\theta/2)\,\mathbf{1} + 2i\sin(\theta/2)\mathbf{I}_x^{(n)}$$

$$= \cos(\theta/2)\begin{bmatrix} 1 & 0 \\ 0 & 1 \end{bmatrix} + 2i\sin(\theta/2)\begin{bmatrix} 0 & \frac{1}{2} \\ \frac{1}{2} & 0 \end{bmatrix}$$

$$= \begin{bmatrix} \cos(\theta/2) & i\sin(\theta/2) \\ i\sin(\theta/2) & \cos(\theta/2) \end{bmatrix} \tag{11.30}$$

In accord with Eq. 11.28, the entire matrix $\mathbf{R} = \exp(i\theta F_x)$ is obtained by taking the product of the matrices of Eq. 11.30 for each spin. This procedure will become clearer in the following sections.

11.3 THE ONE-SPIN SYSTEM

To provide a concrete application of the tools we have developed, let us now apply the general density matrix approach to an ensemble of one-spin systems (with $I = \frac{1}{2}$). To set up the density matrix according to Eqs. 11.1 and 11.5, we need basis functions ϕ_1 and ϕ_2, which are simply α and β. Because these are eigenfunctions of $\mathcal{H}$, the wave functions are simple: $\psi^{(1)} = \alpha$ and $\psi^{(2)} = \beta$, giving the four coefficients needed for Eq. 11.5 the values

$$c_1^{(1)} = 1 \qquad c_2^{(1)} = 0$$
$$c_1^{(2)} = 0 \qquad c_2^{(2)} = 1 \tag{11.31}$$

The probabilities of occurrence, $p^{(1)}$ and $p^{(2)}$, are, at equilibrium, simply the populations of the states as determined by the Boltzmann distribution. If we let $\epsilon = (E_\alpha - E_\beta)/kT$, then from Eq. 2.19 we obtain

$$\frac{p^{(2)}}{p^{(1)}} = e^{-\epsilon} \approx 1 - \epsilon \tag{11.32}$$

for $\epsilon \ll 1$. Thus,

$$p^{(1)} = \frac{1}{2}[1 + \epsilon/2] \qquad p^{(2)} = \frac{1}{2}[1 - \epsilon/2] \tag{11.33}$$

We can then write the matrix elements of ρ (which at equilibrium is the same in the rotating and laboratory frames) as

$$\rho_{11} = \frac{1}{2}[1 + (\epsilon/2)] + 0$$
$$\rho_{22} = 0 + \frac{1}{2}[1 - (\epsilon/2)] \tag{11.34}$$
$$\rho_{12} = \rho_{21} = 0 + 0$$

Then the density matrix at equilibrium is

$$\rho = \frac{1}{2} \begin{bmatrix} 1 + \dfrac{\epsilon}{2} & 0 \\ 0 & 1 - \dfrac{\epsilon}{2} \end{bmatrix}$$

$$= \frac{1}{2} \begin{bmatrix} 1 & 0 \\ 0 & 1 \end{bmatrix} + \frac{\epsilon}{4} \begin{bmatrix} 1 & 0 \\ 0 & -1 \end{bmatrix} \tag{11.35}$$

In Eq. 11.35 the first term is just a constant times the unit matrix **1**. This term is of little interest, as it does not vary with time and it does not contribute to quantities that we wish to measure. It is customary, therefore, to simplify the mathematical expressions by *redefining ρ to represent the density matrix with the unit matrix subtracted out.*

In Section 2.3 we described the effect of the operators I_x, I_y, I_z, I^+ and I^- on the functions α and β. Here we summarize these relations in terms of matrices with α and β as the basis functions. Thus,

$$I_x = \frac{1}{2}\begin{bmatrix} 0 & 1 \\ 1 & 0 \end{bmatrix} \qquad I_y = \frac{1}{2}\begin{bmatrix} 0 & -i \\ i & 0 \end{bmatrix} \qquad I_z = \frac{1}{2}\begin{bmatrix} 1 & 0 \\ 0 & -1 \end{bmatrix}$$

$$I^+ \equiv I_x + iI_y = \begin{bmatrix} 0 & 1 \\ 0 & 0 \end{bmatrix} \qquad I^- \equiv I_x - iI_y = \begin{bmatrix} 0 & 0 \\ 1 & 0 \end{bmatrix} \tag{11.36}$$

As we shall see, it is very helpful to be able to compute the behavior of the spin system from the sort of matrix multiplications that we have already carried out. On the other hand, it is often possible to simplify the algebraic expressions by using the corresponding spin operators. In fact, this is the concept of the product operator formalism that we discuss later. Note that from Eqs. 11.35 and 11.36, the (redefined) density matrix at equilibrium can be written in operator form as

$$\rho = \frac{\epsilon}{2} I_z \tag{11.37}$$

We can use the simple operators in Eq. 11.36 in conjunction with one of the fundamental relations of ρ (Eq. 11.6) to determine the expectation value of the longitudinal magnetization M_z at equilibrium:

$$\langle M_z \rangle = \mathrm{Tr}(M_z\rho) = \mathrm{Tr}\left[(N\gamma\hbar I_z)\left(\frac{\epsilon}{2} I_z\right) \right]$$

$$= \tfrac{1}{2}N\gamma\hbar\epsilon\,\mathrm{Tr}(I_z^2)$$

$$= \tfrac{1}{2}N\gamma\hbar(\gamma\hbar B_0/kT)(\tfrac{1}{4})(2)$$

$$= N\gamma^2\hbar^2 B_0/4kT \tag{11.38}$$

Likewise, substitution of the expressions for M_x and M_y shows that the average values of these quantities at equilibrium are zero, as expected from Fig. 2.3.

It is important to note the difference between the density matrix approach, which deals with an *incoherent* superposition of states, and a linear combination of basis functions, which properly treats a coherent superposition. In Eqs. 2.36 and 2.37, which refer to a single spin, we found that there is a rotating component of magnetization in the xy plane, but the density matrix treatment of an ensemble of spins shows no such component at equilibrium. Likewise, a mixture of many spins, precisely half in state α and half in state β, would be shown by the density matrix to have zero net magnetization, a situation termed "saturation." However, a 1:1 mixture of states α and β for a single spin, as represented by $c_\alpha = c_\beta$ in Eq. 2.31, shows a magnetization in the xy plane, as it should, because the magnetization of a single spin cannot go to zero.

Effect of a 90° rf Pulse

When a strong rf pulse is applied along the x' axis, Eq. 11.26 gives the Hamiltonian operating in the rotating frame, and Eqs. 11.27 and 11.30 show the effect on the density matrix at the end of the pulse period τ. Consider a single spin subject to a 90° pulse. Some care is required in applying these equations to the construction of the necessary matrices. First, note that the matrices must be in the order of the operators of Eq. 11.27, so that the matrix applied on the right of $\boldsymbol{\rho}$ is the same as that given in Eq. 11.30, while the matrix on the left has i replaced by $-i$ in the sine terms. Second, the Larmor relation (Eqs. 2.43 and 2.45) shows that the rotation vector $\boldsymbol{\omega}$ is directed opposite to $\mathbf{B}$. As pointed out in Section 2.11, we must take $\mathbf{B}_1$ along the negative x' or y' axis, so as to make the corresponding rotation positive. These points must be considered in all the applications that we describe. In the present computation, we apply $\mathbf{B}_1$ along the negative x' axis to generate a positive 90° rotation about x'; this is often called simply a 90_x pulse. We then obtain the matrix product

$$
\begin{aligned}
\rho(\tau) &= \frac{\epsilon}{4}
\begin{bmatrix} \cos 45° & -i\sin 45° \\ -i\sin 45° & \cos 45° \end{bmatrix}
\begin{bmatrix} 1 & 0 \\ 0 & -1 \end{bmatrix}
\begin{bmatrix} \cos 45° & i\sin 45° \\ i\sin 45° & \cos 45° \end{bmatrix} \\[2mm]
&= \frac{\epsilon}{4}
\begin{bmatrix} 1/\sqrt{2} & -i/\sqrt{2} \\ -i/\sqrt{2} & 1/\sqrt{2} \end{bmatrix}
\begin{bmatrix} 1 & 0 \\ 0 & -1 \end{bmatrix}
\begin{bmatrix} 1/\sqrt{2} & i/\sqrt{2} \\ i/\sqrt{2} & 1/\sqrt{2} \end{bmatrix} \\[2mm]
&= \frac{\epsilon}{8}
\begin{bmatrix} 1 & -i \\ -i & 1 \end{bmatrix}
\begin{bmatrix} 1 & 0 \\ 0 & -1 \end{bmatrix}
\begin{bmatrix} 1 & i \\ i & 1 \end{bmatrix} \\[2mm]
&= \frac{\epsilon}{4}
\begin{bmatrix} 0 & i \\ -i & 0 \end{bmatrix}
\end{aligned}
\tag{11.39}
$$

Equation 11.39 shows that the 90° pulse has nulled the on–diagonal elements of ρ and has created nonzero off-diagonal elements (coherences). We can also express Eq. 11.39 in the simple algebraic form

$$
\rho(t) = -\frac{\epsilon}{2} I_y
\tag{11.40}
$$

which shows the coherence in the negative y' direction. Macroscopically, this result is consistent with the vector picture approach of Section 2.7, which shows that a 90_x pulse tips the magnetization to the $-y'$ axis, thus reducing M_z to zero.

We can easily verify this result by applying Eq. 11.6 to $\rho(\tau)$:

$$
\begin{aligned}
\langle M_z \rangle &= \mathrm{Tr}(M_z \rho) = \mathrm{Tr}\left(\frac{1}{8} N\gamma\hbar\epsilon
\begin{bmatrix} 1 & 0 \\ 0 & -1 \end{bmatrix}
\begin{bmatrix} 0 & i \\ -i & 0 \end{bmatrix} \right) \\[2mm]
&= \frac{1}{8} N\gamma\hbar\epsilon\, \mathrm{Tr}\left(
\begin{bmatrix} 0 & i \\ i & 0 \end{bmatrix} \right) = 0
\end{aligned}
\tag{11.41}
$$

$$\langle M_x \rangle = \frac{1}{8} N\gamma \hbar \epsilon \operatorname{Tr}\left(\begin{bmatrix} 0 & 1 \\ 1 & 0 \end{bmatrix} \begin{bmatrix} 0 & i \\ -i & 0 \end{bmatrix} \right)$$

$$= \frac{1}{8} N\gamma \hbar \epsilon \operatorname{Tr}\left(\begin{bmatrix} -i & 0 \\ 0 & i \end{bmatrix} \right) = 0 \tag{11.42}$$

$$\langle M_y \rangle = \frac{1}{8} N\gamma \hbar \epsilon \operatorname{Tr}\left(\begin{bmatrix} 0 & -i \\ i & 0 \end{bmatrix} \begin{bmatrix} 0 & i \\ -i & 0 \end{bmatrix} \right)$$

$$= \frac{1}{8} N\gamma \hbar \epsilon \operatorname{Tr}\left(\begin{bmatrix} -1 & 0 \\ 0 & -1 \end{bmatrix} \right) = -\frac{1}{4} N\gamma \hbar \epsilon = -N\gamma^2 \hbar^2 B_0 \tag{11.43}$$

The same result could have been obtained more succinctly by starting with Eq. 11.40 and carrying out the manipulations algebraically (as in Eq. 11.38). The algebraic calculations depend on the fact that the matrices for I_x, I_y, and I_z are traceless and on the cyclic permutation relations $I_r I_s = i I_t$ among I_x, I_y, and I_z.

Effect of Free Precession

We now look at the evolution of the density matrix for the one spin system as the magnetization precesses in the rotating frame. Once more, we apply Eq. 11.16 in the same manner we did in Section 11.2 to take into account the effect of the rotating frame. In this case we obtain

$$\rho(t) = e^{-i\omega F_z t} \rho(\tau) e^{i\omega F_z t} \tag{11.44}$$

Keeping in mind the points noted in setting up the matrices in Eq. 11.39. We obtain the following relation:

$$\rho(t) = \frac{\epsilon}{4} \begin{bmatrix} e^{-i\omega t/2} & 0 \\ 0 & e^{i\omega t/2} \end{bmatrix} \begin{bmatrix} 0 & i \\ -i & 0 \end{bmatrix} \begin{bmatrix} e^{i\omega t/2} & 0 \\ 0 & e^{-i\omega t/2} \end{bmatrix}$$

$$\rho(t) = \frac{\epsilon}{4} \begin{bmatrix} 0 & ie^{-i\omega t} \\ -ie^{i\omega t} & 0 \end{bmatrix} \tag{11.45}$$

The variation of the magnetization with time can be found by applying Eq. 11.6 to each of the components of **M**. $\langle M_z \rangle$ remains zero, because the product matrix $M_z \rho(t)$ is easily shown to be traceless. Rather than computing $\langle M_x \rangle$ and $\langle M_y \rangle$ separately, it is easier and more informative to find $\langle M_{xy} \rangle$, where

$$M_{xy} = M_x + iM_y = N\gamma\hbar(I_x + iI_y) = N\gamma\hbar I_+ \tag{11.46}$$

The complex notation is consistent with our discussion of phase-sensitive detection in Section 3.4, where the real component represents the projection along the x' axis and the imaginary component that along the y' axis.

From Eq. 11.6

$$\langle M_{xy}\rangle = \mathrm{Tr}\,(M_{xy}\rho(t))$$

$$= \frac{1}{4}N\gamma\hbar\epsilon\,\mathrm{Tr}\left(\begin{bmatrix}0 & 1\\0 & 0\end{bmatrix}\begin{bmatrix}0 & ie^{i\omega t}\\-ie^{-i\omega t} & 0\end{bmatrix}\right)$$

$$= \frac{1}{4}N\gamma\hbar\epsilon\,\mathrm{Tr}\left(\begin{bmatrix}-ie^{-i\omega t} & 0\\0 & 0\end{bmatrix}\right)$$

$$= -\frac{1}{4}N\gamma\hbar\epsilon ie^{i\omega t}$$

$$= \frac{1}{4}N\gamma\hbar\epsilon[\sin\omega t - i\cos\omega t] \tag{11.47}$$

As we would expect, Eq. 11.47 describes a magnetization that moves in a circle at angular frequency ω phased so as to lie along $-y'$ at $t = 0$.

We could continue to manipulate this one-spin system with additional pulses and evolution times, but instead we turn to a system of more interest.

11.4 THE TWO-SPIN SYSTEM

Consider now the two-spin system, in which chemical shifts and scalar coupling come into play. In Chapter 6 we discussed the two-spin system in detail, both the weakly coupled AX system and the general case, AB. To illustrate the application of the density matrix, we concentrate first on the AX system and then indicate briefly how the results would be altered for AB. To simplify the notation, we call the nuclei I and S, rather than A and X, and use the common notation in which the spin operators and their components are designated, for example, I_x and S_x, rather than the more cumbersome $I_x(A)$ or I_{xA}. Although the $I-S$ notation is usually applied to heteronuclear spin systems, we use it here to include homonuclear systems (e.g., H−H) as well.

Eigenfunctions and Energy Levels

As we showed in Eq. 6.1, the basis functions are given by

$$\begin{aligned}\phi_1 &= \alpha_I\alpha_S & f_z &= 1\\\phi_2 &= \alpha_I\beta_S & f_z &= 0\\\phi_3 &= \beta_I\alpha_S & f_z &= 0\\\phi_4 &= \beta_I\beta_S & f_z &= -1\end{aligned} \tag{11.48}$$

With only weak spin coupling, these are the eigenfunctions $\psi^{(i)}$, but in the general AB case functions ϕ_2 and ϕ_3 mix. The energy levels for AB are given by

$$E_1 = -\tfrac{1}{2}[\nu_I + \nu_S] + \tfrac{1}{4}J \tag{11.49a}$$

$$E_2 = -\tfrac{1}{2}[(\nu_I - \nu_S)^2 + J^2]^{1/2} - \tfrac{1}{4}J \tag{11.49b}$$

$$E_3 = \tfrac{1}{2}[\nu_I - \nu_S)^2 + J^2]^{1/2} - \tfrac{1}{4}J \tag{11.49c}$$

$$E_4 = \tfrac{1}{2}[\nu_I + \nu_S] + \tfrac{1}{4}J \tag{11.49d}$$

For $|J| \ll |\nu_I - \nu_S|$, J^2 can be ignored in comparison with $(\nu_I - \nu_S)^2$ in Eqs. 11.49b and c, so that the changes in energy levels resulting from weak coupling are only the anticipated first-order effects. Initially we use this approximation, hence can use ϕ_1–ϕ_4 as the eigenfunctions.

Equilibrium Density Matrix

With the approximations used in Eq. 11.33, we can represent the relative populations of the four energy levels in terms of $\epsilon_I \equiv (E_3 - E_1)/kT$ and $\epsilon_S \equiv (E_2 - E_1)/kT$. We already know that the density matrix at equilibrium $\rho(0)$ is diagonal, with elements that describe the excess populations:

$$\rho = \frac{1}{4}\begin{bmatrix} \epsilon_I + \epsilon_S & 0 & 0 & 0 \\ 0 & \epsilon_I - \epsilon_S & 0 & 0 \\ 0 & 0 & -\epsilon_I + \epsilon_S & 0 \\ 0 & 0 & 0 & -\epsilon_I - \epsilon_S \end{bmatrix} \tag{11.50}$$

Effect of a Selective 90° Pulse

The transformation of the density matrix by a 90° pulse and its subsequent evolution as the magnetization precesses freely depend on whether the pulse is applied only to one nucleus or to both I and S. We treat first the situation that would normally occur in a heteronuclear system, where a 90_x pulse is applied only to the I spins—a $90_x^{(I)}$ pulse. This treatment is, of course, also applicable to a homonuclear system subjected to a selective rf pulse.

Equations 11.28 and 11.29 provide the framework, with the matrix manipulations following those in Eq. 11.39, but I_x in Eq. 11.36 must be expressed in the basis functions ϕ_1–ϕ_4, that is, a 4×4 matrix. The elements of I_x may readily be calculated, for example,

$$[I_x]_{11} = \langle \phi_1 | I_x | \phi_1 \rangle = \langle \alpha_I \alpha_S | I_x | \alpha_I \alpha_S \rangle = \langle \alpha_I | I_x | \alpha_I \rangle \langle \alpha_S | \alpha_S \rangle = 0$$

$$[I_x]_{13} = \langle \phi_1 | I_x | \phi_3 \rangle = \langle \alpha_I \alpha_S | I_x | \beta_I \alpha_S \rangle = \langle \alpha_I | I_x | \beta_I \rangle \langle \alpha_S | \alpha_S \rangle = \tfrac{1}{2} \tag{11.51}$$

All 16 elements can be determined in this way to give

$$
I_x = \tfrac{1}{2}
\begin{bmatrix}
0 & 0 & 1 & 0 \\
0 & 0 & 0 & 1 \\
1 & 0 & 0 & 0 \\
0 & 1 & 0 & 0
\end{bmatrix}
\qquad
S_x = \tfrac{1}{2}
\begin{bmatrix}
0 & 1 & 0 & 0 \\
1 & 0 & 0 & 0 \\
0 & 0 & 0 & 1 \\
0 & 0 & 1 & 0
\end{bmatrix}
\tag{11.52}
$$

(Alternative and less tedious methods by use of matrix direct products can also be used when the basis functions are also the eigenfunctions, as shown in Appendix D, where a number of spin matrices are given.) By following the procedure of Eq. 11.39, we obtain the density matrix at the end of the pulse:

$$
\rho(\tau) = \frac{1}{4}
\begin{bmatrix}
\epsilon_S & 0 & i\epsilon_I & 0 \\
0 & -\epsilon_S & 0 & i\epsilon_I \\
-i\epsilon_I & 0 & \epsilon_S & 0 \\
0 & -i\epsilon_I & 0 & -\epsilon_S
\end{bmatrix}
\tag{11.53}
$$

Several important features are shown in this expression for $\rho(\tau)$. First, the S spins are unaffected and remain aligned along the z axis, as indicated by the fact that their populations are shown along the principal diagonal and there are no off-diagonal terms involving S. The I spins, on the other hand, show no terms along the diagonal but do display off-diagonal elements. To see why the off-diagonal terms appear where they do, with many other off-diagonal terms remaining zero, we need to look at the structure of the density matrix.

In Eq. 11.48 we saw that the basis functions for our density matrix are divided into three groups with $f_z = 1$, 0, and -1, respectively. As we saw in Chapter 6, transitions between energy levels $E_1 \leftrightarrow E_2$, $E_3 \leftrightarrow E_4$, $E_1 \leftrightarrow E_3$, and $E_2 \leftrightarrow E_4$ each result in $\Delta f_z = \pm 1$ and are called *single quantum* transitions, while transitions $E_1 \leftrightarrow E_4$ and $E_2 \leftrightarrow E_3$ are termed *double quantum* and *zero quantum* transitions, respectively. The usual selection rules from time-dependent perturbation theory show that only single quantum transitions are permitted in such simple experiments as excitation by a 90° pulse. Moreover, for weakly coupled nuclei, the single quantum transitions each involve only a single type of nucleus, I or S, as indicated in Fig. 6.2.

In Table 11.1 we sketch the form of the density matrix for the two–spin system to show the significance of the elements. P_1–P_4 refer to the populations of the four states, I and S represent single quantum I and S transitions, and Z refers to zero quantum transitions and D to double quantum transitions. We saw in Eq. 11.9 that an off-diagonal element ρ_{mn} is nonzero only if there is a phase coherence between states m and n, and in Eq. 10.19 we saw that ρ_{mn} evolves with a frequency determined by the difference in energies $E_m - E_n$. Thus, these off-diagonal elements represent not only transitions, but single quantum, double quantum, and zero quantum *coherences*, which evolve in free precession at approximate frequencies of ν_I, ν_S, $\nu_I + \nu_S$, and $\nu_I - \nu_S$. In Eq. 11.53 we see that $\rho(\tau)$ has

TABLE 11.1. Structure of the
Density Matrix for a Two-Spin
System

	$\alpha_I\alpha_S$	$\alpha_I\beta_S$	$\beta_I\alpha_S$	$\beta_I\beta_S$
$\alpha_I\alpha_S$	P_1	S	I	D
$\alpha_I\beta_S$	S	P_2	Z	I
$\beta_I\alpha_S$	I	Z	P_3	S
$\beta_I\beta_S$	D	I	S	P_4

single quantum I coherences, as expected from the 90° pulse to the I nuclei, but no S coherences and no double quantum coherences (DQCs) or zero quantum coherences (ZQCs). The I coherences are purely imaginary; hence, by analogy to the one-spin example (Eq. 11.39), the magnetization must lie along the y' axis in the rotating frame. As in the one-spin case, this result could be verified by calculating $\langle M_y \rangle$.

We shall see in Section 11.5 and in Chapter 12 that ZQCs and DQCs can be created by certain pulse sequences and can then be converted to single quantum coherence (SQC) for observation, because ZQC and DQC provide no NMR signal. For reasons that we discuss later, it is often helpful to keep track of the *coherence transfer pathway* by which coherences change in quantum number, or *coherence order*. As we see later, the single quantum coherence initially created has orders $p = \pm 1$, so we must consider both positive and negative coherence orders. These pairs of coherences contain redundant information, and the process of detecting the macroscopic magnetization is normally configured, by convention, to detect the coherence at $p = -1$ rather than that at $+1$. We shall return to a more detailed discussion of coherence transfer pathways and selection of coherence order in Section 11.7.

Evolution of the Density Matrix

As indicated in Eq. 11.19, during the period t after the pulse, each off-diagonal term in ρ evolves at its own frequency in the rotating frame. From Fig. 6.2, we expect two frequencies for the I spins, $\nu_I \pm \frac{1}{2}J$, depending on whether the I spin resides in a molecule with S in state α or β. We can readily see from Table 11.1 that ρ_{13} and ρ_{31} arise from α_S states, while ρ_{24} and ρ_{42} arise from β_S states. Thus

$$\rho(t) = \frac{1}{4}\begin{bmatrix} \epsilon_S & 0 & i\epsilon_I e^{2\pi i(\nu_I - J/2)t} & 0 \\ 0 & -\epsilon_S & 0 & i\epsilon_I e^{2\pi i(\nu_I + J/2)t} \\ -i\epsilon_I e^{-2\pi i(\nu_I - J/2)t} & 0 & \epsilon_S & 0 \\ 0 & -i\epsilon_I e^{-2\pi i(\nu_I + J/2)t} & 0 & -\epsilon_S \end{bmatrix}$$

(11.54)

By calculating $\langle M_{xy} \rangle$, as we did for the one-spin system, we could demonstrate the two precessing components of magnetization that beat with each other as they go in and out of phase.

Effect of a Nonselective 90° Pulse

To see the effect of a 90_x pulse that is applied to both I and S, we can either consider the application of a $90_x^{(S)}$ to the density matrix in Eq. 11.53 or use the procedures outlined in Eqs. 11.29 and 11.30 to form the rotation matrices corresponding to I_x and S_x, then multiply them, as indicated in Eq. 11.28 to obtain

$$R\,(90_x) = \frac{1}{2} \begin{bmatrix} 1 & i & i & -1 \\ i & 1 & -1 & i \\ i & -1 & 1 & i \\ -1 & i & i & 1 \end{bmatrix} \tag{11.55}$$

This matrix and its inverse (formed in this instance by replacing each i by $-i$) applied to the density matrix in Eq. 11.50 gives

$$\rho(\tau) = \frac{1}{4} \begin{bmatrix} 0 & i\epsilon_S & i\epsilon_I & 0 \\ -i\epsilon_S & 0 & 0 & i\epsilon_I \\ -i\epsilon_I & 0 & 0 & i\epsilon_S \\ 0 & -i\epsilon_I & -i\epsilon_S & 0 \end{bmatrix} \tag{11.56}$$

As expected from Table 11.1, application of the nonselective pulse eliminates all terms on the diagonal and generates single quantum I and S coherneces.

By using the matrices for I_y and S_y from Appendix D, we can see that the matrix $\rho(\tau)$ can be expressed as the sum

$$\rho(t) = -\frac{\epsilon_I}{2} I_y - \frac{\epsilon_S}{2} S_y \tag{11.57}$$

with magnetization along the negative y' axis, as we found in Eq. 11.40 for the one-spin system.

The time evolution of ρ follows the pattern of Eq. 11.54 and will not be given here.

Effect of Strong Coupling

Our treatment is directly applicable to a strongly coupled AB system except that the algebra becomes much more tedious. For AB, we showed in Eq. 6.30 that wave functions $\psi_1 = \phi_1$ and $\psi_4 = \phi_4$, but ψ_2 and ψ_3 are formed from a mixture of ϕ_2 and ϕ_3 in a ratio of $Q{:}1$, where Q depends on the value of $|\nu_A - \nu_B|/J$. If the calculation is set up in a basis of the ψ_i, the density matrix looks the same as Eq. 11.50 (with $\epsilon_I = \epsilon_S$). However, the matrix $R(90°)$ based on I_x and S_x as determined in Eq. 11.51, is more complicated than that in Eq. 11.55. After a 90° pulse, the coherences evolve at the frequencies that we discussed for an AB spectrum in Section 6.8. As pointed out by Farrar and Harriman,[113] it is usually easier to carry out the computation in a basis of the ϕ_i, which is formally the same as for the AX system, then transform to the ψ_i basis, using Eq. 6.30, to obtain results that are readily interpretable. Since there is no new insight as a result of a great deal of algebra, we do not discuss strongly coupled systems in further detail.

11.5 INEPT AND RELATED PULSE SEQUENCES

We discussed INEPT in Section 9.5, where we used the vector picture to follow the time dependence of magnetization and grafted on quantum considerations to understand population changes and polarization transfer. With the density matrix, all parts of the process occur in a straightforward manner. It is useful to go through the calculation in some detail, as a number of important points can be illustrated. Moreover, we will examine some modifications of the pulse sequence that provide insight into polarization transfer and multiple quantum methods used in 2D NMR.

Density Matrix Treatment of INEPT

We consider only the two-spin IS system, for which the equilibrium density matrix is given in Eq. 11.50. To simplify the algebraic expressions, we take $I = {}^1\mathrm{H}$, $S = {}^{13}\mathrm{C}$, $\epsilon_S = 1$, and $\epsilon_I = (\gamma_I/\gamma_S)\epsilon_S = 4$, and we drop the normalizing factor. The pulse sequence is illustrated in Fig. 11.1. At time 1, immediately after the 90_x pulse applied only to ${}^1\mathrm{H}$, the density matrix is given in Eq. 11.53, and during the period 1–2, ρ evolves as shown in Eq. 11.54. At time 2, where $t = 1/4J$, the density matrix is

$$\rho(2) = \begin{bmatrix} 1 & 0 & 4ie^A & 0 \\ 0 & -1 & 0 & 4ie^B \\ -4ie^{-A} & 0 & 1 & 0 \\ 0 & -4ie^{-B} & 0 & -1 \end{bmatrix} \qquad (11.58)$$

where

$$A = 2\pi i\left(\nu_I - \frac{J}{2}\right)\frac{1}{4J}$$

$$B = 2\pi i\left(\nu_I + \frac{J}{2}\right)\frac{1}{4J} \qquad (11.59)$$

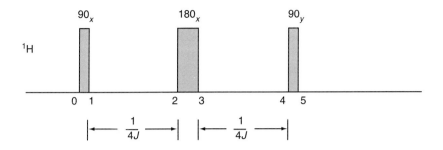

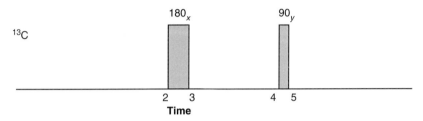

FIGURE 11.1 Pulse sequence for INEPT. See text for discussion of density matrix at each time indicated.

Now we apply a 180_x pulse to both I and S. Following the procedures outlined in Eqs. 11.28–11.30, we construct the appropriate rotation matrix

$$R(180_x) = R^{-1}(180_x) = \begin{bmatrix} 0 & 0 & 0 & -1 \\ 0 & 0 & -1 & 0 \\ 0 & -1 & 0 & 0 \\ -1 & 0 & 0 & 0 \end{bmatrix} \tag{11.60}$$

After the 180° pulses (time 3), the density matrix becomes

$$\rho(3) = \begin{bmatrix} -1 & 0 & -4ie^{-B} & 0 \\ 0 & 1 & 0 & -4ie^{-A} \\ 4ie^{B} & 0 & -1 & 0 \\ 0 & 4ie^{A} & 0 & 1 \end{bmatrix} \tag{11.61}$$

Note that the $180°^{13}$C pulse has interchanged the populations shown on the diagonal, as we would expect. The coherences evolve during the period 3–4 at the same rate indicated in Eq. 11.54, and at time 4 (after another evolution of $1/4J$ s)

$$\rho(4) = \begin{bmatrix} -1 & 0 & -4ie^{A-B} & 0 \\ 0 & 1 & 0 & -4ie^{B-A} \\ 4ie^{B-A} & 0 & -1 & 0 \\ 0 & 4ie^{A-B} & 0 & -1 \end{bmatrix} \tag{11.62}$$

From Eq. 11.59 it is clear that the chemical shift terms cancel, as they should at the echo, but the J terms add to give a multiple of $\pi i/2$ in the exponential, which simplifies (see Appendix C) to the following:

$$\rho(4) = \begin{bmatrix} -1 & 0 & -4 & 0 \\ 0 & 1 & 0 & 4 \\ -4 & 0 & -1 & 0 \\ 0 & 4 & 0 & 1 \end{bmatrix} \tag{11.63}$$

The ^{13}C populations are unchanged from $\rho(3)$, but the two ^1H coherences (ρ_{13} and ρ_{24}) now have opposite signs and are directed along the x' axis (no imaginary term). Calculation of $\langle M_{xy} \rangle$, as in Eq. 11.47, would show no net magnetization.

The next step is application of ^1H and ^{13}C 90° pulses, both along the y' axis. The calculation proceeds in the now familiar way, but we need the matrix for $R(90_y)$, which is formed as in Eq. 11.55, but with the matrix representations of I_y and S_y:

$$R(90_y) = \frac{1}{2} \begin{bmatrix} 1 & 1 & 1 & 1 \\ -1 & 1 & -1 & 1 \\ -1 & -1 & 1 & 1 \\ 1 & -1 & -1 & 1 \end{bmatrix} \tag{11.64}$$

At time 5 the density matrix then becomes

$$\rho(5) = \begin{bmatrix} 0 & 3 & 0 & 0 \\ 3 & 0 & 0 & 0 \\ 0 & 0 & 0 & -5 \\ 0 & 0 & -5 & 0 \end{bmatrix} \tag{11.65}$$

Now we have generated two ^{13}C coherences (ρ_{12} and ρ_{34}) along x' of opposite sign and of considerably enhanced magnitude. In the absence of decoupling, each evolves at its own frequency during the period after time 5, thereby generating two signals of intensity $3:-5$.

It is not necessary that the evolving ^{13}C coherences be detected immediately. As shown in Section 9.6, they can be allowed to precess until they are in phase, then detected while protons are decoupled to provide a single enhanced signal. Alternatively, the entire INEPT sequence can be treated as the preparation period of a 2D experiment. The coherences then evolve during a period t_1 and can be manipulated in various ways by further pulses. One of the most commonly used methods is to apply a second INEPT sequence, without the initial 90_x pulse, after the evolution period to convert the ^{13}C coherences back into ^1H coherences, which can be observed. As we mentioned in Chapter 10, this method, heteronuclear single quantum coherence (HSQC), is widely employed to obtain information on the correlation of chemical shifts of less sensitive nuclei, such as ^{13}C and ^{15}N, with the sensitivity of proton NMR.

The treatment of INEPT by the density matrix is more satisfactory than the development in Section 9.6 in that it follows in a logical way from the initial theory without the need to pull together in an *ad hoc* way certain features from classical and quantum mechanics. However, we have not really gained any new insights into the physics of the processes. Let's now look at some aspects in more detail.

Polarization Transfer

We have seen that the $180°$ ^{13}C pulse reordered the ^{13}C populations, but transfer of proton polarization to the ^{13}C system occurred only when the ^{1}H and ^{13}C $90°$ pulses were applied at time 4. We chose to use a matrix that represented the concurrent application of both pulses. However, these are independent, their Hamiltonians commute, and we could have considered instead the application of the two pulses separately—in either order. Here we examine the effect of applying the ^{1}H pulse.

Using the appropriate rotation operator for a ^{1}H 90_y pulse

$$R^I(90_y) = \frac{\sqrt{2}}{2} \begin{bmatrix} 1 & 0 & 1 & 0 \\ 0 & 1 & 0 & 1 \\ -1 & 0 & 1 & 0 \\ 0 & -1 & 0 & 1 \end{bmatrix} \tag{11.66}$$

we obtain

$$\rho'(5) = \begin{bmatrix} 3 & 0 & 0 & 0 \\ 0 & -3 & 0 & 0 \\ 0 & 0 & -5 & 0 \\ 0 & 0 & 0 & 5 \end{bmatrix} \tag{11.67}$$

We see that the 90_y proton pulse has led to *polarization transfer*, as indicated by the magnitude of the diagonal elements. The role of the ^{13}C $90°$ pulse (which can be applied equally well along x' or y') is then merely to generate the now enhanced ^{13}C coherences that can be detected or allowed to evolve for further manipulation in 1D or 2D experiments.

Multiple Quantum Coherence

Consider, now, the alternative—that at time 4 only a ^{13}C 90_y pulse is applied. Then at time 5

$$\rho''(5) = -\begin{bmatrix} 0 & 1 & 0 & 4 \\ 1 & 0 & 4 & 0 \\ 0 & 4 & 0 & 1 \\ 4 & 0 & 1 & 0 \end{bmatrix} \tag{11.68}$$

Here we have an interesting density matrix. The ^{13}C 90° pulse has generated coherences along the x' axis (ρ_{12} and ρ_{34}), but they have only the normal ^{13}C intensity of 1, with no polarization transfer. However, for the first time we see double quantum and zero quantum coherences (ρ_{14} and ρ_{23}, respectively), each with the proton intensity of 4. If we were now to let $\rho''(5)$ evolve without applying another pulse, we would observe the ordinary, unenhanced ^{13}C signals. The ^{1}H coherences would also evolve at their respective frequencies in the rotating frame (near $\nu_H \pm \nu_C$), but no magnetization is generated that can be observed. (We could confirm this by calculating the value of $\langle M_{xy} \rangle$.)

From Eqs. 6.29 and 6.34 we know that the frequencies of the single quantum transitions include both the chemical shift difference and the coupling constant, and we saw in Eq. 11.54 that the single quantum coherence terms evolve at those frequencies. From Eq. 6.29 we can see that the expression for the double quantum frequency $E_4 - E_1$ would not depend on J, and the difference $E_3 - E_2$ likewise does not depend on J for weakly coupled spins. Thus zero quantum and double quantum coherences evolve as though there were no spin coupling.

Except for being unobservable, these coherences (and antiphase coherences, as well) behave much like magnetizations in that they have relaxation times, which are different from those of single quantum coherences. Multiple quantum coherences can be further manipulated to produce observable magnetization, as we shall see in Chapter 12.

The density matrix approach provides much more information than appears in the simple vector diagrams used in previous chapters. Nevertheless, we can go back to the vector diagram in Fig. 9.8 and see that qualitatively the behavior of the magnetization vectors is in accord with the predictions from the density matrix ρ''. At time 5 the two proton magnetizations are oriented antiparallel, as in Fig. 9.10e. With no z component for ^{13}C magnetization, the two proton magnetizations labeled α and β would now process at the same frequency (as though the spin coupling were "suspended") and no net ^{1}H magnetization would be observed because they remain antiparallel.

11.6 PRODUCT OPERATORS

We have seen that the density matrix can be applied in principle to any spin system, but even for the two-spin system the algebra often becomes very tedious. There are computer programs that permit larger spin systems and more complex experiments to be treated by density matrix procedures. However, it is often difficult to derive an understanding of the processes that occur in intermediate steps. Fortunately, for weakly coupled spin systems an alternative formalism based on *product operators* has become available and is used to describe much of the literature of 2D NMR. We give here the rationale for the method and describe the basic elements.

Density Matrix and Spin Operators

We saw in previous sections that it is often possible to express the density matrix for a particular situation in a more concise operator form. For example, in Eq. 11.37 we saw that the density matrix at equilibrium is given by $(\epsilon/2)I_z$, and in Eq. 11.57 we showed that application of a nonselective 90_x pulse to a two-spin system generated a density matrix that could be expressed as $(\epsilon_I/2)I_y + (\epsilon_S/2)S_y$. To make the expressions even simpler, we can drop the normalizing factors and for heteronuclear systems incorporate the relative magnitudes given by the ϵ's into the spin operators themselves. For this example, we then obtain simply $I_y + S_y$.

Let's now look at the density matrix as it evolves for the pulse sequences that we outlined in Section 11.5. We have illustrated INEPT and related pulse sequences by the two-spin system in which $I = {}^1H$ and $S = {}^{13}C$. The equilibrium density matrix can be considered as the sum of two matrices:

$$\rho(0) = \begin{bmatrix} 5 & 0 & 0 & 0 \\ 0 & 3 & 0 & 0 \\ 0 & 0 & -3 & 0 \\ 0 & 0 & 0 & -5 \end{bmatrix} = \begin{bmatrix} 4 & 0 & 0 & 0 \\ 0 & 4 & 0 & 0 \\ 0 & 0 & -4 & 0 \\ 0 & 0 & 0 & -4 \end{bmatrix} + \begin{bmatrix} 1 & 0 & 0 & 0 \\ 0 & -1 & 0 & 0 \\ 0 & 0 & 1 & 0 \\ 0 & 0 & 0 & -1 \end{bmatrix}$$

$$(11.69)$$

But the first matrix of the sum is just I_z (multiplied by 4 to account for the γ_H/γ_C ratio) and the second is S_z. As indicated earlier, we have also dropped the $\frac{1}{2}$ normalizing factors. (The matrices for a two-spin system are given in Appendix D.) So we can write

$$\rho(0) = I_z + S_z \qquad (11.70)$$

Likewise, after the 90_x pulse applied to ${}^1H$, we have

$$\rho(1) = \begin{bmatrix} 1 & 0 & 4i & 0 \\ 0 & -1 & 0 & 4i \\ -4i & 0 & 1 & 0 \\ 0 & -4i & 0 & -1 \end{bmatrix} = \begin{bmatrix} 0 & 0 & 4i & 0 \\ 0 & 0 & 0 & 4i \\ -4i & 0 & 0 & 0 \\ 0 & -4i & 0 & 0 \end{bmatrix} + \begin{bmatrix} 1 & 0 & 0 & 0 \\ 0 & -1 & 0 & 0 \\ 0 & 0 & 1 & 0 \\ 0 & 0 & 0 & -1 \end{bmatrix}$$

$$(11.71)$$

Once more, it is easy to see that

$$\rho(1) = -I_y + S_z \tag{11.72}$$

Consider now the density matrix at the conclusion of the evolution periods and after the 180_x applied to both ^{1}H and ^{13}C. As shown in Eq. 11.63 (and in the following), the S populations are inverted and can obviously be represented simply by $-S_z$. However, the expressions for the I coherences have some, but not all, signs changed from $\rho(2)$; hence they cannot be represented by I_x or its negative. It turns out that the *product* $I_x S_z$ works:

$$\rho(4) = \begin{bmatrix} -1 & 0 & -4 & 0 \\ 0 & 1 & 0 & 4 \\ -4 & 0 & -1 & 0 \\ 0 & 4 & 0 & 1 \end{bmatrix}$$

$$= -\begin{bmatrix} 0 & 0 & 4 & 0 \\ 0 & 0 & 0 & 4 \\ 4 & 0 & 0 & 0 \\ 0 & 4 & 0 & 0 \end{bmatrix}\begin{bmatrix} 1 & 0 & 0 & 0 \\ 0 & -1 & 0 & 0 \\ 0 & 0 & 1 & 0 \\ 0 & 0 & 0 & -1 \end{bmatrix} - \begin{bmatrix} 1 & 0 & 0 & 0 \\ 0 & -1 & 0 & 0 \\ 0 & 0 & 1 & 0 \\ 0 & 0 & 0 & -1 \end{bmatrix} \tag{11.73}$$

Hence

$$\rho(4) = -I_x S_z - S_z \tag{11.74}$$

This is our first encounter with using this sort of product to represent a density matrix (or part of such a matrix). As we see in the following section, the product of a transverse component (x or y) of the magnetization of one nucleus with the longitudinal component (z) of the other represents *antiphase magnetization*, just as pictured in Figs. 9.10 and 9.11. The vector picture, the density matrix $\rho(4)$, and the product operator are different ways of expressing the same thing. Let's pursue this concept further by examining $\rho(5)$, $\rho'(5)$, and $\rho''(5)$.

We saw in Eq. 11.65 that $\rho(5)$ consists of two ^{13}C coherences of opposite sign, so we expect something similar to the product in Eq. 11.74. Actually, we find

$$\rho(5) = I_z S_x - S_x \tag{11.75}$$

The simple S_z and S_x terms in the last two equations and in Eqs. 11.76 and 11.78 arise from the S coherence (magnetization) that is generated by the 90° S pulse at time 4.

Figure 9.10f gives the vector picture for the situation described by $\rho'(5)$, which resulted from polarization transfer. The antiphase vectors suggest that a product of the sort used to describe antiphase magnetization might be applicable, and in fact, the product $I_z S_z$ is needed. Although pictorially this situation looks like antiphase magnetization, that term is usually reserved for xy components (coherences), and the situation described by $\rho'(5)$ is generally called *longitudinal two-spin order* or *J-order* to characterize the ordered, but non–Boltzmann, distribution in the direction of $\mathbf{B}_0$. (In solids this population distribution is called *dipolar order* because it arises from the interaction of dipolar-coupled spins.) From matrix multiplication, we find that

$$\rho'(5) = I_z S_z - S_z \tag{11.76}$$

Finally, we examine $\rho''(5)$, which we found contains double quantum and zero quantum coherences:

$$\rho''(5) = -\begin{bmatrix} 0 & 1 & 0 & 4 \\ 1 & 0 & 4 & 0 \\ 0 & 4 & 0 & 1 \\ 4 & 0 & 1 & 0 \end{bmatrix} = -\begin{bmatrix} 0 & 0 & 4 & 0 \\ 0 & 0 & 0 & 4 \\ 4 & 0 & 0 & 0 \\ 0 & 4 & 0 & 0 \end{bmatrix}\begin{bmatrix} 0 & 1 & 0 & 0 \\ 1 & 0 & 0 & 0 \\ 0 & 0 & 0 & 1 \\ 0 & 0 & 1 & 0 \end{bmatrix} - \begin{bmatrix} 0 & 1 & 0 & 0 \\ 1 & 0 & 0 & 0 \\ 0 & 0 & 0 & 1 \\ 0 & 0 & 1 & 0 \end{bmatrix}$$

$$\tag{11.77}$$

or

$$\rho''(5) = -I_x S_x - S_x \tag{11.78}$$

It is not immediately obvious why $I_x S_x$ should represent zero and double quantum coherence, but expression of this product in terms of raising and lowering operators makes the realtionship clear. From the definitions in Eq. 2.7, we obtain

$$I_x = \tfrac{1}{2}(I^+ + I^-) \qquad S_x = \tfrac{1}{2}(S^+ + S^-) \tag{11.79}$$

$$I_x S_x = \tfrac{1}{4}[I^+ S^+ + I^- S^- + I^+ S^- + I^- S^+] \tag{11.80}$$

Because the raising and lowering operators change spin quantum number by one, the first two terms on the right account for double quantum coherence, and the last two "flip-flop" terms give zero quantum coherence. Clearly, cross products of any two transverse spin components lead to the same result. Moreover, we can represent two orthogonal forms of pure zero quantum and double quantum coherences by combinations of product operators:

$$\text{ZQC:} \qquad I^+ S^- + I^- S^+ = 2[I_x S_x + I_y S_y] \tag{11.81a}$$

$$I^+ S^- - I^- S^+ = 2i[I_y S_x - I_x S_y] \tag{11.81b}$$

$$\text{DQC:} \qquad I^+ S^+ + I^- S^- = 2[I_x S_x - I_y S_y] \tag{11.82a}$$

$$I^+ S^+ - I^- S^- = 2i[I_x S_y + I_y S_x] \tag{11.82b}$$

Properties of Product Operators

We have seen that several density matrices can be expressed in terms of products and sums of spin operators. It can be shown that for a system of N spin $\tfrac{1}{2}$ nuclei *any* density matrix can be formualted as a linear combination of 4^N spin operators. For the two-spin system that we continue to emphasize, the following 16 operators form a complete set:

I_z, S_z	Longitudinal magnetization
$2I_zS_z$	Longitudinal two-spin order
I_x, I_y, S_x, S_y	Transverse magnetization (single quantum coherence)
$2I_xS_z, 2I_yS_z,$	Transverse I magnetization, antiphase in S
$2I_zS_x, 2I_zS_y$	Transverse S magnetization, antiphase in I
$2I_xS_x, 2I_yS_y, 2I_xS_y, 2I_yS_x$	Zero and double quantum coherence
$\frac{1}{2}E$	Identity operator (unit matrix), needed for completeness

For larger spin systems, a number of three-spin product operators may appear, for example:

$4I_xS_zT_z$	I_x magnetization in antiphase with respect to S and T
$4I_xS_xT_z$	Two-spin coherence in antiphase with respect to T, the "passive" spin
$4I_xS_xT_x$	Three-spin coherence
$4I_zS_zT_z$	Longitudinal three-spin order

The coefficients shown normalize the repetitive factor of $\frac{1}{2}$ in the matrix representation of each individual spin operator and the lack of a $\frac{1}{2}$ factor in the unit matrix. In accord with our practice in the last few pages, we drop the normalizing factors to simplify the terminology in further presentations. This simplification is in accord with the practice in some texts, but many authors retain the normalizing factors. Clearly, for quantitative applications the normalizing factors, as well as the implicit proportionality to γ in the operator, must be taken into account.

The principal benefit from decomposing density matrices into a relatively small number of product operators is that we can determine, once and for all, how each behaves when subject to pulses and to evolution as a result of chemical shifts and spin coupling and tabulate the results. We thus create a product operator algebra that permits us to follow the evolution of a density matrix without having to go through matrix multiplications at each step. Nevertheless, use of this algebra is not trivial. Particular attention needs to be given to signs, and alternative pathways for evolution often arise, which lead to lengthy expressions. Let's look at the principal features:

1. The product operator formalism is normally applied only to weakly coupled spin systems, where independent operators for I and S are meaningful. That means that it is permissible to treat evolution under chemical shifts separately from evolution under spin coupling. It also means that a nonselective pulse can be treated as successive selective pulses affecting only one type of spin. To simplify the notation and to facilitate the handling of the transformation of each product operator, such separations are almost always made.

2. The shorthand scheme for representing the evolution of the operators shows the operator at a particular step in the process, then indicates by an arrow

the type of transformation and gives the resultant operator. For example,

$$I_z \xrightarrow{\ 90_x(I)\ } -I_y$$

3. There are alternative notations for the transformation. Although the one used above to indicate a 90_x pulse applied to nucleus I is clear, it is more common to use an operator (technically, a *superoperator*, which operates on an operator) to designate the transformation. As we would expect, a (super) operator I affects only I components, not S. Also, the pulse flip angle is often specified only if it is other than 90°. For example,

$$I_z \xrightarrow{\ I_x\ } -I_y \qquad I_z \xrightarrow{\ I_x(\theta)\ } I_z \cos \theta - I_y \sin \theta \qquad I_z S_x \xrightarrow{\ I_y\ } I_x S_x$$

Table 11.2 summarizes the effect of various (super)operators, given at the top, on the individual spin operators at the left. The product operators for a two-spin system are included explicitly even though their transformations can readily be deduced from the effects of the superoperators on the individual components of the product operator.

4. Evolution of a coherence based on a chemical shift is usually indicated by a rotation (superoperator) at the precession frequency multiplied by the period during which the chemical shift acts, for example, $\Omega_I \tau$.

5. Evolution based on spin coupling is indicated by the appropriate operator and time. Because the product operator treatment applies solely to weakly coupled systems, the operator is $J_{IS} I_z S_z$, with only the z components considered. As we saw in Section 9.6, we must consider both in-phase and antiphase components of a magnetization (or coherence) as it precesses in the xy plane. In line with Eq. 9.4, we can now use the product operator terminology to note that, for example,

$$I_y \xrightarrow{\ J_{IS} I_z S_z \tau\ } I_y \cos \pi J_{IS} \tau - I_x S_z \sin \pi J_{IS} \tau$$

6. For systems of three or more spins, the treatment in point 5 is applied sequentially for each pair of coupled spins. In each evolution the two spins involved in the coupling are called "active," while the remaining spins, which unaffected in that evolution step are "passive."

7. Evolution of multiple quantum coherence (of order p) is treated as successive evolution of each of the p chemical shifts. As expected, double quantum coherence precesses at $\omega_I + \omega_S$, and zero quantum coherence precesses at $\omega_I - \omega_S$. As we noted in Section 11.5, such coherences are not affected by spin coupling.

8. Because two terms are needed to represent precession during each evolution step and each application of a pulse that is not 90° or 180°, even simple 2D experiments must often be described by many product operator terms. It is often helpful to organize these terms by constructing "trees" that branch at

TABLE 11.2. Conversion Chart for Product Operators[a]

	$90°I_x$	$90°I_y$	$90°S_x$	$90°S_y$	$\Omega t(I_z)$	$\Omega t(S_z)$	JI_zS_zt
I_x	I_x	$-I_z$	I_x	I_x	$I_x \cos \Omega t + I_y \sin \Omega t$	I_x	$I_x \cos \pi Jt + I_yS_z \sin \pi Jt$
I_y	I_z	I_y	I_y	I_y	$I_y \cos \Omega t - I_x \sin \Omega t$	I_y	$I_y \cos \pi Jt - I_xS_z \sin \pi Jt$
I_z	$-I_y$	I_x	I_z	I_z	I_z	I_z	I_z
S_x	S_x	S_x	S_x	$-S_z$	S_x	$S_x \cos \Omega t + S_y \sin \Omega t$	$S_x \cos \pi Jt + I_zS_y \sin \pi Jt$
S_y	S_y	S_y	S_z	S_y	S_y	$S_y \cos \Omega t - S_x \sin \Omega t$	$S_y \cos \pi Jt - I_zS_x \sin \pi Jt$
S_z	S_z	S_z	$-S_y$	S_x	S_z	S_z	S_z
I_zS_z	$-I_yS_z$	I_xS_z	$-I_zS_y$	I_zS_x	I_zS_z	I_zS_z	I_zS_z
I_xS_z	I_xS_z	$-I_zS_z$	$-I_xS_y$	I_xS_x	$I_xS_z \cos \Omega t + I_yS_z \sin \Omega t$	I_xS_z	$I_xS_z \cos \pi Jt + I_y \sin \pi Jt$
I_yS_z	I_zS_z	I_yS_z	$-I_yS_y$	I_yS_x	$I_yS_z \cos \Omega t - I_xS_z \sin \Omega t$	I_yS_z	$I_yS_z \cos \pi Jt - I_x \sin \pi Jt$
I_zS_x	$-I_yS_x$	I_xS_x	I_zS_x	$-I_zS_z$	I_zS_x	$I_zS_x \cos \Omega t + I_zS_y \sin \Omega t$	$I_zS_x \cos \pi Jt + S_y \sin \pi Jt$
I_zS_y	$-I_yS_y$	I_xS_y	I_zS_z	I_zS_y	I_zS_y	$I_zS_y \cos \Omega t - I_zS_x \sin \Omega t$	$I_zS_y \cos \pi Jt - S_x \sin \pi Jt$
I_xS_x	I_xS_x	$-I_zS_x$	I_xS_x	$-I_xS_z$	$I_xS_x \cos \Omega t + I_yS_x \sin \Omega t$	$I_xS_x \cos \Omega t + I_xS_y \sin \Omega t$	I_xS_x
I_xS_y	I_xS_y	$-I_zS_y$	I_xS_z	I_xS_y	$I_xS_y \cos \Omega t + I_yS_y \sin \Omega t$	$I_xS_y \cos \Omega t - I_xS_x \sin \Omega t$	I_xS_y
I_yS_x	I_zS_x	I_yS_x	I_yS_x	$-I_yS_z$	$I_yS_x \cos \Omega t - I_xS_x \sin \Omega t$	$I_yS_x \cos \Omega t + I_yS_y \sin \Omega t$	I_yS_x
I_yS_y	I_zS_y	I_yS_y	I_yS_z	I_yS_y	$I_yS_y \cos \Omega t - I_xS_y \sin \Omega t$	$I_yS_y \cos \Omega t - I_yS_x \sin \Omega t$	I_yS_y

[a] The (super) operator at the top converts each spin operator, at the left, as indicated.

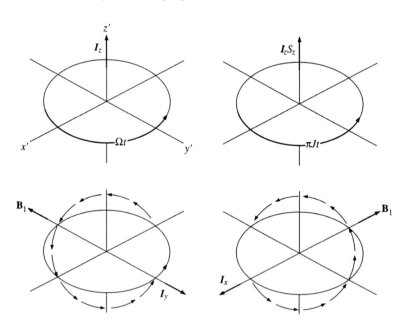

each such step. Many branches ultimately lead to nonobservable terms at the end, and with experience it is possible to truncate these early in the sequential process. Great care must be used in doing so, however, as some terms that can be neglected under ideal conditions cause spurious signals in practical situations.

An Example: INEPT

The entire pulse sequence for INEPT can be summarized in product operator notation as

$$[I_z + S_z] \xrightarrow{90°I_x} \xrightarrow{\Omega_I \frac{1}{4J}} \xrightarrow{\pi J I_z S_z \frac{1}{4J}} \xrightarrow{180°I_x} \xrightarrow{180°S_x} \xrightarrow{\Omega_I \frac{1}{4J}} \xrightarrow{\pi J I_z S_z \frac{1}{4J}} \xrightarrow{90°I_y} \xrightarrow{90°S_y} S(t)$$

$$(11.83)$$

where $S(t)$ indicates acquisition of the S signal. There are four evolution periods, which means that we apparently need to consider a total of 16 different pathways. However, the analysis can be simplified. The 180° I_x (proton) pulse is an essential component in the INEPT experiment because it causes a refocusing of proton chemical shifts, hence makes the sequence equally applicable across the proton spectrum. However, in this analysis it is not necessary to treat this process, as in the end we do not measure any proton signal. Hence the two self-canceling chemical shift evolution steps can be deleted. We can then combine the two spin coupling evolution steps by placing the 180° pulses at the beginning or end of the evolution period. The 180° S_x (^{13}C) pulse is essential because it inverts the S populations, while the 180° I_x pulse reverses the direction of the I magnetization. Thus, we can simplify the sequence and examine the result of each step, as follows:

$$I_z \xrightarrow{90°I_x} -I_y \xrightarrow{180°I_x} I_y \xrightarrow{\pi J I_z S_z \frac{1}{2J}} I_y \cos\frac{\pi}{2} - I_x S_z \sin\frac{\pi}{2} \xrightarrow{90°I_y} I_z S_z \xrightarrow{90°S_y} I_z S_x$$

$$(11.84)$$

$$S_z \xrightarrow{180°S_x} -S_z \xrightarrow{\pi J I_z S_z \frac{1}{2J}} -S_z \xrightarrow{90°S_y} -S_x$$

The initial S polarization is converted by the 90° S_y pulse to magnetization with its normal intensity, but the initial I magnetization (with a relative intensity of 4) has now been converted into antiphase S magnetization. The $I_z S_x$ term indicates that there is S magnetization along the x' axis, that it is antiphase with respect to I, and that its intensity is proportional to the product IS, that is, four times the intensity of normal S magnetization in this example. Both $I_z S_x$ and $-S_x$ evolve during the detection period under the action of the S chemical shift and the spin coupling:

$$I_z S_x \xrightarrow{\pi J I_z S_z t} I_z S_x \cos \pi J t + I_z I_z S_y \sin \pi J t \xrightarrow{\Omega_s t} I_z S_x \cos \pi J t \cos \Omega t$$
$$+ I_z S_y \cos \pi J t \sin \Omega t + S_y \sin \pi J t \cos \Omega t - S_x \sin \pi J t \sin \Omega t$$
$$- S_x \xrightarrow{\pi J I_z S_z t} - S_x \cos \pi J t - I_z S_y \sin \pi J t \xrightarrow{\Omega_s t} - S_x \cos \pi J t \cos \Omega t$$
$$- S_y \cos \pi J t \sin \Omega t - I_z S_y \sin \pi J t \cos \Omega t + I_z S_x \sin \pi J t \sin \Omega t$$

$$(11.85)$$

(We use the fact that $I_z I_z = 1$, ignoring the normalizing factor.) Of the final eight terms, four represent antiphase magnetization, which is not observed, and can thus be deleted. After applying well-known trigonometric identities, we obtain

From I: $S_y[\sin(\Omega + \pi J)t - \sin(\Omega - \pi J)t] - S_x[\cos(\Omega - \pi J)t$
$- \cos(\Omega + \pi J)t]$

From S: $- S_x[\cos(\Omega + \pi J)t + \cos(\Omega - \pi J)t] - S_y[\sin(\Omega + \pi J)t$
$+ \sin(\Omega - \pi J)t]$

$$(11.86)$$

The expressions for S_x and S_y are redundant. They are both detected in quadrature by the normal phase-sensitive detection scheme. Note that from I, where we gain the intensity factor of $\gamma_I / \gamma_S = 4$ for ^{1}H/^{13}C, the α magnetization precessing at $\Omega - \pi J$ has intensity of 4, while from S this gives -1, for an overall intensity of 3. For the β magnetization both components are negative to give a total of -5.

We could have avoided the computation during the detection period by noting that after the final pulses the antiphase magnetization from I resulted from the oppositely directed α and β magnetizations, each of magnitude 4. During the detection period they precess at $\Omega - \pi J$ and $\Omega + \pi J$ respectively, but always with opposite sign because they begin the precession 180° out of phase. Both α and β components of magnetization from S precess with the same phase as the β component from I, hence subtract and add, respectively.

An Example: Solid Echo

As a second example, we look at echoes. We saw in Chapter 9 that a 180° pulse refocuses not only chemical shifts and the effects of magnetic field inhomogeneity but also spin coupling provided that the pulse does not also disturb the spin state of the coupled nucleus (see Fig. 9.2) However, in a homonuclear spin system a nonselective pulse does effect spin states. We found in Chapter 7 that dipolar interactions have the same mathematical from as indirect spin coupling, and it is known that a 180° pulse does not produce an echo in a solid because spin states are disturbed. However, it is possible to obtain a *solid echo* or *dipolar echo* by applying the pulse sequence 90_x, τ, 90_y. It is very difficult to rationalize an echo from this sequence on the basis of vector diagrams, but application of product operators gives a clear demonstration. (We ignore any effect of chemical shifts, which are small compared with the dipolar interaction D. Also, in describing the dipolar interaction we ignore the "flip-flop" term B in Eq. 7.5, so this treatment applies rigorously only to a heteronuclear spin pair. However, the result is also applicable to a homonuclear spin pair.) The pathway proceeds as follows:

$$[I_z + S_z] \xrightarrow{\;90°I_x\;} \xrightarrow{\;90°S_x\;} [I_y + S_y] \xrightarrow{\;\pi D I_z S_z \tau\;}$$

$$(I_y + S_y) \cos \pi D\tau - (I_x S_z + I_z S_x) \sin \pi D\tau \xrightarrow[\;\pi D I_z S_z \tau\;]{\;90°I_y\;\;90°S_y\;}$$

$$(I_y + S_y) \cos \pi D\tau + (I_z S_x + I_x S_z) \sin \pi D\tau \longrightarrow$$

$$(I_y + S_y) \cos \pi D\tau \cos \pi D\tau - (I_x S_z + I_z S_x) \cos \pi D\tau \sin \pi D\tau$$

$$+ (I_z S_x + I_x S_z) \sin \pi D\tau \cos \pi D\tau$$

$$+ (I_z S_x + I_x S_z) \sin \pi D\tau \cos \pi D\tau$$

$$+ (I_z I_z S_y + I_y S_z S_z) \sin \pi D\tau \sin \pi D\tau$$

$$= I_y + S_y \qquad\qquad (11.87)$$

With cancellation of terms and the identities $I_z I_z = 1$ and $\cos^2 \theta + \sin^2 \theta = 1$, the long expression reduces to $[I_y + S_y]$, thus showing an echo at 2τ. (In practice the magnitude of the signal at the echo is somewhat smaller than the initial signal for reasons that we shall not take up.)

These two detailed examples should illustrate the application of the product operator formalism. It is clear that experiments with several pulses and especially with several evolution periods can provide lengthy expressions. In Chapter 12 we indicate the way in which this formalism can be used to explain several 2D NMR experiments but without providing detailed computations.

11.7   COHERENCE TRANSFER PATHWAYS

For INEPT and the dipolar echo, we have just seen examples of coherence transfer pathways, which show how the magnetization goes from its equilibrium orientation along the z axis to precession in the xy plane, where it provides an NMR signal. Because of the doubling of terms during each evolution, as we have seen, there may be a number of possible pathways resulting from a complex sequence of pulses. In a well-designed experiment, most of these do not lead to observable signals. However, we have considered only situations with ideal pulses and straightforward precessions, whereas in practice inhomogeneities in both $\mathbf{B}_0$ and $\mathbf{B}_1$ lead to additional pathways. For example, imperfection in the initial 90° pulse leaves a component of magnetization along z, which we saw in Chapter 10 leads to axial peaks in a simple 2D experiment and often provides additional coherences that can interfere in a more complex experiment. An important aspect in the design of 2D experiments is the provision for eliminating unwanted coherence pathways.

Phase Cycling

We have seen several examples of the cycling of phases of rf pulses and/or of the receiver, with subsequent addition or subtraction of signals, in order to eliminate certain unwanted signals. In Section 3.5 we discussed in detail the commonly used CYCLOPS procedure for avoiding spurious signals as a result of imperfections in phase-sensitive detectors, and we pointed out that this technique serves as a prototype for other applications of phase cycling. In Section 9.5 we noted the use of phase cycling to discriminate against certain signals with X-filters. In 2D and multidimensional NMR experiments phase cycling serves as one of two principal methods for eliminating unwanted signals.

For example, consider NOESY, which was discussed in Chapter 10 and appeared to be quite straightforward. In our current terminology, the following coherence transfer pathway leads to a cross peak:

$$I_z \xrightarrow{90°I_x} -I_y \xrightarrow{\Omega t_1} -I_y \cos \Omega t_1 + I_x \sin \Omega t_1 \xrightarrow{90°I_x} -I_z \cos \Omega t_1 + I_x \sin \Omega t_1$$

$$\xrightarrow{NOE(\tau)} -\lambda S_z + I_x \sin \Omega t_1 \cos \Omega \tau + I_y \sin \Omega t_1 \sin \Omega \tau$$

$$\xrightarrow{90°I_x} \lambda S_y + I_x \sin \Omega t_1 \cos \Omega \tau + I_z \sin \Omega t_1 \sin \Omega \tau \tag{11.88}$$

Here λS_z represents the S magnetization developed by cross relaxation from $I_z \cos \Omega t_1$, and this transforms to λS_y, which precesses during t_2 and is detected. The I_z component shown in the final expression is longitudinal and gives no signal, but the I_x term leads to an unwanted signal. This can be removed by cycling the phase of the second 90° pulse between x and $-x$ and subtracting the resulting signals. The $I_x \sin \Omega t_1$ term is unchanged and is eventually lost in the subtraction, whereas $-I_z \cos \Omega t_1$ changes sign and survives the subtraction.

A second artifact in NOESY (as in other 2D experiments) is the formation of axial peaks (see Section 10.3) from magnetization that has relaxed back to the z axis during the evolution time and is subsequently rotated to the xy plane. We can eliminate axial peaks by cycling the phase of the first 90° pulse between x and $-x$ and again subtracting the alternate signals. The 90_{-x} pulse changes the sign of all subsequent terms in the coherence pathway in Eq. 11.88, but relaxation always give positive I_z, so the sign of axial peaks is unchanged, and subtraction eliminates the resulting axial peak signals.

To suppress both types of artifacts, one phase cycle must be nested within the other to create a four-pulse cycle. In addition, many spectrometers also require the CYCLOPS phase cycle to eliminate artifacts from quadrature detection. In those cases, 16 steps are required overall to complete the cycle. Several 2D experiments require even more phase cycling, thus significantly lengthening the time needed to complete a study and in some instances making it infeasible to use a potentially valuable pulse sequence. Hence, other approaches are used where possible to reduce the number of phase steps, as we describe in the following.

Coherence Order Diagrams

The product operator formalism is normally based on Cartesian coordinates because that simplifies most of the calculations. However, these operators obscure the coherence order p. For example, we found (Eq. 11.80) that products such as $I_x S_x$ represent both zero and double quantum coherence. The raising and lowering operators I^+ and I^- are more descriptive in that (as we saw in Eq. 2.8) these operators connect states differing by ± 1 in quantum number, or coherence order. We can associate I^+ and I^- with $p = +1$ and -1, respectively, and (as indicated in Eq. 11.79) the coherences that we normally deal with, I_x and I_y, each include both $p = \pm 1$. Coherences differing in sign contain partially redundant information, but both are needed to obtain properly phased 2D spectra, in much the same way that both real and imaginary parts of a Fourier transform are needed for phasing.

It is often helpful to summarize in a diagram the changes in coherence level as a function of the discrete periods within the experiment, as illustrated in Fig. 11.2. We begin at equilibrium, where the density matrix (I_z) has only diagonal terms

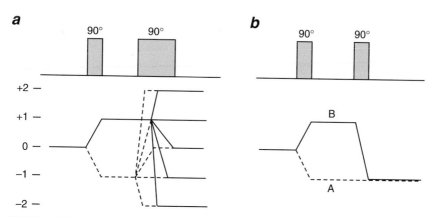

FIGURE 11.2 Coherence pathways for a COSY experiment illustrated in a coherence order diagram. (*a*) All possible pathways resulting from the two 90° pulses. The width of the second pulse has been exaggerated to clarify the diagram. (*b*) The two pathways (A and B) selected from (*a*) that lead to observable magnetization ($p = -1$). Adapted from Günther.[64]

with $p = 0$. As we have seen, pulses can change coherence order; in fact, a pulse of arbitrary angle mixes I_z, I^+, and I^-. An initial 90° pulse leads to coherences of ± 1, and a subsequent 90° pulse operating on I^+ or I^- can generate coherences of various orders, as shown in Fig. 11.2a. A 180° pulse interchanges I^+ with I^-. The coherence order does not change during free precession. For observation, we need magnetization that precesses in the xy plane (i.e., single quantum coherence, $p = \pm 1$). It has become conventional to consider $p = -1$ as the coherence level that is observed with the usual quadrature phase-sensitive detection. Thus, in Fig. 11.2a, most of the pathways shown are irrelevant in terms of observable signal. Usually such diagrams are restricted to paths leading to observable signals, as shown in Fig. 11.2b.

We have seen illustrations of simple phase cycling arrangements designed to minimize artifacts. Coherence level diagrams are often helpful in designing and in understanding such cycles when different coherence orders are present as the density operator evolves. We saw in Eq. 11.19 that coherences evolve at p times the nominal Larmor frequency. The same relation applies to phase shifts as a result of shifting the phase of an excitation pulse. If a pulse phase is changed by $\Delta\psi$, a coherence whose order is changed by that pulse by Δp experiences a phase change

$$\Delta\phi = \Delta p \times \Delta\psi \tag{11.89}$$

Hence, if there are two or more pathways during a particular period that differ in p, it is possible to cycle a pulse and detector phase so as to detect the desired coherence and discriminate against the unwanted coherence. For example, Fig. 11.2b shows the two pathways (A and B) in Fig. 11.2a that lead to observable magnetization at $p = -1$ in a simple COSY experiment. We can discriminate between these by cycling the phase of the first pulse and the receiver in concert, while leaving the phase of the second pulse unchanged, because the first pulse causes a change $\Delta p = +1$ for path A and -1 for path B. With a four-pulse cycle that increments the rf phase by 90° in each step, we obtain the following results for pathways A (the dashed line) and B (the solid line):

	A				B			
Repetition	*1*	*2*	*3*	*4*	*1*	*2*	*3*	*4*
Pulse phase	0°	90°	180°	270°	0°	90°	180°	270°
Signal phase	0°	90°	180°	270°	0°	−90°	−180°	−270°
						=270°	=180°	=90°
Receiver phase	0°	90°	180°	270°	0°	90°	180°	270°
Signal phase	+	+	+	+	+	−	+	−

Summation of the results of the four repetitions gives a signal for path A but none for path B. This approach can be used for more complex pulse sequences to devise effective phase cycles.

Use of Pulsed Field Gradients

An alternative to phase cycling that can be used in many instances to discriminate against particular coherence pathways is the application of pulsed magnetic field gradients. One simple example occurs in NOESY and in other experiments in which only longitudinal magnetization is desired at some step of the pathway. In NOESY a field gradient during the mixing period τ rapidly removes the transverse magnetization (the I_x term in the coherence pathway given in Eq. 11.88) and prevents unwanted signal from this source.

On the other hand, the reversibility of dephasing in the presence of a field gradient can be turned to advantage when coherences of different orders are present during some step of an experiment. When a gradient is applied, the magnetic field at a particular point in the sample is augmented by a value ΔB and the precession frequency is increased by $p\Delta\Omega$ for a coherence of order p in a molecule at that point in the sample. For example, consider an experiment in which coherences with $p = 1$ and 2 precess during period A, with both converted to single quantum coherence during period B. If a field gradient is applied during A for a period τ, the double quantum coherence dephases twice as fast as the single quantum coherence and thus accumulates twice the phase at τ. If a reversed field gradient of the same magnitude is now applied during B for a period 2τ, the DQC will rephase at 2τ, but the SQC will not.

Many variants of this scheme are possible and can most readily be devised by considering coherence order diagrams. The requirement for rephasing of a coherence is that

$$\gamma_1 p_1 G_1 \tau_1 = -\gamma_2 p_2 G_2 \tau_2 \tag{11.90}$$

where the magnetogyric ratio γ is needed to account for coherence transfer between different nuclides. The strength and direction of the field gradients G may be varied with τ held constant, or the value of τ can be changed while the magnitude of G is held constant. The signs of γ, p, and G must all be taken into account. When applicable, the use of field gradients is usually the preferred method for discriminating between coherence orders. The alternative of (often extensive) phase cycling lengthens the experiment time and allows spectrometer instabilities to become more intrusive.

Gradients can be applied along one of the Cartesian axes or, with more sophisticated apparatus, at the magic angle. One consequence of a gradient applied only along x, y, or z is that an anisotropic environment is created, in which long-range intermolecular dipolar couplings may not completely average to zero. Although such interactions fall off as $1/R^3$, the number of molecules at distance R increases as R^2. Strong signals have been observed that arise from *intermolecular* double quantum coherences that are created and rendered observable by a suitable pair of gradient pulses. Although there are potential applications for this phenomenon, the appearance of such peaks in normal high resolution NMR represents an artifact, which can be avoided by use of magic angle gradients.[48]

11.8 ADDITIONAL READING AND RESOURCES

A large number of NMR books provide various levels of treatment of the topics covered in this chapter. *Density Matrix Theory and Its Applications in NMR Spectroscopy* by T. C. Farrar and J. E. Harriman[113] gives a clear exposition of this subject. *2D NMR Density Matrix and Product Operator Treatment* by G. D. Mateescu and A. Valeriu[114] provides an introductory perspective with several examples worked out in detail. *Spin Choreography* by Ray Freeman[106] gives a very good treatment of the product operator formalism, along with discussions of many other aspects of 1D and 2D pulsed NMR methods. *Coherence and NMR* by M. Munowitz[115] and articles by Malcolm Levitt and Thomas Mareci in *Pulse Methods in 1D and 2D Liquid-Phase NMR* edited by Wallace Brey[116] provide very useful insights into the nature of coherence and the formalisms covered in this chapter.

11.9 PROBLEMS

11.1 Fill in the details of the derivation of Eq. 11.12 from 11.11.

11.1 Verify Eq. 11.16*a* by differentiating to obtain Eq. 11.15.

11.3 Verify the derivation of Eqs. 11.29 and 11.30 by repeating the algebra in each step.

11.4 Carry out the matrix multiplications to convert $\rho(0)$ in Eq. 11.50 to $\rho(\tau)$ in Eq. 11.53.

11.5 Derive the matrix for a nonselective 90° pulse in Eq. 11.55, and apply it to derive $\rho(\tau)$ in Eq. 11.56.

11.6 Show that $\rho(4)$ in Eq. 11.62 reduces to $\rho(4)$ as expressed in Eq. 11.63, and show how $\rho(4)$ transforms to $\rho(5)$ in Eq. 11.65.

11.7 Use the density matrix treatment to verify the results shown in Fig. 9.2 for application of a 180° pulse to (a) I only, (b) S only, or (c) both I and S. Begin with the equilibrium density matrix, Eq. 11.50, apply a nonselective 90° pulse to obtain ρ as given in Eq. 11.56, consider the evolution through τ, and then treat *a*, *b*, and *c* separately.

11.8 Summarize the results of Problem 11.7 in terms of product operators.

11.9 Verify that the multiple quantum coherence expressed in the density matrix of Eq. 11.68 does not give rise to observable ^1H magnetization.

CHAPTER 12

Selected 1D, 2D, and 3D Experiments: A Further Look

Now that we have available the more powerful theoretical approaches of the density matrix and product operator formalisms to augment the still very useful vector picture, we can examine the mechanisms of some common NMR methods in more detail. In this chapter we discuss some one-dimensional techniques but concentrate on 2D experiments. We see that some of the 2D experiments can be extended to three or four dimensions to provide additional correlations and to spread out the crowded 2D peaks that sometimes arise from large molecules.

12.1 SPECTRAL EDITING

The ^{13}C spectrum of an organic molecule arises from ^{13}C nuclei coupled through one bond to zero, one, two, or three protons. It can be very helpful to ascertain for each line in a proton-decoupled ^{13}C spectrum whether it arises from a CH_3, CH_2, CH, or quaternary carbon. In our discussion of INEPT in Section 9.7, we noted that each of the first three moieties has different precession behavior and refocuses at a specific time, hence can serve as a means of distinguishing among these groups. We describe here three methods that are particularly useful for such editing of ^{13}C spectra (or for other nuclei, such as ^{15}N or ^{29}Si), all of which are based on this concept.

Each method employs a spin echo to refocus chemical shifts, so we can direct our attention to the precession of the coherences for the separate spin components in a frame rotating at the chemical shift frequency. The magnetizations for ^{13}C in CH precess at $\pm J/2$; in CH$_2$ the central component remains fixed in the rotating frame, while the other components precess at $\pm J$; and CH$_3$ has two components precessing at $\pm J/2$ and two at $\pm 3J/2$.

APT

Attached proton test (APT) can be carried out in several ways, one of which is depicted in Fig. 12.1. It can be readily understood with only the vector picture. The 180° pulse refocuses chemical shifts, but spin coupling evolution occurs only during the period of $1/J$ after the first 180° pulse, when the decoupler is turned off. At the end of that period the total signals S are given by

CH: $\quad S = \frac{1}{2}[\cos 2\pi(J/2)(1/J) + \cos 2\pi(-J/2)(1/J)] = \cos \pi = -1$

CH$_2$: $\quad S = \frac{1}{4}[2 + \cos 2\pi(J)(1/J) + \cos 2\pi(-J)(1/J)]$
$\qquad\quad = \frac{1}{4}[2 + \cos 2\pi] = 1$

$$\text{(12.1)}$$

CH$_3$: $\quad S = \frac{1}{8}[3 \cos 2\pi(J/2)(1/J) + 3 \cos 2\pi(-J/2)(1/J)$
$\qquad\qquad + \cos 2\pi(3J/2)(1/J) + \cos 2\pi(-3J/2)(1/J)] = -1$

C: $\quad S = 1$

The initial pulse can be selected to conform to the Ernst angle to improve S/N, rather than being restricted to 90°. In that case a second 180° pulse is needed to

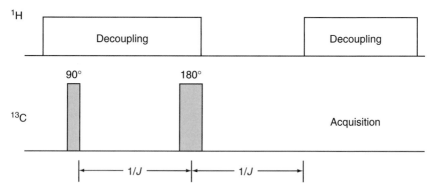

FIGURE 12.1 Pulse sequence for attached proton test (APT). The initial pulse width can be set to conform to the Ernst angle in order to optimize sensitivity.

eliminate interference from magnetization left along z by the initial pulse, because the combination of two 180° pulses leaves this magnetization unchanged.

APT is simple to use and gives signals for quaternary carbons, as well as hydrogen-bound carbons (in contrast to some other methods discussed subsequently). Discrimination between CH and CH_2 is clear, but the editing is only partial, as indicated, and does not distinguish CH from CH_3 or CH_2 from quaternary C. The signals of the hydrogen-bound carbons are enhanced by the NOE, which improves sensitivity, whereas those for quaternary carbons are usually weaker because of generally lower NOE and longer T_1.

INEPT

As we saw in Section 9.7, INEPT is based on excitation of ^1H and polarization transfer to ^{13}C via one-bond $H-^{13}C$ coupling, so quaternary carbons do not appear in INEPT spectra, but the other signals are considerably enhanced. In the simple version of INEPT without decoupling, we saw that the two signals for a CH group have relative intensities of $1: -1$ after subtracting the ordinary ^{13}C signal. An analysis similar to that for APT shows that CH_2 gives a triplet of intensities $1:0:-1$, and CH_3 shows a quartet with two positive and two negative lines. The relative intensities depend on several factors and are often nearly equal, as illustrated in Fig. 12.2. A coupled INEPT spectrum can thus help identify CH_n groups but suffers from the disadvantage of spreading the signal from each carbon among several components that often overlap with those from other carbon nuclei. As we saw in Section 9.7, it is possible to obtain a decoupled INEPT spectrum by allowing the components to refocus during an additional precession period 2Δ, where $\Delta = 1/4J$ for $^{13}C-H$. For CH_2 and CH_3, optimum refocusing occurs at other values of Δ, so repetition of the experiment with different values of Δ permits spectral editing. However, there are distortions in relative intensities resulting from the ratios given previously and from variations in the values of J (for example, in sp^2 versus sp^3 hybridized carbon atoms). We turn now to another method that avoids these problems.

DEPT

An alternative pulse sequence that provides the same multiplicities as INEPT but with intensity ratios that follow the binomial theorem is DEPT (*distortionless enhancement by polarization transfer*). The pulse sequence, depicted in Fig. 12.3a, can be used, like refocused INEPT, for sensitivity enhancement but is usually employed as an editing technique. The three evolution periods τ are chosen to approximate $1/2J$, but the length of the pulse labeled "θ" can be varied. As we show in the following, CH has maximum intensity at $\theta = 90°$; CH_2 has zero

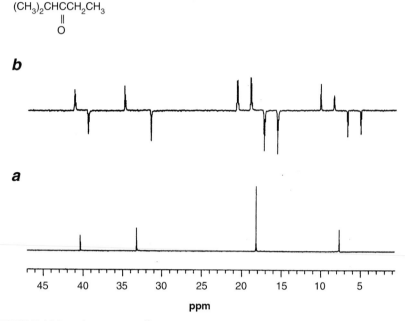

FIGURE 12.2 (*a*) ¹H-decoupled ¹³C spectrum of 2-methyl-3-pentanone in CDCl₃. The carbonyl resonance is not shown. (*b*) ¹H coupled ¹³C INEPT spectrum of the same sample. Note the CH doublet centered about 40.5 ppm, the CH₂ "triplet" with the center line missing at 33 ppm, and the two CH₃ quartets centered near 18 and 8 ppm. Courtesy of Herman J. C. Yeh (National Institutes of Health).

intensity at 90° but maximum positive and negative intensity at 45° and 135°, respectively; CH₃ also has zero intensity at 90° and maximum positive intensity near both 45° and 135°. Manipulation of the data obtained at each of these three values of θ permits clean separation of a normal decoupled ¹³C spectrum into three separate decoupled subspectra for CH₃, CH₂, and CH carbons, as illustrated in Fig. 12.3*b*. DEPT is less sensitive to the setting of precession time τ than is INEPT, so that differences in values of J can be better tolerated, and the lack of intensity distortion means that meaningful intensities are obtained for the subspectra. As in INEPT, quaternary carbons do not appear in ordinary DEPT spectra, but a modification described later, DEPTQ, also displays signals from quaternary carbons.

Unlike APT and INEPT, DEPT cannot be explained with classical vector diagrams, but the mechanism for DEPT can be summarized in a coherence pathway with product operators. As in INEPT, the 180° I pulse at time 2 (indicated in Fig. 12.3*a*) refocuses proton chemical shifts at time 4, so we need consider only evolution from spin coupling. Also, the 180° S pulse refocuses ¹³C chemical shifts

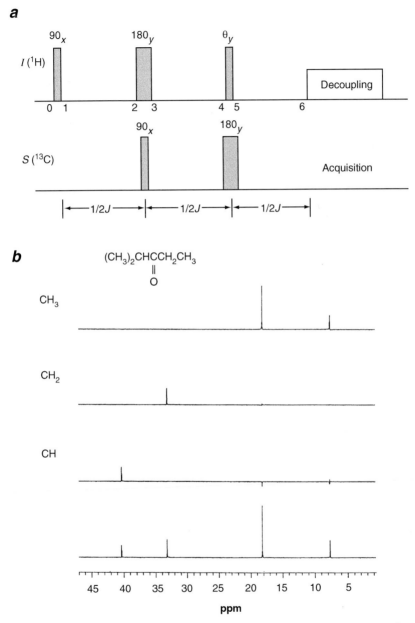

FIGURE 12.3 (*a*) Pulse sequence for DEPT. Times indicated (0–6) are discussed in the text. (*b*) Example of DEPT applied to 2-methyl-3-pentanone, showing edited spectra from CH$_3$, CH$_2$, and CH carbons, as well as the decoupled ^{13}C spectrum of the aliphatic carbons. Spectra courtesy of Herman J. C. Yeh (National Institutes of Health).

at time 6. Initially we take up an ^{1}H–^{13}C system (I and S, respectively) and obtain the following pathway:

$$I_z \xrightarrow{\;90°I_x\;} -I_y \xrightarrow{\;\pi JI_zS_z\frac{1}{2J}\;} I_xS_z \xrightarrow{\;90°S_x\;} -I_xS_y \xrightarrow{\;\pi JI_zS_z\frac{1}{2J}\;} -I_xS_y \xrightarrow{\;90°I_y\;} I_zS_y \xrightarrow{\;\pi JI_zS_z\frac{1}{2J}\;} -S_x \qquad (12.2)$$

Evolution of I_y under the influence of spin coupling for a period $1/2J$ generates pure antiphase I magnetization, just as in INEPT, and the 90_x S pulse converts it to double quantum and zero quantum coherence. As we saw in Chapter 11, the coupling has no effect on the multiple quantum coherence during the second evolution period, and the 90_y I pulse converts it to S magnetization that is antiphase in I. Further evolution for a period $1/2J$ converts this entirely to observable S coherence, but with four times the intensity because polarization is transferred from I. During the acquisition period broadband decoupling eliminates splitting from ^{13}C–H couplings.

If the I pulse at time 4 is θ, rather than 90°, the only change in the preceding pathway is to convert the final S_x to $S_x \sin\theta$. Some multiple quantum coherence remains but is not observed. For ^{13}CH$_2$ and ^{13}CH$_3$ systems, the pathway is much the same. The ^{13}C can be considered as being coupled to one of the magnetically equivalent protons as described, and the other protons behave as "passive" spins, each of which can be shown to contribute an additional modulation of $\cos\theta$. Thus, the general expression becomes

$$\text{Intensity } (I_nS) = n\sin\theta\cos^{n-1}\theta \qquad (12.3)$$

This expression gives maximum intensities at the angles listed before.

We also need to mention a second pathway, because the 90° S pulse converts equilibrium S_z to S_y coherence, which (as in INEPT) would add to that produced before. Normally DEPT is carried out by phase cycling the S pulse (90_x, 90_{-x}) to cancel magnetization produced in this pathway. However, a modification called DEPTQ has been introduced[117] that utilizes this pathway to generate a signal also from quaternary (Q) carbon nuclei by adding a 90° S pulse at time $\tau = 1/2J$ *before* the initial I pulse. This initial pulse excites all ^{13}C nuclei directly, rather than via coherence transfer from ^{1}H. Several additional changes in the sequence and phase cycling (which we do not describe here) are required to combine the directly excited S coherence with that transferred from I.

12.2 DOUBLE QUANTUM FILTERING EXPERIMENTS

The ability to create and manipulate double quantum coherence is an essential component in a number of 1D and 2D NMR experiments, as we have already seen in several examples. In other experiments, we can eliminate certain unwanted signals by choosing only coherence pathways that go through a $p = \pm 2$ step. We describe here the simple procedure needed for a *double quantum filter* (DQF) and return to consideration of INADEQUATE (introduced in Section 10.2), which depends on use of a ^{13}C DQF. In Section 12.3 we shall see a further example of the use of a DQF.

Double Quantum Filters

Double quantum coherence (DQC) can be produced in a pair of coupled spins by a 90°, 2τ, 90° sequence, where the pulses are nonselective, and $\tau = 1/4J$. The coherence pathway is:

$$
I_z + S_z \xrightarrow{90°I_x, \, 90°S_x} -(I_y + S_y) \xrightarrow{\pi J I_z S_z (2\tau)} I_x S_z + I_z S_x \xrightarrow{90°I_x, \, 90°S_x} -(I_x S_y + I_y S_x) \tag{12.4}
$$

As we saw in Eq. 11.82b, the last expression represents DQC. Chemical shifts are often refocused by inserting 180° I and S pulses at time τ, the midpoint in the evolution period, so the DQC-generating sequence may appear as 90°, τ, 180°, τ, 90°. In a 2D experiment, the DQC thus produced is allowed to evolve and is eventually converted to observable magnetization whereas in a 1D experiment the DQC is often converted almost immediately to observable magnetization.

This procedure can be used as a filter if there is another pathway by which unwanted observable magnetization is produced but never goes through the DQC step. By phase cycling and/or by imposition of suitable magnetic field gradients (as discussed in Section 11.7), it is usually feasible to discriminate very effectively between the coherence pathways and to eliminate virtually all traces of unwanted signal. We turn now to one important example.

INADEQUATE

As we saw in Section 10.2, INADEQUATE is designed to establish correlations between coupled ^{13}C nuclei and thus elucidate the carbon framework of an organic molecule. This experiment can be carried out in either a 1D or 2D form. In the one-dimensional manifestation, INADEQUATE is designed simply to observe ^{13}C–^{13}C coupling. In a molecule that is highly and uniformly enriched, such an observation could be made directly with a simple 90° pulse. However, at the natural abundance of 1.1%, the peak ^{13}C signal from molecules containing only one ^{13}C is at least 200 times as large as the desired signal from coupled nuclei, so a method is needed to discriminate against it. The double quantum

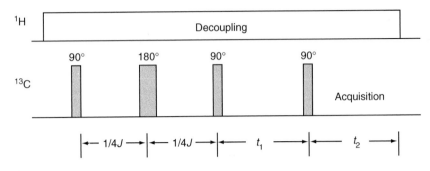

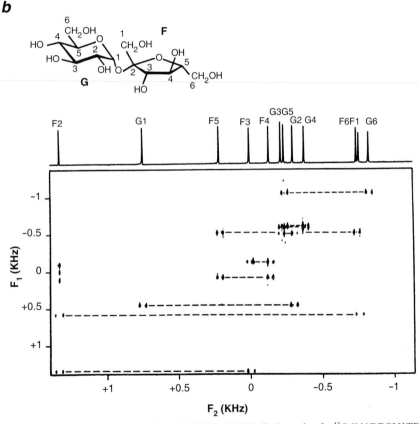

FIGURE 12.4 (*a*) Pulse sequence for 2D INADEQUATE. (*b*) Example of a ^{13}C INADEQUATE spectrum at 50 MHz from sucrose. The spectra in the F_2 dimension consist of AX (or AB) patterns from $^1J(^{13}C-^{13}C)$ couplings. Under the conditions used, smaller long-range $^{13}C-^{13}C$ couplings do not appear, and proton decoupling during both t_1 and t_2 removes ^{13}CH splitting. The resonances are separated along F_1 by double quantum frequencies. Most of the assignments are clear, as indicated on a normal decoupled ^{13}C spectrum at the top of the spectral plot. The overlap of some resonances from the glucose (G) ring can be removed, even at 50 MHz, by plotting on an expanded vertical scale. Spectrum courtesy of Ad Bax (National Institutes of Health).

filter provides exactly what is needed, because a ^{13}C that is not coupled to another ^{13}C cannot provide double quantum coherence, hence can be eliminated by proper phasing.

The pulse sequence for 2D INADEQUATE is given in Fig. 12.4. We recognize the familiar 90°, τ, 180°, τ sequence, with $\tau = 1/4J_{CC}$, which refocuses the ^{13}C chemical shifts and leaves the magnetization in pure antiphase for conversion by the next 90° pulse to DQC. The only difference between the 1D and 2D versions is that for 1D the variable evolution time t_1 is replaced by a very short fixed period (a few microseconds). The DQC generated by pulse 3 is thus almost immediately transformed by pulse 4 into antiphase magnetization that develops into observable signal during the acquisition period. Broadband proton decoupling is provided throughout the entire pulse sequence and detection period, so ^{1}H couplings can be ignored.

The 1D version of INADEQUATE provides excellent suppression of ^{13}C signals from noncoupled nuclei, and the ^{13}C$-^{13}$C couplings provide useful but limited information because many one-bond C–C couplings are of similar magnitude, hence do not provide specific information on the carbon framework of the molecule. On the other hand, 2D INADEQUATE provides very specific information, as illustrated in Fig. 12.4b, because the DQC precesses during the t_1 period and acquires phase information in the same way that single quantum coherences (magnetizations) do in other 2D experiments that we have examined. Thus, the 2D spectrum shows the DQ frequency along the F_1 axis (of course, with no splitting from spin coupling) and the double-quantum filtered ^{13}C coupled spectrum along F_2.

The key to suppressing coherence pathways that do not include the DQF is suitable manipulation of the phases of the pulses and data acquisition. Phase cycles are employed that are similar in concept to those used in CYCLOPS (Section 3.5) and COSY (Section 11.7) but usually have many more steps. For INADEQUATE, the DQC evolves during t_1 at twice the normal ^{13}C frequency and experiences twice the phase shift of an SQC. If the first three pulses, which generate the DQC, are cycled in concert through 0, $\pi/2$, π, and $3\pi/2$ phases, the DQC emerges 180° out of phase (i.e., changed in sign) at $\pi/2$ and $3\pi/2$, but unchanged at π. So if signals are alternately added and subtracted for these four phases, the DQC will coherently add. On the other hand, an SQC from uncoupled ^{13}C nuclei experiences only a 90° phase change each time, so that signals at 0 and π cancel on addition, as do signals at $\pi/2$ and $3\pi/2$. Thus, this cycle eliminates the unwanted SQC.

In practice, additional phase cycling is needed in order to generate quadrature signals that can be processed as discussed in Section 10.3 to produce properly phased line shapes. Because of the DQC precession frequency, phase increments of $\pi/4$ are needed to provide the same sort of data usually obtained with $\pi/2$ phase shifts. Hence, a second set of phase-cycled repetitions is needed, with data stored in a separate location. Overall, the following phase cycling can be carried out:

Location I			Location II		
Pulses 1,2,3	*Pulse 4*	*Receiver*	*Pulses 1,2,3*	*Pulse 4*	*Receiver*
x	x	Add	$x + \pi/4$	x	Add
y	x	Subtract	$y + \pi/4$	x	Subtract
$-x$	x	Add	$-x + \pi/4$	x	Add
$-y$	x	Subtract	$-y + \pi/4$	x	Subtract

If CYCLOPS is used to eliminate artifacts in quadrature detection, this eight-step cycle must then be nested within CYCLOPS to give a 32-step cycle overall. In this simple treatment we have not taken into account the effect of pulse imperfections, which generate additional coherence pathways from coherences that were found to vanish in the preceding analyses, so that further phase cycling is often necessary.

The large number of phase cycling steps might appear to provide an unacceptable lengthening of the experiment, but in practice many repetitions are needed in any event in order to enhance signal/noise by time averaging, as only 0.01% of the molecules are contributing to each signal. Nevertheless, the use of pulsed field gradients provides an alternative method for coherence selection, as discussed in Section 11.7.

Proton-Detected INADEQUATE

Because the principal shortcoming of INADEQUATE is low S/N, a method that enhances the signal from ^{13}C (or other rare, insensitive nucleus such as ^{15}N) by transfer of polarization from protons is beneficial for observing $^{13}C-^{13}C$ moieties in which at least one of these nuclei is coupled to a proton. The technique uses an INEPT sequence to transfer polarization from ^1H to ^{13}C, followed by INADEQUATE, and finally a reverse INEPT sequence to transfer polarization back to ^1H for detection. This experiment places even more demands on suppression of unwanted coherences, as the ^1H signal from the 99% of molecules with no ^{13}C must also be eliminated. We shall not describe the experimental details or the coherence pathway except to note that it again depends on ^{13}C DQC, but with ^1H also in antiphase (terms of the sort $I_x S_y T_z$, where I and S again represent ^{13}C and T is ^1H). With phase cycling, pulsed field gradients, and a selective ^1H "trim" pulse to remove unwanted coherences, this experiment can improve S/N by an order of magnitude.

Other Uses of DQF

Double quantum filters can be used as a component of many different 2D, 3D, and 4D experiments. In some applications, COSY benefits from DQF, as we discuss in the next section.

12.3   COSY

COSY was the first 2D experiment attempted and in many ways serves as the prototype, as we discussed in Chapter 10. Now that we can apply product operator formalism, let's return to a further consideration of the mechanism for COSY and look at some of the factors that make it a more complex technique than was apparent in our initial treatment.

Coherence Pathways

We noted in Chapter 10 that the pulse sequence for COSY is just $90°$, t_1, $90°$, t_2. In Section 11.6 we found that each precession must be described by two terms, so that evolution as a result of chemical shifts and spin coupling can lead to a rapid escalation in the number of coherence pathways, which can be conveniently grouped as "trees." Table 12.1 shows the pathways that result from an I spin whose initial magnetization along z (1) evolves to produce diagonal peaks in the COSY spectrum, (2) transfers to S to produce cross peaks, (3) leads to antiphase I or S magnetization that is undetectable, and (4) generates zero and double quantum coherence, which is also undetectable. Table 12.1 actually displays only half the pathways, because spin S is also rotated into the xy plane by the initial $90°$ pulse, and there is a completely equivalent pathway for S that leads to diagonal peaks centered at Ω_S and to complementary cross peaks symmetrically placed with respect to the diagonal.

For the four pathways in Table 12.1 that lead to observable I signals and the four that arise from S, we can use the procedure in Section 11.7 to write expressions for the evolution during t_2:

$$
\begin{aligned}
&(1) \quad I_x \sin \Omega_S t_1 \sin \pi J t_1 \cos \Omega_I t_2 \sin \pi J t_2 \\
&(2) \quad I_x \sin \Omega_I t_1 \cos \pi J t_1 \cos \Omega_I t_2 \cos \pi J t_2 \\
&(3) \quad I_y \sin \Omega_S t_1 \sin \pi J t_1 \sin \Omega_I t_2 \sin \pi J t_2 \\
&(4) \quad I_y \sin \Omega_I t_1 \cos \pi J t_1 \sin \Omega_I t_2 \cos \pi J t_2 \\
&(5) \quad S_x \sin \Omega_I t_1 \sin \pi J t_1 \cos \Omega_S t_2 \sin \pi J t_2 \\
&(6) \quad S_x \sin \Omega_S t_1 \cos \pi J t_1 \cos \Omega_S t_2 \cos \pi J t_2 \\
&(7) \quad S_y \sin \Omega_I t_1 \sin \pi J t_1 \sin \Omega_S t_2 \sin \pi J t_2 \\
&(8) \quad S_y \sin \Omega_S t_1 \cos \pi J t_1 \sin \Omega_S t_2 \cos \pi J t_2
\end{aligned}
\qquad (12.5)
$$

Each expression has a modulation by J in both t_1 and t_2, so we expect the spectrum to show frequencies related to J in both frequency dimensions. Expressions (2) and (4) are modulated by Ω_I during both t_1 and t_2, so these lead to diagonal peaks at or near Ω_I, and expressions (6) and (8) lead to diagonal peaks near Ω_S. The other four expressions are modulated by Ω_I during one time period and Ω_S during the other period, so they give cross peaks. Note that the precessions during t_1 represent an amplitude modulation of the signal obtained during t_2. Also,

Table 12.1 Coherence Pathways in a COSY Experiment[a, b]

$$I_z \xrightarrow{I_x} -I_y \xrightarrow{\Omega I_z}$$

Upper branch: $I_x \xrightarrow{JI_zS_z}$

$$I_yS_x \xrightarrow{I_x} I_zS_x \xrightarrow{S_x} -I_zS_y \xrightarrow{\Omega I_z} -I_zS_y \xrightarrow{\Omega S_z}$$
$$I_zS_x \xrightarrow{JI_zS_z} S_y^c \quad , \quad I_zS_x$$
$$-I_zS_y \xrightarrow{JI_zS_z} S_x \quad , \quad -I_zS_y$$

$$I_x \xrightarrow{I_x} I_x \xrightarrow{S_x} I_x \xrightarrow{\Omega I_z}$$
$$I_y \xrightarrow{\Omega S_z} I_y \xrightarrow{JI_zS_z} -I_xS_z \quad , \quad I_y$$
$$I_x \xrightarrow{\Omega S_z} I_x \xrightarrow{JI_zS_z} I_yS_z \quad , \quad I_x$$

Lower branch: $-I_y \xrightarrow{JI_zS_z}$

$$I_xS_z \xrightarrow{I_x} I_xS_z \xrightarrow{S_x} -I_xS_y \xrightarrow{\Omega I_z}$$
$$-I_yS_y \xrightarrow{\Omega S_z} I_yS_x \xrightarrow{JI_zS_z} I_yS_x$$
$$-I_yS_y \xrightarrow{JI_zS_z} -I_yS_y$$
$$-I_xS_y \xrightarrow{\Omega S_z} I_xS_x \xrightarrow{JI_zS_z} I_xS_x$$
$$-I_xS_y \xrightarrow{JI_zS_z} -I_xS_y$$

$$-I_y \xrightarrow{I_x} -I_z \xrightarrow{S_x} -I_z \xrightarrow{\Omega I_z} -I_z \xrightarrow{\Omega S_z} -I_z \xrightarrow{JI_zS_z} -I_z$$

[a] The pathways originating in I_z are shown. An equivalent set of pathways originates in S_z.

[b] The rotation operators I_x and S_x represent 90° rotations; ΩI_z and ΩS_z represent evolution of I and S, each under the influence of its chemical shift; and JI_zS_z represents evolution under the influence of spin coupling. For each evolution, the upper branch implies multiplication by the appropriate sine term, and the lower branch implies multiplication by the appropriate cosine term.

[c] Of the 13 final coherences, I_x and I_y give diagonal peaks, and S_x and S_y give cross peaks. The remaining coherences are unobservable—antiphase magnetization, double quantum coherence, and longitudinal magnetization.

Adapted from Freeman.[106]

the J modulation is a maximum for $t_1 = 1/2J$, which is a consideration in setting the evolution time range. As we noted in Section 10.2, for small values of J the pulse sequence can be modified to include time for the J modulation to develop.

We can picture the peaks more easily if we use trigonometric identities to obtain for expressions (3) and (4), that is, one diagonal and one cross peak, the following:

$$(3) \quad \tfrac{1}{4}I_y[\cos(\Omega_S - \pi J)t_1 - \cos(\Omega_S + \pi J)t_1]$$
$$\times [\cos(\Omega_I - \pi J)t_2 - \cos(\Omega_I + \pi J)t_2]$$
$$= \tfrac{1}{4}I_y[\cos(\Omega_S - \pi J)t_1 \times \cos(\Omega_I - \pi J)t_2$$
$$- \cos(\Omega_S - \pi J)t_1 \times \cos(\Omega_I + \pi J)t_2$$
$$- \cos(\Omega_S + \pi J)t_1 \times \cos(\Omega_I - \pi J)t_2$$
$$+ \cos(\Omega_S + \pi J)t_1 \times \cos(\Omega_I + \pi J)t_2] \quad (12.6)$$

(4) $\frac{1}{4}I_y[\sin(\Omega_I + \pi J)t_1 + \sin(\Omega_I - \pi J)t_1]$
  $\times [\sin(\Omega_I + \pi J)t_2 + \sin(\Omega_I - \pi J)t_2]$

$= \frac{1}{4}I_y[\sin(\Omega_I + \pi J)t_1 \times \sin(\Omega_I + \pi J)t_2$

  $+ \sin(\Omega_I + \pi J)t_1 \times \sin(\Omega_I - \pi J)t_2$

  $+ \sin(\Omega_I - \pi J)t_1 \times \sin(\Omega_I + \pi J)t_2$

  $+ \sin(\Omega_I - \pi J)t_1 \times \sin(\Omega_I - \pi J)t_2]$ (12.7)

Consider first the diagonal signal (4). There are four peaks, all of the same sign, centered at (Ω_I, Ω_I). Two of the peaks lie on the diagonal, separated in both ω_I and ω_S directions by $2\pi J$ (or by J when converted to frequency in Hz), while the others are symmetrically placed off the diagonal by $2\pi J$, as illustrated in Fig. 12.5. All functions in both dimensions are sines, which represent dispersion mode, so each of the four peaks has positive and negative lobes, as shown.

Expression (3) also shows four peaks, but these are centered at (Ω_S, Ω_I), as shown in Fig. 12.5. These peaks arise from coherences that precess near Ω_S during t_1 and near Ω_I during t_2. Again, the J spacings are apparent. Here the functions are all cosines, hence represent absorption mode, but the four peaks have both positive and negative signs, as indicated in Fig. 12.5.

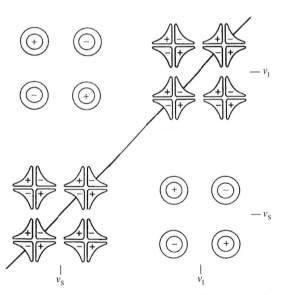

FIGURE 12.5 Schematic representation of the diagonal and off-diagonal peaks in a COSY spectrum arising from an I–S spin system. The diagonal peaks are in dispersion mode, and the off-diagonal peaks are absorption mode.

Expressions (7) and (8) are of the same form as (3) and (4) and each gives four peaks centered at (Ω_S, Ω_I) and at (Ω_S, Ω_S), respectively. The remaining four expressions provide redundant information, but with a $\pi/2$ phase shift.

Detection and Spectral Display

Our description of a COSY spectrum provided the essential features but was based on consideration of only the y' component, thus ignoring the practical aspects of quadrature detection. As we saw in Section 10.3, a $\pi/2$ phase shift in the second pulse can be used to produce a separate data set modulated by sin $\Omega_1 t_1$, rather than cos $\Omega_1 t_1$, and with quadrature detection these two data sets can be manipulated in various ways. The phase cycling for COSY illustrated in Section 11.8 leads to an N-type (echo) spectrum, which provides narrower lines but still with phase twist. Display of the absolute value (Section 10.3) is of some help in converting both diagonal and cross peaks to only positive signals, but the dispersion tails are still present to reduce resolution and peak S/N. Alternatively (and preferably), the TPPI and/or States method can be used to obtain absorption mode in both dimensions for the cross peaks. The diagonal peaks are in dispersion mode and the interesting cross peaks are in the preferred absorption mode, but with alternating signs, as illustrated in Fig. 12.5. (Passive couplings give in–phase peaks, which is often helpful in interpretation.) Adequate digital resolution must be used to prevent cancellation of cross peaks, especially for small values of J.

Because we can adjust phasing in both dimensions, we can convert the diagonal peaks to absorption mode, but only at the expense of converting the cross peaks to dispersion, which is quite undesirable.

DQF-COSY

In many instances strong singlet resonances and, especially, resonances from solvents such as H_2O can interfere significantly with a large portion of the COSY spectrum and create dynamic range problems. By replacing the second 90° pulse of the COSY sequence by two very closely spaced 90° pulses, as shown in Fig. 12.6a, we can insert a double quantum filter (see Section 12.2). Because the magnetically equivalent nuclei in H_2O or CH_3 groups do not display their internal spin coupling, they behave as a pseudospin of 1 or 3/2 and do not develop DQC, hence do not generate observable signal in the final step. DQF-COSY is widely used, especially for studying molecules in aqueous solutions, where it is often combined with selective water presaturation or other methods for reducing the H_2O signal (see Section 9.5). The gain in reducing unwanted signals is, however, accompanied by a 50% loss of sensitivity relative to COSY itself.

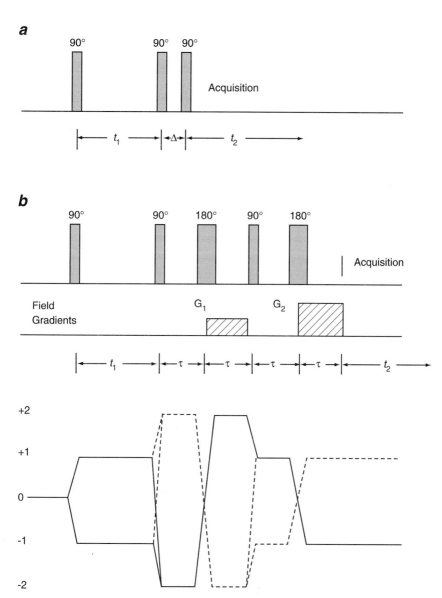

FIGURE 12.6 (*a*) Pulse sequence for double quantum filtered COSY. The period Δ, during which the DQC evolves, is normally quite short (~4 μs). Phase cycling is employed with all pulses. (*b*) Pulse sequence for DQF COSY with pulsed field gradients. Gradients G_1 and G_2, with relative amplitudes 1:2, are both applied for the same time τ (typically ~2 ms). The 180° pulses and additional τ periods are designed to refocus coherences that dephase during the relatively long gradient pulses. (*c*) Coherence pathways shown on a coherence order diagram for DQF-COSY with pulsed field gradients. Only coherence at +2 during G_1 is refocused by G_2 at the −1 level needed for detection, so the pathway given by the solid line is selected.

By following the coherence pathway described in Section 12.2, we find that there are again eight observable signals, which are similar in form to those in Eq. 12.5 but somewhat different in detail. For example, Eq. 12.6 describing the cross peaks becomes:

$$\frac{1}{8} I_y [\sin(\Omega_S + \pi J)t_1 \times \sin(\Omega_I + \pi J)t_2$$
$$- \sin(\Omega_S + \pi J)t_1 \times \sin(\Omega_I - \pi J)t_2$$
$$- \sin(\Omega_S - \pi J)t_1 \times \sin(\Omega_I + \pi J)t_2$$
$$+ \sin(\Omega_S - \pi J)t_1 \times \sin(\Omega_I - \pi J)t_2] \tag{12.8}$$

and Eq. 12.7 describing the diagonal peaks remains the same except for multiplication by $\frac{1}{8}$ rather than $\frac{1}{4}$. At first glance this might appear to be a step backward, because now both diagonal and cross peaks are in dispersion mode. However, the phasing can be adjusted to put *all* signals in absorption (with positive and negative peaks as before). The elimination of dispersion tails from diagonal peaks can be helpful in observing nearby cross peaks.

We have seen in Section 11.7 that pulsed field gradients can be very effective in distinguishing between orders of coherence. DQF-COSY benefits from use of gradients to reduce phase cycling and optimize coherence selection. The rf pulse sequence, shown in Fig. 12.6*b*, differs from the DQF-COSY sequence in Fig. 12.6*a* in the addition of two 180° pulses to compensate for phase problems introduced by the finite length (a few milliseconds) of the pulsed field gradients. G_1 acts to dephase the $p = +2$ coherence, and G_2 (with twice the strength) refocuses the $p = -1$ coherence that has been designated in the desired path shown by the solid line, whereas all other coherences are simply dephased by G_2 and give no signal.

ω_1-Decoupled COSY

The COSY pulse sequence can be modified to provide a 2D spectrum that is decoupled in the ω_1 dimension while retaining coupling in ω_2. The simplification in ω_1 can be helpful in interpreting crowded spectra, as the chemical shifts are clearly revealed. Although this experiment is not widely used as such, it serves as a prototype for a building block in other more complex 2D and 3D experiments.

The pulse sequence is shown in Fig. 12.7. This sequence represents a *constant time experiment* and differs from those that we have seen in that the total evolution time comprises two variable periods. As t_1 is incremented, one period increases but the other decreases to retain a constant overall evolution time, Δ. The 180° pulse refocuses chemical shifts at the end of the t_1 period, but they then precess as

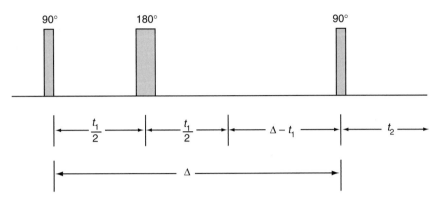

FIGURE 12.7 Pulse sequence for ω_1-decoupled COSY. The period Δ is a fixed time, while t_1 is incremented as usual in a 2D experiment.

usual during the period $\Delta - t_1$, which serves as the effective evolution period. Because the 180° pulse affects both I and S, evolution from J coupling is unaffected by the pulse and simply precesses as usual (at $\pm J/2$) during the entire period Δ. However, as Δ is fixed, there is no J modulation as t_1 is incremented and no coupling is thus shown in the ω_1 dimension.

Other Variants of COSY

The wide utility of COSY has spawned a number of other variations and improvements for specific purposes. For example, reducing the angle of the mixing pulse from 90° to 45° (sometimes called COSY-45) results in a change in the relative intensities of the cross peaks that depends on the relative signs of J, a feature that can often be helpful in distinguishing between $^2J_{HH}$ and $^3J_{HH}$ in a CH_2—CH_2 fragment (see Section 5.4). Alternatively, the dependence on pulse angle can be used to distinguish between connected and nonconnected transitions. ECOSY (*exclusive COSY*) and an alternative PECOSY (*primitive ECOSY*) provide simplified cross peak patterns by restricting coherence transfer to connected transitions.

COSY can be combined with a z filter (see Section 9.5) into z-COSY, which is occasionally used to produce absorption phase spectra. A 90°, τ, 90° sequence replaces the second 90° pulse in the usual COSY sequence, just as in DQF-COSY (see Fig. 12.6a). In this case, however, pulse 2 rotates in-phase magnetization to the z axis, where it is stored for the short τ period and restored to the xy plane by pulse 3. Suitable phase cycling or imposition of a pulsed field gradient eliminates the coherences remaining in the xy plane during τ, so that the final magnetization arises solely from the polarization stored along z, hence is in phase and can produce pure absorption mode in both diagonal and cross peaks. The period τ must be kept short to avoid NOE, but even so there is a 50% loss of signal, as in DQF-COSY.

A variation of COSY called RELAY has been used to relay coherence from one coupled set of spins to a second set. RELAY employs a pulse sequence of 90°, τ, 180°, τ, 90° in place of the usual 90° mixing pulse. Although this method can be useful, it has largely been supplanted by TOCSY-HOHAHA, as described in Section 10.2.

12.4 HETERONUCLEAR CORRELATION BY INDIRECT DETECTION

As we saw in previous chapters, correlation of proton NMR signals with those of another nuclide S, such as ^{13}C, is usually carried out by starting with ^1H polarization (I), allowing it to evolve during t_1, and transferring polarization to S, for example, via an INEPT-type sequence. If the S magnetization then precesses during the detection period, the correlation can be established, and the S signal is enhanced by a factor of γ_I/γ_S as a result of polarization transfer. However, much larger signals can be obtained if the polarization is transferred back to ^1H and detected there with proton sensitivity, rather than detected as S magnetization, a gain of an additional factor of $(\gamma_I/\gamma_S)^{3/2}$. Table 12.2 indicates the very significant gains obtainable from this "round trip" transfer of polarization with ^1H detection. Because of this sensitivity gain, most heteronuclear correlations are now carried out by indirect detection with modern instruments that have probes designed for such studies. To some extent even older instruments can be used, with the proton decoupler coil serving as a (less efficient) detection coil.

Other than the availability of equipment, the only significant drawback to indirect detection is that the strong ^1H signals from isotopomers containing ^{12}C or ^{14}N must be eliminated. In fact, in some instances in which sensitivity is not a problem and the ^1H spectrum has many lines, it may be preferable to use

TABLE 12.2 Gain in Sensitivity by Indirect Detection

Transfer	Formula	Relative sensitivity	
		$I = {}^1H; S = {}^{13}C$	$I = {}^1H; S = {}^{15}N$
S	$\gamma_S^{5/2}$	1	1
$I \rightarrow S$	$\gamma_I\gamma_S^{3/2}$	4	10
$S \rightarrow I$	$\gamma_S\gamma_I^{3/2}$	8	30
$I \rightarrow S \rightarrow I$	$\gamma_I^{5/2}$	32	300

HETCOR. However, in general, indirect detection methods are preferred, with suitable techniques to suppress the unwanted ^1H signals. For example, a BIRD pulse sequence can be applied in the preparation period. As we saw in Section 9.5, this sequence can be configured to invert $^{12}C-H$ magnetization and leave $^{13}C-H$ magnetization along $+z$. The 90° pulse ending the preparation period is then timed to occur as the $^{12}C-H$ magnetization is relaxing through zero.

We discussed several indirect detection methods in Section 10.2 and now amplify on some aspects.

HSQC

The pulse sequence for the heteronuclear single quantum coherence (HSQC) experiment is shown in Fig. 12.8 (As we noted, it is often preceded by a BIRD sequence or other means of suppressing unwanted signal.) The initial pulses are just those of an INEPT sequence, which we have examined in detail, and the last portion, as we shall see, is a reverse INEPT sequence, in which polarization is transferred from S back to I. At time 5, as we saw in Chapter 11, I polarization has been transferred to S, which begins the evolution period in antiphase relative to I (as I_zS_x) and precesses at $\Omega_S \pm \pi J$, as described in Section 11.6. However, the 180° I pulse at the midpoint of the evolution period causes the coupling (but not the S chemical shift) to refocus at the end of t_1. Thus, ignoring the coupling terms, the coherence transfer pathway starting at time 5 is

$$I_zS_x \xrightarrow{\Omega_St_1} I_zS_x \cos \Omega_St_1 + I_zS_y \sin \Omega_St_1 \xrightarrow{90°I_y\,90°S_x} I_xS_x \cos \Omega_St_1 + I_xS_z \sin \Omega_St_1$$

$$(12.9)$$

Thus, at time 9 we have a double quantum coherence (I_xS_x), which will not generate any signal, and antiphase I magnetization (I_xS_z) multiplied by the sine

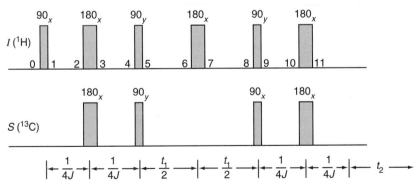

FIGURE 12.8 Pulse sequence for the heteronuclear single quantum coherence experiment. See text for discussion of the state of the spin system at the times indicated.

of the phase angle accumulated during the evolution period. Before detection, this antiphase term evolves through a period of $1/2J$ in order to generate only in-phase magnetization (just the reverse of the initial $1/2J$ period). Again, a 180° I pulse at the center of this period refocuses chemical shifts at the beginning of the detection period, while the simultaneous 180° S pulse ensures that the coupling does not refocus.

If broadband decoupling is applied to the S spins (e.g., ^{13}C), a proton spectrum decoupled from ^{13}C is obtained in dimension ω_2. We have ignored the H−H couplings, which also evolve during the experiment, as they are usually a small

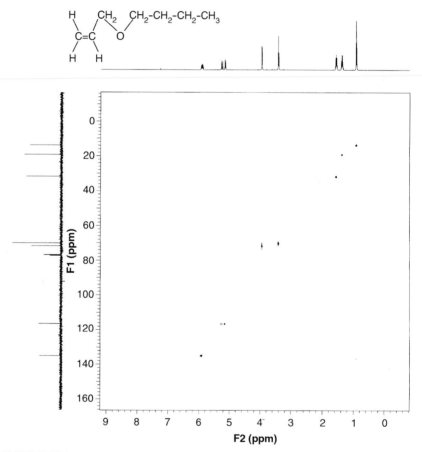

FIGURE 12.9 Example of heteronuclear single quantum coherence (HSQC) applied to allylbutyl ether (300 MHz). The correlations of ^{1}H and ^{13}C chemical shifts are clearly shown. Note the similarity to Fig. 10.10, which displays a HETCOR spectrum. For a sample of this sort, where signal/noise ratio is no problem, there is little to choose between the two techniques, but HSQC is inherently much more sensitive.

perturbation. However, they appear in the proton spectrum, as shown in Fig. 12.9, which depicts a typical HSQC spectrum.

HMQC

An alternative to HSQC is heteronuclear multiple quantum coherence (HMQC), which can be generated in a very similar manner, as indicated in Fig. 12.10. This begins in the same way as INEPT in generating antiphase proton magnetization during the period $1/2J$, but without the need for the 180° pulses of INEPT. As we saw in Section 11.5 and Eq. 11.68, after the application of the 90° S pulse, the density operator at time 3 has zero and double quantum coherence, which evolve at frequencies $\omega_S - \omega_I$ and $\omega_S + \omega_I$, respectively, during the t_1 period, after which another 90° S pulse at time 6 reconverts this coherence to antiphase proton magnetization. During t_1 the coupling does not affect evolution of ZQC or DQC. The 180° I pulse at time 4 interchanges DQC and ZQC, thus eliminating the proton (I) frequency at the end of t_1 and leading to a spectrum with only Ω_S in the ω_1 dimension. The final period of $\Delta = 1/2J$ allows the antiphase I magnetization to precess and to arrive in phase at time 8 for detection while broadband S decoupling is applied. Again, H–H couplings appear in the ω_2 dimension. In addition, such couplings also appear (often unresolved) in the ω_1 dimension, because they evolve during t_1 even though IS coupling is absent in the IS multiple quantum coherence.

A number of variations of both HSQC and HMQC have been developed. Both methods are widely used and are of comparable value. The HMQC experiment uses fewer pulses, hence can be somewhat shorter and is less dependent

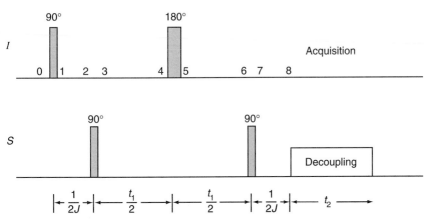

FIGURE 12.10 Pulse sequence for the heteronuclear multiple quantum coherence experiment. See text for discussion of the state of the spin system at the times indicated.

on experimental aberrations. However, the unresolved H−H couplings in the ω_1 dimension broaden the lines, thus reducing resolution and peak signal/noise.

As usual, suitable phase cycling and/or use of pulsed field gradients is critical to avoid the detection of undesired coherences.

HMBC

As we noted in Section 10.2, HMBC is a slightly modified version of HMQC that is designed to emphasize couplings through more than one heteronuclear bond (e.g., ^{13}C—C—C—H). One pulse sequence for HMBC, as illustrated in Fig. 12.11, is identical to that in Fig. 12.10 with the addition of a 90° pulse and adjustment of timing and phase cycling.

As in HMQC, $\Delta = 1/(2\ ^1J_{CH})$, so that antiphase magnetization for one-bond coupling is present at time 2, is converted into ZQC and DQC by the first S pulse and evolves during period Δ'. However, magnetization representing the smaller values of $^nJ_{CH}$ evolves at a rate of about a factor of 20 more slowly, hence the first S pulse has little effect on it, and it continues to evolve during $\Delta' \approx 1/(2\ ^nJ_{CH})$, reaching antiphase only at time 4, where the second S pulse converts it to ZQC and DQC. Thus, information on the smaller couplings behaves according to our discussion of HMQC and produces observable signals. The unwanted coherence from the one-bond coupling is eliminated by cycling the phase of the first S pulse between x and $−x$ and adding the final signals. The sign of the ZQC and DQC for the one-bond coupling is thus reversed by the phase

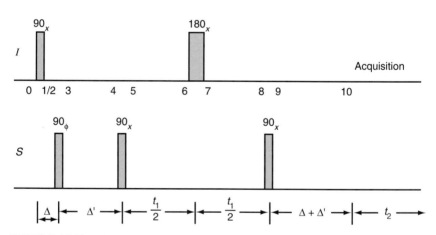

FIGURE 12.11 Pulse sequence for the heteronuclear multiple bond correlation experiment. $\Delta = 1/(2^1J)$ and $\Delta' = 1/(2^nJ)$, where ^1J and ^nJ are spin couplings between I and S through one and n bonds, respectively. The first S pulse, marked 90_ϕ, is cycled through $+x$ and $−x$. See text for discussion of the state of the spin system at the times indicated.

cycling and eliminated, while the desired coherences from n-bond couplings are unaffected by the phase cycling and coherently add.

The use of an additional 90° pulse with timing and phase cycling as given here acts as a *low-pass filter*, because it eliminates the high frequency component from the larger ^1J. This technique is also used in other complex pulse sequences as a filter.

12.5   THREE- AND FOUR-DIMENSIONAL NMR

The two reasons for extending NMR studies beyond three dimensions are the same as those for going from one to two dimensions: (1) to spread out crowded resonances and (2) to correlate resonances. 3D and 4D experiments have been carried out almost exclusively to interpret the spectra of macromolecules, principally large proteins. We return in Chapter 13 to a discussion of these applications, but here we give a brief description of some types of 3D and 4D experiments.

Spectral Editing in the Third Dimension

Figure 12.12*a* gives a good illustration of the need for going to a third dimension to facilitate the interpretation of a crowded 2D spectrum. The NOESY spectrum of a uniformly ^{15}N-enriched protein, staphylococcal nuclease, has so many cross peaks that interpretation is virtually impossible. However, it is possible to use ^{15}N chemical shifts to edit this spectrum, as indicated in Fig. 12.12*b* and *c* in a three-dimensional experiment. With the ^{15}N enrichment, NOESY can be combined with a heteronuclear correlation experiment, in this case HMQC, but HSQC could also be used. A 3D pulse sequence can be obtained from two separate 2D experiments by deleting the detection period of one experiment and the preparation period of the other to obtain two evolution periods (t_1 and t_2) and one detection period (t_3). In principle, the two 2D components can be placed in either order. For the NOESY-HMQC experiment, either order works well, but in some instances coherence transfer proceeds more efficiently with a particular arrangement of the component experiments. We look first at the NOESY-HMQC sequence, for which a pulse sequence is given in Fig. 12.13. The three types of spins are designated I and S (as usual), both of which are ^{1}H in the current example, and T, which is ^{15}N in this case.

The coherence pathway leading to NOESY cross peaks was illustrated in Eq. 11.88, where we showed that with appropriate phase cycling we had only a term λS_y (with λ measuring the extent of NOE polarization transfer). Double quantum and zero quantum coherences are then generated by the pathway

$$\lambda S_y \xrightarrow{\pi J_{ST} S_z T_z \frac{1}{2J}} \lambda S_x T_z \xrightarrow{90° T_x} -\lambda S_x T_y \tag{12.10}$$

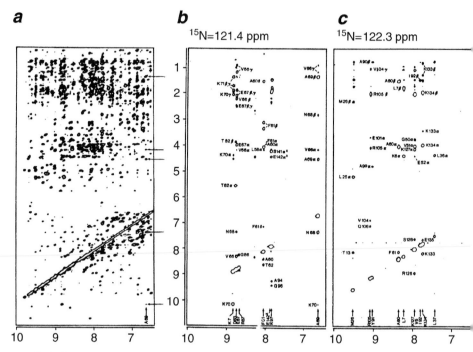

a

b
^{15}N=121.4 ppm

c
^{15}N=122.3 ppm

FIGURE 12.12 (*a*) NOESY spectrum (500 MHz) of staphylococcal nuclease, uniformly enriched in ^{15}N. (*b*) Two adjacent planes from a NOESY-HMQC experiment with staphylococcal nuclease, separated by ^{15}N chemical shift difference of 0.9 ppm. Adapted from Marion *et al.*[118]

As we have seen, the DQC and ZQC evolve at frequencies $\Omega_S \pm \Omega_T$. The 180° proton (I,S) pulse, which is a standard part of the HMQC sequence, refocuses proton chemical shifts that evolve during t_2 and interchanges ZQC and DQC. The final 90° T (^{15}N) pulse converts $S_x T_y$ to $S_x T_z$, and this antiphase S magnetization evolves during the final $1/2J$ period to in–phase S magnetization, which is detected while T is broadband decoupled.

Application of NOESY-HMQC to the nuclease sample replaces the 2D NOESY spectrum of Fig. 12.12*a* by a 3D spectrum that can be displayed in a cube but is more easily interpreted as a set of planes, as indicated schematically in Fig. 12.14. Two of the NOESY planes obtained in the nuclease experiment are illustrated in Fig. 12.12*b*. Clearly, the 3D experiment is successful in "editing" the uninterpretable spectrum of Fig. 12.12*a* into manageable pieces. Moreover, each NOESY plane is labeled by the chemical shift of the ^{15}N that is coupled to one of the protons, so additional useful information may be available if that ^{15}N chemical shift can be related to structural features in the molecule. Note that a proton not coupled to any ^{15}N generates an axial peak, rather than a cross peak, in the 3D spectrum, but phase cycling is used to remove axial peaks. Also, displays other

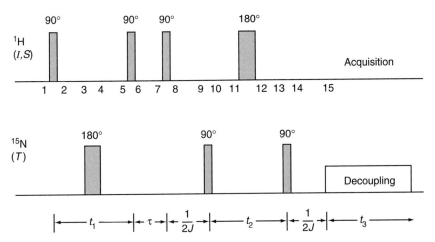

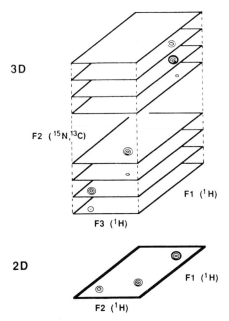

FIGURE 12.13 Pulse sequence for the three-dimensional experiment NOESY-HMQC. The first three 90° ¹H pulses constitute the usual NOESY sequence, with mixing time τ. The 180° ¹⁵N pulse at time 3 removes the effect of ¹H−¹⁵N couplings. The ¹H and ¹⁵N pulses at times 7, 9, 11, and 13 constitute the HMQC sequence, with the pulse at time 7 serving as part of both sequences.

FIGURE 12.14 Schematic representation of the data from a three-dimensional NMR experiment shown as a set of planes, as compared with data from a 2D experiment in a single plane.

than the NOESY planes may be prepared by manipulation of the 3D data; for example, an ^{1}H–^{15}N plane or a diagonal plane may be selected for specific purposes.

This type of experiment may be extended to four dimensions with suitable samples. For example, a protein that is uniformly enriched in both ^{15}N and ^{13}C may be studied by the 4D experiment HMQC-NOESY-HMQC, with the pulse sequence shown in Fig. 12.15. Here the ^{1}H–^{13}C HMQC component comes first. The in-phase proton magnetization resulting at the end of that component goes through the usual NOESY sequence before being subjected to the ^{1}H–^{15}N HMQC in a repeat of the procedure described in the preceding paragraphs. Analysis of the data again is carried out by selecting appropriate planes from the 4D data set.

In principle, the procedure can be extended to additional dimensions. However, as we pointed out in Section 10.1, a 4D experiment takes 4–5 days of data accumulation, even when phase cycling is minimized by optimum use of pulsed field gradients. Expansion beyond four dimensions with adequate spectral resolution to cover complex multiline spectra would require exorbitant amounts of

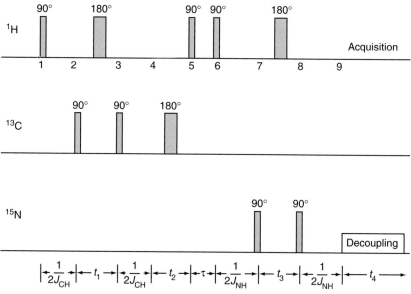

FIGURE 12.15 Pulse sequence for a four-dimensional NMR experiment, HMQC-NOESY-HMQC. The initial HMQC experiment correlates ^{1}H and ^{13}C frequencies, and the final HMQC experiment correlates ^{1}H and ^{15}N frequencies. The period τ is the usual NOESY mixing period. The initial ^{1}H magnetization is converted to ^{13}C multiple quantum coherence, which evolves during t_1 and is then converted back to ^{1}H coherence for evolution during t_2. After the NOESY mixing period, the ^{1}H coherence is converted to ^{15}N MQC, which evolves during t_3 and is converted once more to ^{1}H coherence for detection during t_4.

instrument time and would be subject to additional instrumental instabilities. In fact, instrument time can often be used more effectively to perform a series of separate 3D experiments that provide complementary structural information, rather than carrying out a single 4D experiment.

Correlation by Triple Resonance Experiments

Many different types of 3D and 4D experiments can be devised. We described briefly one class, in which two different kinds of 2D experiments are joined, primarily to use the third dimension to clarify the presentation of spectral data. With the ability of most modern NMR spectrometers to provide precise frequency and phase control for several radio frequencies simultaneously and to apply them efficiently in the probe, a class of 3D and 4D triple resonance experiments has become feasible. These experiments usually use two HSQC and/or HMQC sequences to transfer magnetization in the path $I \rightarrow S \rightarrow T \rightarrow S \rightarrow I$. I is almost always ^1H; S and T are ^{13}C and ^{15}N in proteins, the type of molecule in which these experiments are most often utilized. In addition, ^{31}P may be involved in nucleic acids, another frequent subject for these methods, and many other nuclides can be used in other applications.

As an example, we describe one of many similar experiments devised for assigning resonances in proteins, a subject that we take up in more detail in Chapter 13. This particular experiment is designed to correlate the frequencies within the ^1H—^{15}N—^{13}C=O portion of a peptide group and is appropriately called simply *HNCO*. The basic pulse sequence for HNCO is shown in Fig. 12.16. To simplify the notation, instead of I, S, and T, we identify the active spins as H, N, and C, and use K to denote the spin of α-^{13}C. In a peptide chain, one α carbon is bonded and spin coupled to the nitrogen and another α carbon is bonded and coupled to the carbonyl carbon atom. With recombinant DNA methods, the protein is uniformly and highly enriched in both ^{13}C and ^{15}N, so all of these spins need be considered.

By now, we can readily recognize the ^1H–^{15}N INEPT sequence at the beginning of the experiment, which starts an HSQC sequence that generates antiphase ^{15}N magnetization (H_zN_y) at the beginning of the first evolution period t_1. During t_1 the ^{15}N magnetization precesses under the influence of the various nitrogen chemical shifts in different peptide groups, but each ^{15}N also experiences several one-bond couplings: $^1J_{NH}$, $^1J_{NC}$, and $^1J_{NK}$. (Several two- and three-bond couplings also occur, but we ignore these at present because they are considerably smaller than the one-bond couplings.) The H, C, and K 180° pulses in the middle of t_1 refocus these modulations at the end of t_1 and effectively decouple these spins from N.

Next comes a fixed period $\Delta = 1/2J_{NC}$, during which N is influenced by its chemical shift and all couplings except that from K, which is removed by the

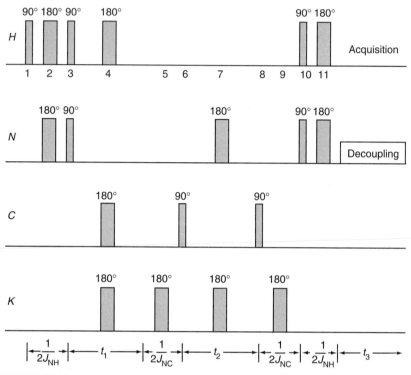

FIGURE 12.16 Pulse sequence for the triple resonance 3D NMR experiment HNCO. H and N denote ^1H and ^{15}N, C denotes $^{13}C=O$, and K denotes $^{13}C_\alpha$. Pulses at times 1, 2, and 3 constitute an INEPT sequence that transfers coherence from H to N, where it precesses during t_1. Pulses at times 6, 7, and 8 represent an HMQC sequence that creates multiple quantum coherence in C (where it precesses during t_2) and transfers coherence back to N. Pulses 10 and 11 are an inverse INEPT sequence that transfers coherence back to H for detection during t_3. The other 180° pulses refocus heteronuclear spin couplings. Note that coherence is not transferred to spin K.

180° K pulse in the middle of the Δ period. However, the N 180° pulse in the middle to the later t_2 period will eventually refocus the N chemical shifts and J_{NH}, so we can concentrate on the effect of $^1J_{NC}$ during the Δ period. As we have seen previously, evolution of N magnetization for a period $1/2J_{NC}$ develops N magnetization that is antiphase in C. Thus, the initial term H_zN_y becomes $H_zN_xC_z$, and the first C 90° pulse converts this to DQC and ZQC, as we have seen for HMQC:

$$H_zN_y \xrightarrow{\pi J_{NC}N_zC_z \frac{1}{2J_{NC}}} -H_zN_xC_z \xrightarrow{90°C_x} H_zN_xC_y \qquad (12.11)$$

During t_2 the DQC and ZQC precessions are tagged by the C frequencies. The final 90° C pulse converts the multiple quantum coherence to antiphase

N magnetization, which evolves to in-phase N magnetization at the end of the second Δ period. Finally, a reverse INEPT sequence converts this to H magnetization that is antiphase in N, and it evolves during the period $\tau = 1/2J_{NH}$ to in-phase H magnetization for detection. A summary of this last portion of the coherence pathway is

$$H_z N_x C_y \xrightarrow{90°C_x} H_z N_x C_z \xrightarrow{\pi J_{NC} N_z C_z \frac{1}{2J_{NC}}} H_z N_y \xrightarrow{90°H_x, 90°N_x} -H_y N_z \xrightarrow{\pi J_{NH} N_z H_z \frac{1}{2J_{NH}}} H_x \qquad (12.12)$$

Although the general course of the experiment is clear, there are many important aspects that are disguised in the coherence pathway and our brief discussion. We have not accounted for all couplings, which cause additional splittings or line broadenings. We have not considered the rates at which the various coherence transfers occur and the effect of relaxation during the evolution periods. Finally, we have not specified the phase cycling needed to eliminate undesired coherence pathways.

In practice, HNCO is now carried out by a somewhat more complex pulse sequence than that given in Fig. 12.16 in order to improve its efficiency. Pulsed field gradients are added to aid coherence pathway selection; an INEPT transfer from N to K replaces the multiple quantum coherence step; and the N evolution is carried out with a constant time experiment.

12.6 ADDITIONAL READING AND RESOURCES

In addition to the reading suggestions given Chapter 10, a large number of NMR books give very good descriptions of many 1D, 2D, and 3D experiments, almost always in terms of product operators and coherence transfer pathways. *Principles of NMR in One and Two Dimensions* by Richard Ernst *et al.*[27] and *Modern Techniques in High-Resolution FT-NMR* by N. Chandrakumar and S. Subramanian[119] give the theoretical background and apply it to many kinds of experiments. *Spin Choreography* by Ray Freeman[106] gives very readable descriptions of all the important 2D NMR methods, often approached from more than one perspective. *Protein NMR Spectroscopy: Principles and Practice* by J. Cavanagh *et al.*[120] provides a considerable amount of information on general NMR methods, in spite of its more restrictive title. A review article by H. Kessler, M. Gehrke, and C. Griesinger[121] provides an excellent pedagogical presentation and a very good description of the basic 2D experiments.

Books listed in previous chapters by van de Ven,[109] Günther,[64] and Brey[116] contain good descriptions of many 2D experiments. *150 and More Basic NMR Experiments—A Practical Course* by S. Braun, *et al.*,[41] described in Chapter 3, provides excellent summaries of pulse sequences, applications, and product operator descriptions of many 2D and 3D experiments. The *Encyclopedia of NMR*[1] includes 16 articles specifically devoted to more detailed exposition of most of the commonly used 2D and 3D methods.

12.7 PROBLEMS

12.1 Use a coherence transfer pathway for a noncoupled ^{13}C similar to that in Eq. 12.4 or prepare a suitable vector diagram to show its coherence state after the evolution period of Fig. 12.4. Verify that the phase cycling procedure described for INADEQUATE cancels the single quantum coherence from this ^{13}C.

12.2 Verify the derivation of Eq. 12.6 and 12.7 by using trigonometric identities from Appendix C.

12.3 Use a coherence pathway or vector description to show that the pulse sequence in Fig. 12.7 leads to a signal at the end of the evolution period that shows no modulation from spin coupling.

12.4 Expand the coherence pathway of Eq. 12.9 to include the relevant steps at the beginning and end of the path that results from the HSQC pulse sequence in Fig. 12.8.

12.5 Write a coherence pathway for the four-dimensional HMQC-NOESY-HMQC experiment, based on the pulse sequence in Fig. 12.15.

CHAPTER 13

Elucidation of Molecular Structure and Macromolecular Conformation

Although NMR has applications in many scientific and technological areas, *high resolution* NMR is used mainly as a valuable tool in the elucidation of molecular structure, as a quantitative analytical technique, and as a means for understanding molecular reactions. Organic chemistry has long used NMR as one of the two principal instrumental methods (mass spectrometry is the other) for determining the structure of molecules of ever-increasing complexity. Inorganic chemistry also benefits from the availability of multinuclear spectrometers that can be used to study virtually any element, as well as the organic ligands that characterize many inorganic complexes.

NMR methods have also been used extensively to determine the configuration and conformation of both moderate-size molecules and synthetic polymers, whose primary molecular structure is already known. During the past decade high resolution NMR, particularly employing 2D and 3D methods, has become one of only two methods (x-ray crystallography is the other) that can be used to determine precise three-dimensional structures of biopolymers—proteins, nucleic acids, and their cocomplexes—and NMR alone provides the structure in solution, rather than in the solid state.

There are now a very large number of books aimed at techniques for studying molecular structure, both in small molecules and in polymers and biopolymers.

In this chapter we provide only an overview of the approach to such problems and give a few examples to indicate the scope of these important applications.

13.1 ORGANIC STRUCTURE ELUCIDATION

In problems of structure elucidation an NMR spectrum may provide useful, even vital data, but it is seldon the sole piece of information available. A knowledge of the source of the compound or its method of synthesis is frequently the single most important fact. In addition, the interpretation of the NMR spectrum is carried out with concurrent knowledge of other physical properties, such as elemental analysis from combustion or mass spectral studies, the molecular weight, and the presence or absence of structural features, as indicated by infrared or ultraviolet spectra or by chemical tests. Obviously, the procedure used for analyzing the NMR spectrum is highly dependent on such ancillary knowledge.

The best way to gain proficiency in structure elucidation by NMR is to work through a number of examples. This is outside the scope of this book, but a number of books (some listed at the end of the chapter) provide good step-by-step procedures for tackling various kinds of structural problems.

Features of ¹H Spectra

Two-dimensional NMR provides powerful tecniques to aid interpretation, but the starting point is a simple, one-dimensional proton NMR spectrum, with careful integration to ascertain the relative numbers of protons in different lines or multiplets. In some instances one or two good ¹H NMR spectra may be sufficient to solve the problem with little expenditure of instrument time. In other instances, where only minute amounts of sample are available, it may not be feasible to obtain any NMR data other than a simple ¹H spectrum. However, as we pointed out in Chapter 3, with modern instrumentation and microprobes, it is usually possible to use indirect detection methods to obtain correlations with less sensitive nuclei, such as ¹³C and ¹⁵N, even with quite small amounts of sample.

Before attempting to interpret an NMR spectrum, it is wise to ascertain whether it has been obtained under suitable experimental conditions so that it is meaningful. The symmetry and sharpness of the line due to TMS (or other reference) provides an indication of magnetic field homogeneity and general spectrometer performance. Most spectrometers designed for routine studies are sufficiently automated that reasonable sets of instrument parameters are selected, but it is possible to misset parameters for acquisition or data processing and obtain distorted spectra.

Lines that are clearly due to impurities, to the solvent itself, or to a small amount of undeuterated solvent can usually be identified. Water is often present in solvents, its resonance frequency depending on the extent of hydrogen bonding to the solvent or the sample and on the concentration of water.

An examination of the relative areas of the NMR lines or multiplets (resolved or unresolved) is often the best starting point for the interpretation of the spectrum. If the total number of protons in the molecule is known, the total area can be equated to it, and the number of hydrogen atoms in each portion of the spectrum established. The opposite procedure of assigning the smallest area to one or two protons and comparing other areas with this one is sometimes helpful, but it should be used with caution because appreciable error can be introduced in this way. Occasionally, lines so broad that they are unobservable in the spectrum itself can be detected in the integral trace.

The positions of strong, relatively sharp lines can usually be correlated with expected chemical shifts. This correlation, together with the area measurements, frequently permits the establishment of a number of methyl and methylene groups. In organic solvents such as $CDCl_3$ exchangeable protons can usually be observed as single lines and can often be identified by addition of a drop of D_2O to the sample and resultant disappearance of peaks.

Figure 13.1 shows a spectrum typical of a steroid at two different magnetic fields. At low field, the protons of the many CH and CH_2 groups in the condensed ring system are so nearly chemically equivalent that they give rise to a broad, almost featureless hump, but even at low field the angular and side-chain methyl groups show very pronounced sharp lines, the positions of which can provide valuable information on molecular structure. With modern, high field spectrometers these methyl lines are just as informative, but the lines from alicyclic protons are sufficiently resolved that 2D methods permit a vast additional amount of information to be obtained from their analysis.

The approximate centers of all simple multiplets, broad peaks, and unresolved multiplets can also usually be correlated with functional groups. The *absence* of lines in characteristic regions often furnishes important data.

First-order splitting in multiplets can easily be identified, and values of J deduced directly from the splittings. As noted in Chapter 6, the first-order criterion $\nu_A - \nu_X \gg J_{AX}$ is often not strictly obeyed, resulting in a distortion or "slanting" of the expected first-order intensity distribution toward the center of the entire pattern (e.g., Fig. 6.13). The number of components, their relative intensities, the value of J, and the area of the multiplet together provide valuable information on molecular structure. Commonly occurring, nearly first-order patterns, such as that due to CH_3CH_2X where X is an electronegative substituent, should be recognized immediately. Other patterns that are actually not first order, such as those due to the magnetically nonequivalent protons in *para*-disubstituted benzene rings, are also characteristic and should be easily identified, as indicated in Section 6.17. When necessary, other non−first-order patterns can often be analyzed quickly by simulation programs or by procedures given in Appendix B.

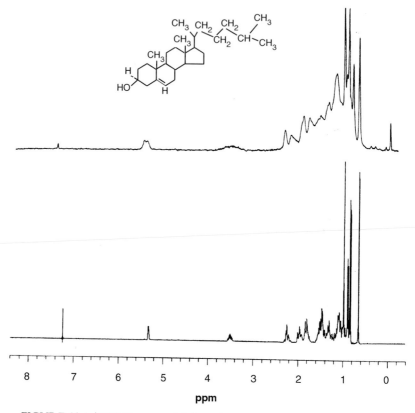

FIGURE 13.1 ^{1}H NMR spectra of cholesterol at 60 MHz (top) and 300 MHz (bottom).

In simple cases, the magnitudes of coupling constants are often definitive in establishing the relative positions of substituents. For example, Fig. 13.2 shows that the three aromatic protons of 2,4-dinitrophenol give rise to a spectrum that is almost first order in appearance, even at 60 MHz. The magnitudes of the splittings suggest that two protons *ortho* to each other give rise to the peaks in the regions of 450 and 510 Hz and that the latter proton is *meta* to the one resulting in the lines near 530 Hz. From the known effects of electron-withdrawing and electron-donating substituents (Chapter 4) it is clear that the most deshielded protons must be adjacent to the NO_2 groups.

When magnetic nuclei other than protons are present, it should be recalled that some values of J might be as large as many proton chemical shifts. For example, in Fig. 13.3, $^2J_{HF} = 48$ Hz, accounting for the widely spaced 1:3:3:1 quartets due to the CH that is coupled to both the fluorine and the adjacent methyl group. Because $^3J_{HF} = 21$ Hz and $^3J_{HH} = 7$ Hz, the CH_3 resonance is a doublet of doublets.

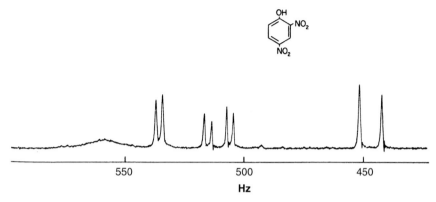

FIGURE 13.2 ¹H NMR spectrum at 60 MHz of 2,4-dinitrophenol in CDCl₃.

Features of ¹³C Spectra

Directly recorded one-dimensional ¹³C NMR spectra are routinely obtained for structure elucidation of organic compounds except in instances where a very limited amount of sample precludes direct ¹³C detection. It is usually worthwhile to spend enough instrument time to obtain a ¹³C spectrum with good signal/noise ratio and with sufficient delay times between pulse repetitions to ensure that quaternary carbons with long T_1 are clearly observed.

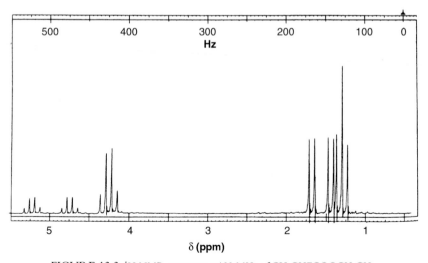

FIGURE 13.3 ¹H NMR spectrum at 100 MHz of CH₃CHFCOOCH₂CH₃.

Most of the general comments in Section 13.1 apply, *mutatis mutandis*, to the interpretation of ^{13}C spectra. Almost all ^{13}C spectra are obtained initially with complete proton decoupling, so that (in the absence of ^{19}F, ^{31}P, or other nuclei that might couple to carbon) the spectrum consists of a single line for each chemically different carbon atom in the molecule. As we have seen, ^{13}C spectra are usually obtained under conditions where the areas of chemically shifted lines are *not* proportional to the numbers of carbon nuclei contributing to the lines, so integration of peaks is usually of little value. With the structural information already derived from the ^1H spectrum, plus ancillary data on the sample, the ^{13}C chemical shifts may provide sufficient results to provide a good idea of the molecular structure and guide the selection of additional correlation experiments. As pointed out in Chapter 4, large databases of both ^1H and ^{13}C chemical shifts are available and can often be very helpful in interpreting the one-dimensional spectra.

Selective decoupling experiments are sometimes still useful in correlating specific peaks via spin coupling and deducing bonding networks, and continuous wave NOE can occasionally be helpful in establishing spatial relationships. Generally, however, correlations and their implications for molecular structure are best carried out via 2D techniques.

Spectral Editing

We saw in Section 12.1 that APT and DEPT provide straightforward methods for differentiating among CH_3, CH_2, CH, and quaternary carbons. This information is almost essential in interpreting the ^{13}C spectrum, and in some instances it can best be obtained with APT or DEPT, where ^{13}C is observed, especially when the proton spectrum is crowded. If HETCOR and/or indirect detection of ^{13}C is needed, these one-dimensional editing techniques provide somewhat redundant information but may still be helpful.

13.2 APPLICATION OF SOME USEFUL 2D METHODS

As we have seen, there are many 2D methods and variants that can provide data critical to structure determination. There is no general way to decide which ones are "best" for solving a particular problem, as various kinds of information may be needed. Also, some experiments may be less feasible than others because of instrumental capabilities, amount of sample available, and time requirements. We summarize here only a few of the major factors involved in selecting such 2D methods.

Proton–Proton Correlation

Except for the simplest spectra, we almost always need to establish the nuclei responsible for observed couplings, hence to develop structural information on at least individual molecular fragments. COSY (and its various modifications) is probably the most popular 2D homonuclear correlation experiment and is often the one performed first, after obtaining the one-dimensional spectra. Although DQF-COSY suffers a factor of 2 loss in sensitivity, it is usually preferred because it provides a cleaner diagonal in the plot by reducing or eliminating interfering peaks from solvent resonances and from some intense singlets in the sample.

Long-range coupling patterns are often very helpful also. The parameters for COSY can be adjusted, as we have seen, to permit correlation through smaller couplings, but the optimum experiment is usually HOHAHA/TOCSY. In fact, this experiment can be carried out with parameters set to limit the isotropic mixing to spins that have large coupling and thus to replace COSY. The real value of HOHAHA, however, is in identifying, sets of nuclei that are pairwise coupled, hence must be within one isolated part of the molecule. It is widely applied to identify monomeric units in a peptide, oligonucleotide or oligosaccharide, as illustrated in Fig. 13.4.

Heteronuclear Correlations

Because of the sensitivity gain from the indirect detection experiments, either HMQC or HSQC is usually the method of choice for establishing correlations between protons and ^{13}C (as well as other heteronuclei, such as ^{15}N or ^{31}P). HET-COR, the heteronuclear analog of COSY, must be used when the available (somewhat older) instrument does not have good indirect detection capability, but use of HETCOR is declining.

As we saw in Chapters 10 and 12, all of these correlation experiments are used primarily to establish one-bond connectivities, but parameters may be adjusted to permit coherence to build up from longer range couplings. In many instances HMBC is a key technique, as it provides correlations between a proton and carbons that are two or three bonds removed (even via a nitrogen, oxygen, or other nonmagnetic nuclide). This information, along with the one-bond correlations, ties molecular fragments together.

Other Useful 2D Experiments

As we have been, INADEQUATE provides a very effective method for establishing ^{13}C—^{13}C one-bond couplings and can be valuable in establishing the carbon

FIGURE 13.4 ^{1}H TOCSY (HOHAHA) spectrum (500 MHz) of sucrose in D$_2$O. The lower 2D spectrum is a typical TOCSY spectrum with a long contact time, 100 ms. Correlations among protons in the glucose ring are shown in the region 3.4–3.8 ppm and at 5.4 ppm (the anomeric proton), while those in the fructose ring fortuitously fall in a largely separate region, 3.8–4.2 ppm.

framework of the molecule provided it is not broken by nitrogen, oxygen, or other nonmagnetic nuclides. INADEQUATE is used relatively infrequently because of its low sensitivity, but with continuing improvements in instrumentation and the availability of proton-detected INADEQUATE (Section 12.2), we might expect to see increased use of this method.

We have seen that NOESY provides information on internuclear (principally interproton) distances. For many organic molecules (as distinguished from macromolecules such as proteins and nucleic acids) "structure elucidation" often involves only the establishment of the structural formula and bonding scheme. However, where ambiguities in configuration or preferred conformation remain to be settled, NOESY is often crucial for establishing stereochemistry.

13.3   STRUCTURE AND CONFIGURATION OF POLYMERS

Even when the structures of the individual monomer units in a polymer are known, the determination of their sequence and of the geometric arrangement, configuration, and conformation of the entire polymer presents challenging problems. We comment on only a few aspects here.

The NMR spectrum of a synthetic homopolymer may be very simple if the monomeric unit repeats regularly. On the other hand, irregularities, such as head-to-head junctions mixed with head-to-tail junctions, in such cases as vinyl polymers, for example, introduce additional lines that can often be valuable in structure elucidation.

A synthetic copolymer provides additional degrees of freedom in the arrangement of the repeating units. For example, the spectrum of a copolymer of vinylidine chloride and isobutylene, shown in Fig. 13.5, indicates that various tetrad sequences (sequences of four monomer units) display significantly different spectra. Copolymers composed of more than two monomer types, including biopolymers, have much more complex spectra, as we discuss later.

When the repeating unit possesses a center of asymmetry, further complexity is introduced into the spectrum. For example, in a vinyl polymer

$$\left[\begin{array}{c} CH_3 \\ | \\ -C-CH_2- \\ | \\ X \end{array} \right]_n$$

Note that the information provided here is similar to that found in the INADEQUATE spectrum of sucrose (Fig. 12.4b), but the sensitivity of TOCSY is much greater. The upper 2D spectrum shows a TOCSY spectrum with a short contact time, 10 ms, where the cross peaks are restricted largely to protons that are directly coupled (as in COSY). Courtesy of Daron Freedberg (Food and Drug Administration).

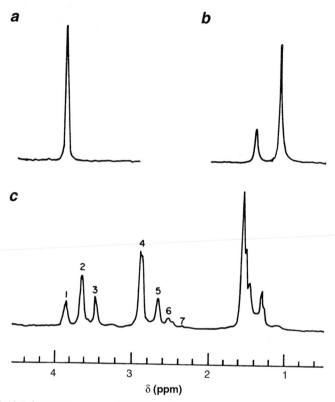

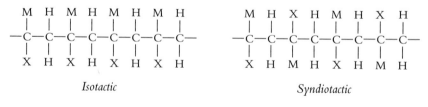

FIGURE 13.5 ^{1}H NMR spectrum (60 MHz) of (*a*) polyvinylidene chloride, A; (*b*) polyisobutylene, B; and (*c*) a copolymer of 70% A and 30% B. Peaks can be assigned to various tetrad sequences: (1) AAAA, (2) AAAB, (3) BAAB, (4) AABA, (5) BABA, (6) AABB, (7) BABB. From Bovey.[122]

the CH_3 and CH_2 chemical shifts are strongly dependent on the relative configuration (handedness) of adjacent monomeric units. An *isotactic* sequence is one in which all monomer units have the same configuration (*ddd* or *lll*); a *syndiotactic* sequence is one in which the configurations alternate (e.g., *dldl*); a *heterotactic* sequence is one in which a more nearly random configurational arrangement occurs (e.g., *ddld*). We can picture the sequences as follows:

$$\begin{array}{cccccccc} M & H & M & H & X & H & M & H \\ | & | & | & | & | & | & | & | \\ -C & -C & -C & -C & -C & -C & -C & -C- \\ | & | & | & | & | & | & | & | \\ X & H & X & H & M & H & X & H \end{array}$$

Heterotactic

In the syndiotactic sequence the two methylene protons adjacent to an M—C—X group are in the same environment and are chemically equivalent, whereas in the isotactic sequence they are chemically nonequivalent and give rise to an AB spectrum. An example of the spectrum of a vinyl polymer, polymethyl methacrylate, is shown in Fig. 13.6. Note that the CH_2 resonance around 2.1 ppm is a singlet in *a* but an AB quartet in *b*.

Our description of the expected methylene spectrum depends only on the configuration of two adjacent monomer units, a *dyad*. The environment of a given methyl group, on the other hand, depends on the relative configurations of both of the neighboring M—C—X groups, hence on a *triad* sequence. The three lines

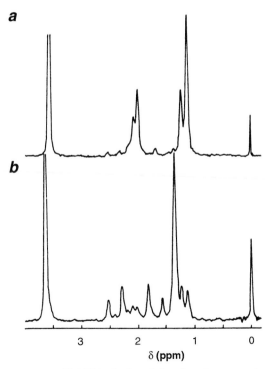

FIGURE 13.6 ^{1}H NMR spectra (60 MHz) of polymethyl methacrylate prepared under conditions is which the sample in *a* is primarily syndiotactic, and that in *b* is primarily isotactic. From Bovey.[122]

for the methyl resonances at $\delta \approx$ 1.3, 1.5, and 1.7 ppm in Fig. 13.6 are attributable to the syndiotactic, heterotactic, and isotactic triad sequences, respectively. In many instances additional effects from more distant groups can be discerned, and longer sequences must be considered.

13.4 THREE-DIMENSIONAL STRUCTURE OF BIOPOLYMERS

Molecules of biological interest, including small proteins, nucleic acids, carbohydrates, and lipids, have long been subject to study by NMR. In fact, NMR has been a principal technique for elucidating the structure of monosaccharides, and 2D experiments are increasingly being used to unravel more complex oligosaccharides, as illustrated previously. The principal contributions of NMR to the structure of lipids and phospholipids relate to the determination of conformation, often under conditions of highly restricted motion, which requires application of methods used for studying solids and liquid crystals described in Chapter 7.

For proteins and nucleic acids, NMR has never been a method of choice for structure elucidation in the sense of determining the identity or sequence of monomer units, because chemical and biotechnology techniques are simpler and far superior. However, NMR has been and remains of great value in providing detailed information on secondary and tertiary structure in biopolymers and oligomers. For example, the hydrogen bonding scheme in polynucleotides (as illustrated in Fig. 4.14) and other interactions such as base stacking and intercalation of complexed molecules can best be studied by NMR. In proteins, NMR has long been used to probe internal structural elements and to discriminate among titratable groups, because it was recognized that the local environments of otherwise chemically identical amino acid residues markedly alter their chemical shifts.

Our concern in this section, however, is not the application to biopolymers of methods that are equally applicable to smaller molecules. Rather, we discuss here a totally different approach to the determination of precise three-dimensional structures of these molecules, in which NMR data play a key role. We illustrate the concept with proteins, which have yielded particularly useful information, but the general approach can also be used with nucleic acids and with complexes of a protein and a nucleic acid.

There are three aspects to consider. First, we summarize briefly the underlying computational framework needed and the general strategy used in the structure determination. Second, we cover the use of 2D, 3D, and 4D methods to permit the sequential assignment of peaks to specific amino acids. Finally, we describe the use of nuclear Overhauser enhancements and spin coupling constants to provide restraints on interproton distances and bond angles, and we indicate how dipolar coupling and chemical shifts can sometimes add further information on molecular conformation.

Computational Strategy

The overall approach to determining the structure of a protein is to use computational power to take into account concurrently (1) the known sequence of the amino acids in the protein; (2) the known molecular structure of each of those amino acid residues, including bond distances and angles; (3) the known planar structure of the peptide group; (4) internuclear distances and interresidue bond angles, as determined from NMR data; (5) correlations of chemical shifts and structural features; and (6) minimization of energy and avoidance of unreasonable atomic contacts. There are a number of ways to handle the computations and to derive the molecular structure, but all of them depend critically on the data supplied by NMR.

All of the computational methods begin with structures generated by taking into account the known structures of peptides (1–3 above). Some methods use approximate protein structures generated by using a portion of the NMR data and the principles of *distance geometry*, which converts internuclear distances into a compatible structure of N atoms in ordinary three-dimensional space. Other approaches begin with an extended polypeptide chain or a random coil, as generated by the computer program.

The initial structure is then subjected to a classical mechanical treatment that takes into account NMR data, along with potential functions that relate the energy to bond distances, various dihedral bond angles, and interatomic interactions (e.g., van der Waals forces, electrostatic effects). The most popular of the computational methods, *simulated annealing*, does just what the name implies. As the computer program subjects the initial (often random coil) structure to an energy minimization calculation that includes all the preceding restraints, it repeatedly simulates the effect of raising the temperature to allow molecular motions in order to avoid any local or "false" energy minima and slowly lowers the simulated temperature to allow the structure to achieve a global energy minimum. This whole process is repeated with different initial structures. Provided there is a sufficiently large number of restraints (as we discuss later) and they are mutually consistent, a set of 30–40 converged structures is obtained. Figure 13.7 shows pictorially the result of a single simulated annealing computation that starts from a random coil, along with a set of structures generated by repeating the process.

The average of these converged structures is taken as "the" protein structure, whose precision can be assessed by the deviations of the individual structures from the average. The "quality" of the final structure can be described in terms of this root mean square deviation, for both the peptide backbone and side chains, and to some extent by the extent to which it conforms to limitations of dihedral bond angles and interatomic contacts anticipated from thousands of previously known structures (the "Ramachandran plot"). By all criteria, NMR structures of proteins that are determined in this way are comparable to structures determined by x-ray crystallography. In addition, NMR methods can be applied to evaluate the

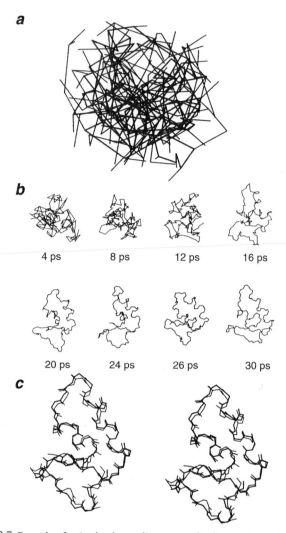

a

b

4 ps 8 ps 12 ps 16 ps

20 ps 24 ps 26 ps 30 ps

c

FIGURE 13.7 Example of a simulated annealing process for the protein crambin. (*a*) Constituent atoms of the protein in a random arrangement. (*b*) Computed structures obtained as the folding process is simulated for times given in picoseconds. (*c*) Average converged structure obtained with 20 repetitions of the simulated annealing process, as compared with the structure determined by x-ray crystallography (stereo view). Courtesy of G. Marius Clore (National Institutes of Health).

flexibility of the structure, because (as we saw in Chapter 8) relaxation times and NOEs are very sensitive to internal motions, as well as overall molecular tumbling.

With this general overview of the process used for structure determination, we now look at the way in which NMR data are employed.

Spectral Assignment

A protein subject to NMR analysis may have 100–200 amino acid residues, which provide a ^{1}H NMR spectrum of many hundreds of lines. Because the amino acid sequence can be assumed to have been determined previously by non-NMR methods, the first step in the NMR study is to assign each line in the spectrum to a specific moiety (NH, α-CH, side chain CH$_3$, etc.) of a specific amino acid residue. Without the 2D methods that we have discussed, it would be virtually impossible to make such assignments. For relatively small proteins (∼50–100 residues) it is often possible to use "conventional" homonuclear 2D methods, such as COSY and HOHAHA, to define some bonding paths and to supplement these results with NOE data for residues that are very close in space as a result of secondary structural elements such as α helices. However, for proteins of moderate size such techniques are insufficient, and special methods had to be developed and now constitute the standard method of making sequential assignments.

These techniques depend on the availability of uniformly highly enriched ^{13}C and ^{15}N proteins, which can usually be prepared by recombinant DNA methods. In the following discussion we assume that the protein is enriched to 95% or more in both ^{13}C and ^{15}N. In addition, for proteins larger than about 200 residues, it becomes important for some experiments to reduce line broadening from H–H and ^{13}C–H dipolar interactions by replacing most or all carbon-bonded protons in the protein by deuterium (also by rDNA methods). This substitution provides substantial line narrowing, because $1/T_2$ varies as γ^2, as we saw in Chapter 8. For other experiments, such as measurement of H–H NOEs, the presence of protons is, of course, essential.

The assignment techniques use the triple resonance experiments discussed in Section 12.5. Assignments of resonances along the peptide backbone are based on the fact that large one-bond couplings occur, as illustrated in Fig. 13.8, and are relatively independent of the particular amino acid. A number of experiments are summarized in Fig. 13.9, which highlights the portions of the protein involved in the correlation for each type of experiment. The first listed is HNCO, which we found in Section 12.5 to provide correlations among the peptide ^{15}N, peptide H, and adjacent ^{13}C in the carbonyl group. HNCA provides similar correlations to the α(A) carbon, and HN(CO)CA furnishes a correlation through four bonds between the α carbon and the nitrogen and hydrogen in the adjacent amino acid residue. Those involving "CB" and "HB" extend correlations beyond the immediate peptide chain to β carbons and hydrogens, and the HCCH-type experiments yield correlations between pairs of side chain hydrogens and carbons.

Each of these experiments has unique features, but the general concept follows our discussion of HNCO—proton magnetization is transferred to other nuclei, where it is tagged by their frequencies during evolution periods and is

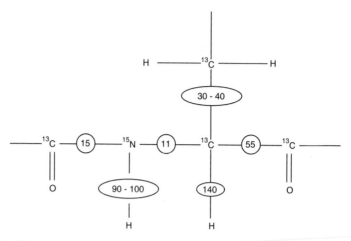

FIGURE 13.8 Typical values (Hz) of one-bond coupling constants in a peptide group are shown in circles.

eventually returned to ^1H for detection. Because the experiments largely provide complementary information (with some useful redundancies), the assignment of all lines in a large protein requires a great deal of instrument time, and each experiment must be optimized to minimize the number of repetitions required. Use of pulsed field gradients in lieu of extensive phase cycling is often of particular value.

Data analysis can be partially automated but is far from trivial, as indicated in Fig. 13.10, where several planes from a number of different experiments [HNCO, HNCA, HCACO, HCA(CO)N, and ^{15}N HOHAHA-HMQC] are displayed for the protein calmodulin. The illustration gives small sections of planes that are relevant for making the sequential connectivities between just two residues, lysine-21 and asparagine-22. Some of the correlations are indicated in the figure, but we shall not discuss the details. In practice, few such displays are prepared, as computer programs have been developed to select the peaks, sort them into groups of correlated chemical shifts, and position the fragments iteratively to obtain a consistent and complete assignment for the entire polypeptide backbone.

Assignment of protons in side chains is also critical; HCCH-COSY and HCCH-HOHAHA are widely used to facilitate these assignments. Because each amino acid has an asymmetric center, stereospecific assignments for methylene and methyl protons are obtained whenever feasible.

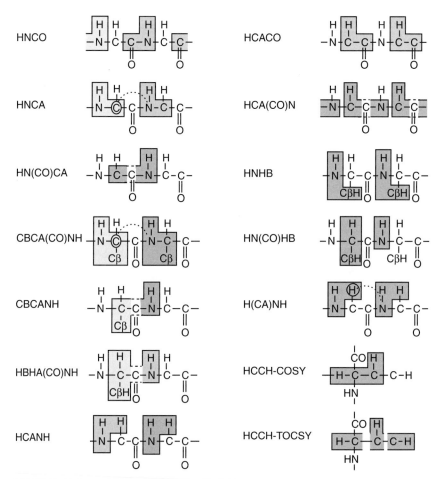

FIGURE 13.9 Summary of some useful 3D double resonance and triple resonance experiments that are used for assigning resonances in proteins. The highlighting indicates the correlations obtained in each experiment. Courtesy of G. Marius Clore (National Institutes of Health).

NMR and Structure Determination

The spectral assignment process, complex as it is, is only a prelude to the use of NMR data to assist in the final determination of the three-dimensional structure of the protein. Here, the most important experiment is NOESY, often in conjunction with HMQC or other methods for spreading out the 2D NOESY peaks in three or four dimensions. For larger proteins 4D experiments are essential to spread the NOESY peaks according to both ^{13}C and ^{15}N chemical shifts. We

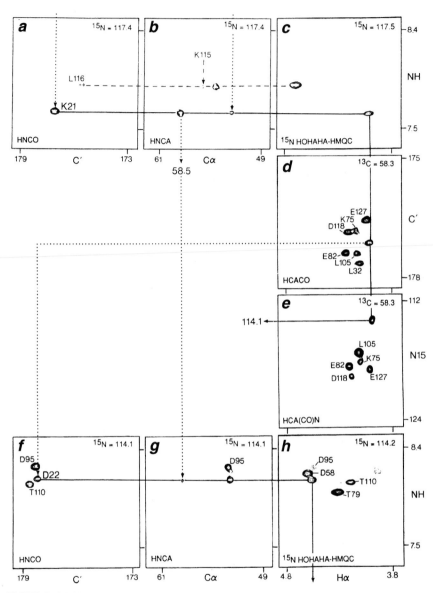

FIGURE 13.10 Illustration of correlations found in a number of different experiments for the protein calmodulin. From Ikura et al.[123]

saw an example in Fig. 12.11 of the importance of the additional dimension(s) in obtaining interpretable results.

This distance information is augmented with data on the conformation of sections of the polypeptide chain, obtained from a suitably parameterized Karplus equation (Section 5.4) and a large number of coupling constants obtained in multidimensional NMR experiments. Because lines in proteins are often broad relative to the magnitude of H−H couplings, the values of J are most often determined not from observation of line splittings as in small molecules, but from an analysis of the intensities of cross peaks. We have seen that the efficiency of coherence transfer and the generation of cross peaks depends in most experiments on the magnitudes of such couplings.

Both distance and angular data are formally put into the simulated annealing computation by including potential functions that are minimized for distances or angles within ranges that are compatible with experimental data. Neither NOEs nor the Karplus relation provides exact distances or angles, but with a large number of precise spectral data, the compatible ranges can be made quite narrow.

The process of determination of a three-dimensional structure proceeds in stages, as additional NMR data are supplied at each stage of refinement. For example, NOE results from protons that are very close to each other, along with the estimates of dihedral angles, provide mainly information on secondary structural elements, such as α helices and β sheets. Additional information may also be supplied on hydrogen bonding (from proton exchange rates as measured by NMR) and on proximity and orientation of C=O groups and aromatic rings (from neighbor anisotropy and ring current effects, as discussed in Section 4.5). Also, ^{13}C chemical shifts, which have been found to correlate with secondary structure, are fed into the computation for further refinement.

The long-range distance information that ultimately defines the overall protein structure comes primarily from NOESY experiments. However, because of the $1/r^6$ dependence, the long-range structural information is limited, and there are often regions of the protein that are poorly defined because of limited NOE data. An additional restraint is sometimes available in the direct dipolar interactions, which as we saw in Chapter 7 vary with $1/r^3$. For molecules tumbling rapidly *and randomly*, we also learned in Chapter 7 that such interactions average to zero. However, the random tumbling can be overcome in paramagnetic molecules (e.g., heme proteins), where the overall magnetic susceptibility is high enough to permit small, selective orientation in the magnetic field (see Section 7.13). Also, proteins can usually be dissolved at low concentration in lyotropic liquid crystal systems that orient in the magnetic field. Two particularly useful systems are diamagnetic disk-shaped bicelles and rod-shaped viruses. In line with the discussion in Section 7.13, the solute (protein) molecules become slightly ordered because even globular proteins are not exactly spherical. The ordering parameter is (fortunately) very small ($\sim 10^{-3}$), which results in small and interpretable dipolar splittings in the resonances, $\sim 10-20$ Hz for an ^{15}N−H dipole and $\sim 20-40$ Hz for a ^{13}C−H dipole. Because the orientation of the dipolar tensor is not determined by symmetry, a number of factors must be considered in interpreting the data in detail. The use of dipolar couplings in the structure calculation often results in an appreciable improvement in the quality of the structure, particularly in regions where few NOEs are available. For example, in multidomain proteins and in complexes between proteins and nucleic acids, dipolar couplings provide information on the relative orientations of the components that is not otherwise available.

An example of a protein structure determined by use of these NMR methods is shown in Fig. 13.11. Cyanovirin-N is a protein of molecular weight 11,000, which is a potent inactivator of the human immunodeficiency virus (HIV). Although this is a modest-size protein with 101 amino acid residues, over 2500 separate NMR restraints were used in the structural calculation. As indicated in Table 13.1, about half the restraints involve distances estimated from NOE data, and 334 dipolar couplings (primarily one-bond couplings) also provide distance restraints. There are a large number of angular restraints from spin couplings and from NOESY and ROESY data, and both ^{1}H and ^{13}C chemical shifts are also used in the target function minimized in the simulated annealing process.

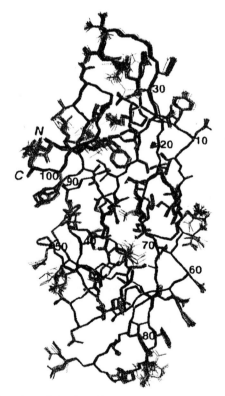

FIGURE 13.11 Illustration of the three-dimensional structure of the protein cyanovirin-N, as determined by NMR methods. From Bewley et al.[124]

TABLE 13.1 Structural Data for Cyanovirin-N

Distance restraints from NOEs	1241
Dihedral angle restraints	334
$^3J_{HNC\alpha}$ couplings	81
^{13}C chemical shifts	157
^1H chemical shifts	362
Dipolar couplings	334
Coordinate precision	
Backbone	0.15 Å
All nonhydrogen atoms	0.45 Å

13.5 ADDITIONAL READING AND RESOURCES

As indicated previously, there are now a large number of very good books directed primarily toward the use of NMR in structure elucidation of organic molecules. Some relatively recent books are *Modern NMR Spectroscopy—A Guide for Chemists* by Jeremy Sanders and Brian Hunter,[125] *Modern NMR Techniques for Chemistry Research* by Andrew Derome,[126] and *One and Two-Dimensional NMR Spectroscopy* by H. Friebolin.[127] Several other books, including *Structure Elucidation by NMR in Organic Chemistry* by Eberhard Breitmaier,[128] *Modern NMR Spectroscopy—A Workbook of Chemical Problems* by Jeremy Sanders et al.,[129] and *Two-Dimensional NMR Methods for Establishing Molecular Connectivity* by Gary Martin and Andrew Zektzer,[130] provide good discussions of the strategies and tactics for organic structure elucidation and include a large number of practical structural problems with their step-by-step solutions.

Application of NMR to three-dimensional structure determination is covered in several books, including *NMR of Proteins and Nucleic Acids* by Kurt Wüthrich,[60] *NMR of Proteins* edited by G. M. Clore and A. M. Gronenborn,[131] *Biomolecular NMR Spectroscopy* by Jeremy Evans,[132] and *Protein NMR Spectroscopy* by John Cavanagh et al.[120]

The *Encyclopedia of NMR*[1] contains a very large number of articles on biological applications of NMR, including discussions of the methodology used for three-dimensional structure determination, along with presentations on individual biopolymers.

As indicated in Chapter 4, BioMagResBank (www.bmrb.wisc.edu) is largely devoted to a database of tens of thousands of ^1H, ^{13}C, and ^{15}N chemical shifts in proteins.

CHAPTER 14

NMR Imaging and Spatially Localized Spectroscopy

During the past two decades, NMR has become widely known to the general public, not as a tool for the elucidation of molecular structure but as a technique for medical diagnosis. In this application, NMR is usually known as *magnetic resonance imaging* and abbreviated MRI. The details of the development of MRI methodology and its use in medicine are outside the scope of this book, but in this chapter we outline the basic principles of the use of NMR to produce two-dimensional and three-dimensional images. We shall find that many of the pulse techniques that were described in previous chapters are used to create images that show both structure and function in living samples (animal and human) and in inanimate objects, such as solid polymers and composite materials. In addition, we see how imaging techniques can be used to permit the observation of ordinary NMR spectra within a localized volume of interest in both living and inanimate samples.

14.1 USE OF MAGNETIC FIELD GRADIENTS TO PRODUCE IMAGES

In Section 3.3 we saw that carefully designed electrical coils can be used to generate magnetic field gradients in various directions and that the strength of each gradient can be used to compensate for inhomogeneities in the applied magnetic field—a process that we called "shimming." For NMR imaging we again obtain a homogeneous magnetic field across the sample but then deliberately introduce a linear magnetic field gradient in one or more directions. We continue to take the main magnetic field B_0 along the z axis, and the gradients G represent changes in the value of B_0 in various directions:

$$G_x = dB_0/dx \qquad G_y = dB_0/dy \qquad G_z = dB_0/dz \qquad (14.1)$$

Suppose that a gradient G_x is imposed along the x axis in an otherwise uniform magnetic field B_0. From the Larmor equation (2.42), the resonance frequency of a sample at a specific value of x is then

$$\omega = \gamma(B_0 + G_x x) \qquad (14.2)$$

The resonance frequency thus becomes a measure of the location of the sample along the x axis. For example, if we place a sample tube containing two capillaries of water in an ordinary NMR spectrometer and introduce the gradient G_x, apply a 90° rf pulse, and Fourier transform the FID, we obtain a "spectrum" of the sort shown in Fig. 14.1. The two broad lines show the relative positions of the capillaries in the x direction and constitute a one-dimensional "image" of the overall sample. Several important points can be deduced from Fig. 14.1 and Eq. 14.2:

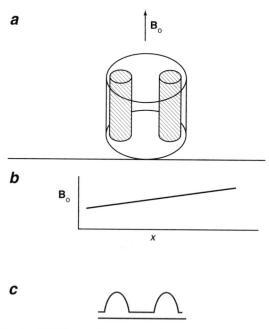

FIGURE 14.1 Simulated NMR signal from two capillaries of water in a normal sample tube, with an imposed magnetic field gradient in the x direction, transverse to the sample tube axis. (a) Sample tube arrangement. (b) Variation of B_0 across the sample. (c) NMR signal.

1. We saw in Section 3.7 that in ordinary NMR spectroscopy the rate of data acquisition determines the maximum spectral range. For imaging it is clear from Eq. 14.2 that for a given value of G_x the spectral width determines the maximum range, F_x, that can be observed in the x direction. F_x is called the *field of view* (usually abbreviated as FOV) in the x direction.

2. For a resonance that produces a sharp spectral line in a homogeneous field (such as water), the width of the response in frequency units is determined by the magnitude of the gradient imposed and the actual spatial width of the capillary. Thus, to provide a desired *spatial resolution*, d_x, in the x direction, an adequate gradient must be used and data must be acquired long enough, as also discussed in Section 3.7. The ratio F_x/d_x then gives N_x, the number of data points needed (in imaging jargon, the number of *pixels*), in complete analogy to the considerations in Section 3.7. In practice, there are limits to both N and G, which impose limits on spatial resolution.

3. Equation 14.2 applies to a sample that contains only a single spectral line. When chemical shifts are taken into account, it becomes

$$\omega_k = \gamma(B_0 + G_x x)(1 - \sigma_k) \tag{14.3}$$

If, instead of water, our sample were a compound with n chemically shifted resonance lines, Fig. 14.1 would consist of n overlapping one-dimensional images. Although the chemical shift can be turned to advantage in some applications, as described later, for a simple image this *chemical shift artifact*, as it is called in imaging jargon, must be taken into account. In many instances the size of a pixel (as expressed in frequency units) is large enough to encompass all spectral lines in the subject to be imaged. We return to the effect of chemical shifts in Section 14.5.

14.2   USE OF 2D NMR METHODS IN IMAGING

Although we were able to illustrate several important aspects of imaging in the one-dimensional example in Fig. 14.1, actual objects of interest are three-dimensional, and all three dimensions must be taken into account. For visualization, however, two-dimensional images are often displayed and analyzed. There are several ways in which such images can be obtained, but the most generally applicable is the use of the 2D (or 3D) NMR methods that we have discussed in detail. The basic 2D experiment was illustrated in Fig. 10.1. For imaging, a gradient G_x is applied during the evolution period, while during the detection period G_x is turned off and a gradient G_y is applied. Let us see how the use of these two orthogonal gradients permits all points in the field of view in the xy plane to be observed and imaged.

Figure 10.2 illustrates the behavior of a nuclear spin magnetization following a 90° pulse, where the phase angle ϕ accumulated by the end of the evolution period is the product of t_1 and Ω. If a magnetic field gradient G_x is applied during

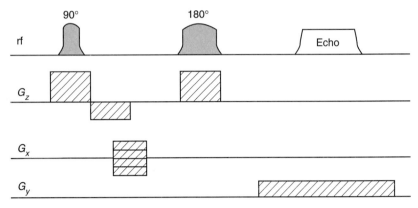

FIGURE 14.2 Pulse sequence for a two-dimensional NMR image in the xy plane, with slice selection along z. The value of the phase-encode gradient G_x is altered in successive repetitions, as indicated. The rf pulses are frequency selective and define the slice while gradient G_z is applied. Within one value of G_x there may be several repetitions with different values of frequency to obtain data from multiple slices. Data are acquired from the spin echo while the readout gradient G_y is applied.

the evolution period, the phase angle then provides information on the value of x, according to Eq. 14.2. (In imaging, this is called the *phase encode* direction.) During detection, gradient G_y is applied along y (the *frequency encode*, or *read* direction) so that a signal at a particular resonance frequency indicates the presence of sample at a particular value of y. Repetition of the experiment as the value of t_1 is incremented, as in a standard 2D experiment, could thus provide the data needed to define the xy plane over the region of interest. However, in practice an improvement is made when the 2D pulse sequence is applied to imaging. For normal 2D spectroscopy, the value of Ω is fixed by molecular parameters, whereas for imaging we can specify the value of G_x, hence alter the value of Ω and the product Ωt_1 while keeping t_1 fixed. Thus, the 2D sequence typically used for imaging is shown in Fig. 14.2, where the multiple values shown for G_x indicate that this gradient is symmetrically incremented in successive repetitions over the range $\pm(G_x)_{\max}$. The advantage of changing G_x and keeping t_1 fixed is that the same period is allowed in each repetition for transverse relaxation to occur, an important consideration in avoiding distortion in the image. Thus a two-dimensional Fourier transform provides the frequencies, hence the location of signals in the xy plane.

The rf portion of Fig. 14.2 shows a $90°$, τ, $180°$ spin echo pulse sequence, rather than a simple $90°$ pulse. All imaging studies employ either a spin echo sequence or a gradient echo to avoid acquisition of data during the FID, which decays rapidly in the presence of a magnetic field gradient. Instead, data acquisition occurs during the echo, when the rf circuitry is not subject to aberrations from having just been switched, as in the FID. If a spin echo is obtained, chemical shifts are refocused in the phase encoding direction, which may also have some advantages.

Slice Selection

As we know, the 2D pulse sequence can easily be extended to three dimensions. In a very few instances true 3D images are obtained in this way, but the number of repetitions is very large for reasonable resolution (as we see subsequently). A more efficient method, as illustrated in Fig. 14.2, is to select a small region along the z axis by applying a gradient G_z and a *selective* 90° pulse to excite only a range of nuclei in a particular region. This *slice selection* procedure is the technique used for almost all imaging. Repetition of the whole procedure with an rf pulse at a different frequency then selects another slice. The improvement in efficiency over a true 3D method results from the fact that it is unnecessary to wait for relaxation to restore the system to equilibrium between the repetitions because nuclei in the second slice are not affected by the first selective pulse. Thus, within one overall repetition period T_R, we can obtain data from a large number of slices, the thickness of each determined by the characteristics of the selective pulse. As we discussed in Section 9.6, a long rectangular pulse has the frequency profile of a sinc function; hence, in practice, the rf amplitude of the selective pulse is tailored to provide a sharp (ideally rectangular) frequency profile.

A more complete analysis of the physics involved in this imaging technique shows that it is necessary to augment the gradient in the slice selection (z) direction with a gradient of half the intensity and opposite polarity in order to compensate for unwanted dephasing effects. We shall not describe the details here, but the compensating gradient is included in Figure 14.2.

Repetition Time

For good resolution and a large field of view, the time to obtain a human image with conventional methods can be of the order of minutes. As in 2D NMR spectroscopy, it is important to minimize the use of instrument time, but for imaging of living subjects, there is even more incentive to shorten the overall scan time to minimize motion artifacts and the patient's discomfort. We saw in Chapter 3 that use of the Ernst flip angle (Eq. 3.26), rather than a 90° pulse, optimizes signal/noise ratio in a given total experiment time. In imaging of biological tissues that have different values of T_1, it is not possible to optimize the flip angle for all tissues. Nevertheless, because $T_1 \gg T_2$ for most tissue, the use of a small flip angle can decrease the repetition time by a factor of 10 in the FLASH (*fast low angle shot*) method and its later modifications.

14.3 k SPACE; ECHO PLANAR IMAGING

Although the 2D methods that we have described are very widely used and are adequate for many purposes, other approaches are also used that have advantages in certain applications. The object of the study is, of course, to obtain an image that faithfully reproduces the features over (usually) a planar field of view. In 2D NMR spectroscopy we wish to obtain all frequencies within a specified two-dimensional range (less than the Nyquist frequency), but we obtain the data as a function of time, which is the Fourier conjugate of frequency. In imaging, the data are, of course, acquired during real time. However, they should be thought of as being a function, not of time, but of the Fourier conjugate of a spatial variable, which is usually given the symbol k, the spatial frequency vector, with the units radians/cm (rather than radians/s for ω):

$$k(t) = \int_0^t \mathbf{G}(t')\, dt' \tag{14.4}$$

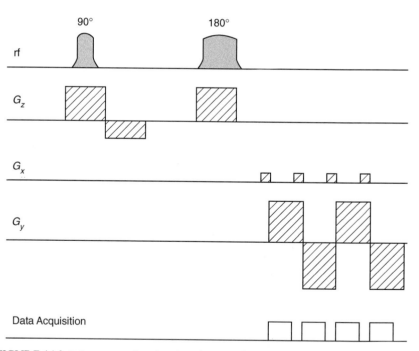

FIGURE 14.3 Pulse sequence for echo-planar imaging. Slice selection is carried out as in Fig. 14.2. The readout gradient G_y is rapidly switched in direction in synchrony with application of a pulsed phase-encode gradient G_x, as indicated. Data are acquired while G_y is turned on but G_x is off.

$\mathbf{G}(t')$ is the time-dependent spatial encoding gradient in a vector format. There are many ways of scanning k space, ranging from the standard 2D procedure to permutations of the sequence for acquiring 2D data, use of gradients in various angular directions (*projection-reconstruction* method), spiral encoding, and finally *echo planar imaging* (EPI).

We saw in Section 9.2 that an echo can be produced without a 180° pulse by simply reversing the direction of a magnetic field gradient (a *gradient-recalled echo*), and we explored the use of such a gradient. EPI uses a sequence of such rapidly reversed field gradients to produce a series of echoes, just as a series of 180° pulses produces echoes in a Carr−Purcell pulse sequence. However, for imaging, a phase encoding gradient is applied before each strong bipolar gradient in order to generate information from a single, continuous back-and-forth rectilinear trajectory that covers the entire desired planar region of k space. The entire pulse and gradient sequence is shown in Fig. 14.3. (Slice selection is carried out by a selective pulse, as in the conventional 2D imaging method.)

An echo planar image is acquired in tens of milliseconds, rather than in the 1−2 minutes required for conventional imaging. It can therefore be used more readily to acquire images where the sample is subject to motion. Humans and animals have regular motions, such as respiration and heartbeat, and irregular motions, such as abdominal contractions. Such motion artifacts can seriously degrade the quality of conventional images, but the time scale for EPI is sufficiently short that such motions are "frozen" in time. The principal disadvantage of EPI is the additional complexity and expense of the rapidly reversed gradients.

14.4 FACTORS AFFECTING IMAGE CONTRAST

The information in an image is perceived by the human eye in terms of features that contrast in some way from their surroundings. In a 2D NMR spectrum we observe peaks at positions where NMR resonances occur and nothing in between if we have set the noise threshold to a satisfactory level. An analogous situation occurs if we image an NMR-active material dispersed in an NMR-inactive material, such as the ^{1}H resonance of water droplets in air or in sandstone, or the ^{19}F resonance of a fluorine-containing substance injected into an animal. However, most samples of interest are not so simple. In particular, all biological tissues are composed largely of water, which is the principal component that is imaged. Although the amounts of water in tissues vary, using the intensity of the resonance alone would provide poor contrast in most instances between different types of tissue. Other factors, such as relaxation times, fluid flow, and diffusion coefficients, vary from one type of material or tissue to another and can provide additional contrast features.

Relaxation Times

As we saw in Chapter 8, relaxation is determined by the rate of molecular motion and by interactions such as chemical exchange. Water in the complex milieu of a biological tissue has restricted mobility in the vicinity of large macromolecules and even larger supramolecular organelles, and because of rapid exchange, the relaxation time of bulk intracellular water can be significantly altered from its value in a liquid. Whereas $T_1 \approx 3$ s and $T_2 \approx 2$ s in pure liquid water, in biological tissues $T_1 \approx 0.2-1$ s and T_2 is often an order of magnitude shorter. Both T_1 and T_2 in tissues are frequency dependent, and the values of both relaxation times vary substantially from one type of tissue to another, while water in extracellular fluids has T_1 ranging from about 1 to 3 s. In general, relaxation times need not be measured in practical imaging applications. However, from a very large number of empirical observations during laboratory and clinical studies, it is known that image contrast can be improved in specific instances by emphasizing differences in T_1 and/or T_2 between tissue components. From our discussion in Sections 2.9 and 9.2, it is easy to see how such differences might be optimized.

If it is desired to emphasize T_1 differences, one easy choice is to repeat the pulse sequence with a short repetition time T_R, so that magnetizations are unable to relax completely back to equilibrium. Thus signals from water molecules with long T_1 are reduced in intensity. Alternatively, an inversion recovery sequence (180°, τ, 90°) can be applied instead of just a 90° pulse to excite the resonance. With careful selection of the value of τ, as indicated in Fig. 14.4, T_1 differences can be emphasized.

In many instances, it has been found that pathological conditions can most easily be observed by emphasizing differences in T_2. Because images are normally obtained by use of a spin echo sequence (90°, τ, 180°), adjustment of τ permits

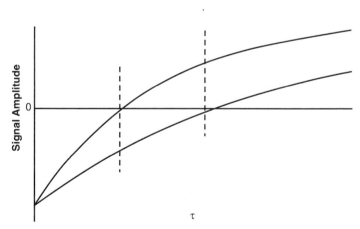

FIGURE 14.4 Recovery of magnetization to its equilibrium value after inversion in a 180°, τ, 90° pulse sequence. The relative intensities of two components with different values of T_1 can be varied by choice of value of τ.

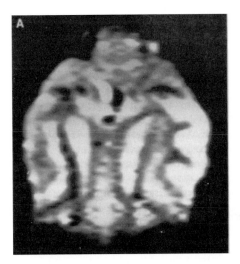
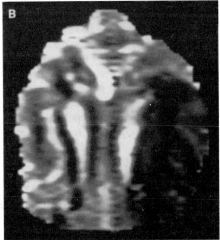

FIGURE 14.5 Transverse images of a cat brain following occlusion of the middle cerebral artery, which causes damage to the right side of the brain. (A) Conventional image, showing little difference between the two sides of the brain. (B) Diffusion image from a pulsed field gradient experiment, showing infarcted area, which has little diffusion. From Moonen et al.[133]

discrimination on the basis of T_2, in a manner quite analogous to the illustration in Fig. 14.4. (The imaging literature uses the time of the echo, called TE $= 2\tau$, rather than τ itself.)

Diffusion

In certain pathological conditions contrast by emphasizing relaxation times is inconclusive for diagnostic purposes, but the diffusion of water in the affected tissue is significantly altered. We saw in Section 9.3 that the CPMG pulse sequence can be used to measure diffusion coefficients, according to Eq. 9.2. Such a sequence can be readily incorporated into an imaging study to produce an image based on the diffusion coefficient. An example is given in Fig. 14.5, which shows the greatly enhanced contrast obtained in the brain of an experimental animal subjected to a simulated stroke.

Flow

We saw in Section 9.2 that the use of a bipolar magnetic field gradient permits the measurement of bulk flow. This technique, again, is readily adaptable to NMR imaging and can be used to discriminate between the flowing water in blood and the static water in tissues.

Magnetic Susceptibility

In *in vivo* NMR imaging it is often feasible to alter the magnetic susceptibility of the blood by injecting a paramagnetic material, such as a chelate of gadolinium, which can alter susceptibility by approximately 1 ppm. The local magnetic field in the capillaries is then different from that of the surrounding tissue. Although the field difference is very small, it acts across the extremely thin capillary wall to generate a magnetic field *gradient* that may be more than 100 G/cm—a far larger value than imposed externally. This large gradient causes nuclear spins in the nearby surrounding tissue to dephase very rapidly, thus shortening T_2^* significantly. Because individual molecules move rapidly through the gradient, refocusing effect of the observing spin echo sequence is negligible, and a suitable pulse sequence using gradient recalled echoes can be used to accentuate the local inhomogeneity effect. Thus the paramagnetic agent acts indirectly to increase the contrast between the tissue near the capillaries and that farther away. One example of the use of a gadolinium *contrast agent* is the visualization of certain tumors, which have a reduced blood supply, against adjacent normal tissue.

In addition to exogenous contrast agents supplied to the blood, the blood itself may serve as a contrast agent. Blood cells come from the lungs saturated with oxygen in the form of oxyhemoglobin, which is diamagnetic, but as the oxygen is removed for cell metabolism, the (deoxy)hemoglobin thus formed is paramagnetic. Hence, the level of oxygen in the blood in the vicinity of a rapidly metabolizing cell is altered, and by the susceptibility mechanism that we described, nuclei in the nearby metabolizing tissue may be identified. This technique, called *b*lood *o*xygenation *l*evel *d*ependent (BOLD) contrast, is used particularly to identify active regions in the brain associated with some particular sensory, motor, or cognitive function. NMR imaging is thus used to measure biological function, as well as anatomy, an area called *functional magnetic resonance imaging* (fMRI).

14.5 CHEMICAL SHIFT IMAGING AND *IN VIVO* SPECTROSCOPY

Most imaging in humans, animals, and other living biological systems is done by ^{1}H NMR, in which the principal constituent observed is water, with lipids providing a smaller but nonnegligible signal. For the studies we have described the parameters are usually set to include both water and lipid in a single pixel, but there are techniques that can easily distinguish between the two. Of more interest, however, are the many small molecules in cellular systems that participate in metabolic function and that can be studied by ^{1}H, ^{13}C, or ^{31}P NMR. The procedure for obtaining their spectra in a localized environment consists of a slice selection along the z axis as before, then phase encoding in both the

x and y directions, and finally recording the FID in the *absence* of a magnetic field gradient. Thus a normal NMR spectrum is obtained for each pixel. Because of the low (millimolar) concentrations, the pixel size is much larger than used for normal high resolution imaging in order to obtain adequate signal. As usual in ^{1}H NMR spectroscopy, it is necessary to suppress the water signal, and in biological systems also the lipid signals, which are much stronger than the compounds of interest. As we saw in Section 9.6, there are a number of methods for suppression of a solvent line, and several can be adapted for use in *in vivo* spectroscopy.

In vivo NMR spectra are often displayed in the normal fashion of signal versus frequency. However, an alternative method of displaying such chemical shift data is to select the frequency at the peak of a particular component and to show the 2D *chemical shift image*. Such a presentation is often useful in investigating the spatial variation in concentration of a particular substance.

14.6 NMR IMAGING IN SOLIDS

NMR imaging is increasingly used to study structures and processes in solid materials. In some instances, the material being imaged is a liquid that penetrates a solid substance—for example, water and oil in sandstone. Here the methods are similar to those that we have already described for imaging water in biological materials, with parameters adjusted according to the nature and size of the samples.

By applying the methods that we discussed in Chapter 7 for narrowing NMR lines of solids, it is possible to obtain an image of a solid material itself. A description of the experiments is outside the scope of this book. However, the whole range of methods described in Chapter 7 can be used, including multiple pulse line narrowing, magic angle spinning (synchronized with a rotating rf field), use of gradient echo techniques, generation of an rf field that rotates at the magic angle with a stationary sample, and multiple quantum approaches. Utilizing the line-narrowing methods in conjunction with strong magnetic field gradients presents a technical challenge, but very useful results are obtained. As in medical imaging, contrast often depends on differences in T_2 or $T_{1\rho}$ among various components in a complex solid material.

An alternative approach is to accept the broad lines characteristic of a solid but to use an extremely large gradient, such as that found in the stray magnetic field near a large superconducting magnet (*stray field imaging*, STRAFI). For example, with a 4.7 T magnet a gradient of about 40 tesla/m can be obtained, which is two orders of magnitude larger than used for medical imaging. Because the gradient is fixed in space, the sample must be moved and data obtained on a single point at a time. STRAFI is thus very inefficient but can produce useful images of solid substances not amenable to study in other ways.

14.7 ADDITIONAL READING AND RESOURCES

Because NMR imaging constitutes a very broad area from human studies to animals, microscopy, and solids, there are many specialized books in the field. We mention two that include good discussions of many aspects of the techniques: *NMR in Physiology and Biomedicine* edited by Robert Gillies[134] and *Principles of NMR Microscopy* by Paul Callaghan.[135] About 20% of the articles in the *Encyclopedia of NMR* relate to NMR imaging and to *in vivo* spectroscopy. All aspects of the technology, as well as many applications to biomedicine and to studies of solid materials, are included.

Properties of Common Nuclear Spins

Isotope	Spin	Resonance frequency[a]	Relative sensitivity[b]	Natural abundance[c]	Spin of γ	Quadrupole moment[d]
^1n	$\frac{1}{2}$	68.506	32.2	—	−	
^{1}H	$\frac{1}{2}$	100.0000	100	99.985	+	
^{2}H	1	15.3506	0.97	0.015	+	0.286
^{3}H	$\frac{1}{2}$	106.6640	121	—	+	
^{3}He	$\frac{1}{2}$	76.1790	44.2	0.0001	−	
^{6}Li	1	14.7161	0.85	7.4	+	−0.082
^{7}Li	$\frac{3}{2}$	38.8638	29.3	92.6	+	−4.01
^{9}Be	$\frac{3}{2}$	14.0518	1.39	100	−	5.23
^{10}B	3	10.7437	1.99	19.6	+	8.46
^{11}B	$\frac{3}{2}$	32.0840	16.5	80.4	+	4.06
^{13}C	$\frac{1}{2}$	25.1450	1.59	1.11	+	
^{14}N	1	7.2263	0.10	99.6	+	2.01
^{15}N	$\frac{1}{2}$	10.1368	0.10	0.4	−	
^{17}O	$\frac{5}{2}$	13.5564	2.91	0.037	−	−2.56
^{19}F	$\frac{1}{2}$	94.0940	83.3	100	+	
^{21}Ne	$\frac{3}{2}$	7.899	0.25	0.26	−	10.16
^{23}Na	$\frac{3}{2}$	26.4518	9.25	100	+	10.06
^{25}Mg	$\frac{5}{2}$	6.1216	0.27	10.1	−	19.94
^{27}Al	$\frac{5}{2}$	26.0569	20.6	100	+	14.03
^{29}Si	$\frac{1}{2}$	19.8672	0.78	4.70	−	
^{31}P	$\frac{1}{2}$	40.4807	6.63	100	+	

(Continues)

Isotope	Spin	Resonance frequency[a]	Relative sensitivity[b]	Natural abundance[c]	Spin of γ	Quadrupole moment[d]
^{33}S	$^3/_2$	7.6701	0.23	0.76	+	−6.78
^{35}Cl	$^3/_2$	9.7979	0.47	75.5	+	−8.11
^{37}Cl	$^3/_2$	8.1558	0.27	24.5	+	−6.39
^{39}K	$^3/_2$	4.6664	0.05	93.1	+	5.9
^{41}K	$^3/_2$	2.5613	0.008	6.9	+	7.2
^{43}Ca	$^7/_2$	6.7300	0.64	0.15	−	−4.09
^{45}Sc	$^7/_2$	24.2917	30.1	100	+	−22
^{47}Ti	$^5/_2$	5.6376	0.21	7.28	−	29
^{49}Ti	$^7/_2$	5.6391	0.38	5.51	−	24
^{51}V	$^7/_2$	26.3030	38.2	99.8	+	−5.2
^{53}Cr	$^3/_2$	5.6525	0.09	9.55	−	−15
^{55}Mn	$^5/_2$	24.7891	17.5	100	+	33
^{57}Fe	$^1/_2$	3.2378	0.003	2.19	+	
^{59}Co	$^7/_2$	23.7271	27.7	100	+	40.4
^{61}Ni	$^3/_2$	8.9361	0.36	1.19	−	16.2
^{63}Cu	$^3/_2$	26.5154	9.31	69.1	+	−21.1
^{65}Cu	$^3/_2$	28.4037	11.4	30.9	+	−19.5
^{67}Zn	$^5/_2$	6.2568	0.29	4.11	+	15
^{69}Ga	$^3/_2$	24.0013	6.91	60.4	+	16.8
^{71}Ga	$^3/_2$	30.4966	14.2	39.6	+	10.6
^{73}Ge	$^9/_2$	3.4883	0.14	7.76	−	−17.3
^{75}As	$^3/_2$	17.1227	2.51	100	+	31.4
^{77}Se	$^1/_2$	19.0715	0.69	7.58	+	
^{79}Br	$^3/_2$	25.0545	7.86	50.5	+	33.1
^{81}Br	$^3/_2$	27.0070	9.85	49.5	+	27.6
^{83}Kr	$^9/_2$	3.8476	0.19	11.6	−	25.3
^{85}Rb	$^5/_2$	9.6552	1.05	72.2	+	22.8
^{87}Rb	$^3/_2$	32.7212	17.5	27.9	+	13.2
^{87}Sr	$^9/_2$	4.3338	0.27	7.02	−	33.5
^{89}Y	$^1/_2$	4.9002	0.012	100	−	
^{91}Zr	$^5/_2$	9.2963	0.95	11.2	−	−20.6
^{93}Nb	$^9/_2$	24.4762	48.2	100	+	−32
^{95}Mo	$^5/_2$	6.5169	0.32	15.7	−	2.2
^{97}Mo	$^5/_2$	6.6537	0.34	9.5	−	25.5
^{99}Tc	$^9/_2$	22.5083	37.6	—	+	30
^{99}Ru	$^3/_2$	4.6052	0.019	12.7	−	7.9
^{101}Ru	$^5/_2$	5.1614	0.14	17.1	−	45.7

(*Continues*)

Isotope	Spin	Resonance frequency[a]	Relative sensitivity[b]	Natural abundance[c]	Spin of γ	Quadrupole moment[d]
^{103}Rh	$\frac{1}{2}$	3.1723	0.003	100	−	
^{105}Pd	$\frac{5}{2}$	4.5761	0.11	22.2	−	66.0
^{107}Ag	$\frac{1}{2}$	4.0479	0.007	51.8	−	
^{109}Ag	$\frac{1}{2}$	4.6536	0.010	48.2	−	
^{111}Cd	$\frac{1}{2}$	21.2155	0.95	12.8	−	
^{113}Cd	$\frac{1}{2}$	22.1932	1.09	12.3	−	
^{113}In	$\frac{9}{2}$	21.8657	34.5	4.3	+	79.9
^{115}In	$\frac{9}{2}$	21.9125	34.7	95.7	+	81.0
^{117}Sn	$\frac{1}{2}$	35.6323	4.52	7.61	−	
^{119}Sn	$\frac{1}{2}$	37.2906	5.18	8.58	−	
^{121}Sb	$\frac{5}{2}$	23.9306	16.0	57.3	+	−36
^{123}Sb	$\frac{7}{2}$	12.9589	4.57	42.7	+	−49
^{123}Te	$\frac{1}{2}$	26.1698	1.8	0.87	−	
^{125}Te	$\frac{1}{2}$	31.5498	3.15	6.99	−	
^{127}I	$\frac{5}{2}$	20.0086	9.34	100	+	−78
^{129}Xe	$\frac{1}{2}$	27.8089	2.12	26.4	−	
^{131}Xe	$\frac{3}{2}$	8.2443	0.28	21.2	+	−12
^{133}Cs	$\frac{7}{2}$	13.1162	4.74	100	+	−0.37
^{135}Ba	$\frac{3}{2}$	9.9344	0.49	6.59	+	16
^{137}Ba	$\frac{3}{2}$	11.1129	0.69	11.3	+	24.5
^{139}La	$\frac{7}{2}$	14.1256	5.92	99.9	+	20
^{141}Pr	$\frac{5}{2}$	30.62	29.3	100	+	−5.89
^{143}Nd	$\frac{7}{2}$	5.45	0.34	12.2	−	−63
^{145}Nd	$\frac{7}{2}$	3.36	0.079	8.3	−	−33
^{147}Sm	$\frac{7}{2}$	4.17	0.15	15.0	−	−25.9
^{149}Sm	$\frac{7}{2}$	3.44	0.075	13.8	−	7.5
^{151}Eu	$\frac{5}{2}$	24.86	17.8	47.8	+	90.3
^{153}Eu	$\frac{5}{2}$	10.98	1.53	52.2	+	241
^{155}Gd	$\frac{3}{2}$	3.07	0.028	14.7	−	130
^{157}Gd	$\frac{3}{2}$	4.03	0.054	15.7	−	136
^{159}Tb	$\frac{3}{2}$	24.04	5.83	100	+	143
^{161}Dy	$\frac{5}{2}$	3.44	0.042	18.9	−	251
^{163}Dy	$\frac{5}{2}$	4.82	0.112	25.0	+	265
^{165}Ho	$\frac{7}{2}$	21.34	18.1	100	+	349
^{167}Er	$\frac{7}{2}$	2.88	0.051	22.9	−	357
^{169}Tm	$\frac{1}{2}$	8.29	0.057	100	−	
^{171}Yb	$\frac{1}{2}$	17.4993	0.55	14.3	+	

(*Continues*)

Isotope	Spin	Resonance frequency[a]	Relative sensitivity[b]	Natural abundance[c]	Spin of γ	Quadrupole moment[d]
^{173}Yb	$^5/_2$	4.87	0.133	16.1	−	280
^{175}Lu	$^7/_2$	11.42	3.12	97.4	+	349
^{177}Hf	$^7/_2$	4.042	0.064	18.5	+	337
^{179}Hf	$^9/_2$	2.54	0.022	13.8	−	379
^{181}Ta	$^7/_2$	11.9896	3.60	100	+	328
^{183}W	$^1/_2$	4.1664	0.007	14.4	+	
^{185}Re	$^5/_2$	22.5246	13.3	37.1	+	218
^{187}Os	$^1/_2$	2.2823	0.001	1.64	+	
^{189}Os	$^3/_2$	7.7654	0.23	16.1	+	85.6
^{191}Ir	$^3/_2$	1.7195	0.003	37.3	+	81.6
^{193}Ir	$^3/_2$	1.8725	0.003	62.7	+	75.1
^{195}Pt	$^1/_2$	21.4144	0.99	33.8	+	
^{197}Au	$^3/_2$	1.7540	0.003	100	+	54.7
^{199}Hg	$^1/_2$	17.9103	0.57	16.8	+	
^{201}Hg	$^3/_2$	6.6114	0.14	13.2	−	38.5
^{203}Tl	$^1/_2$	57.1232	18.7	29.5	+	
^{205}Tl	$^1/_2$	57.6338	19.2	70.5	+	
^{207}Pb	$^1/_2$	20.9206	0.92	22.6	+	
^{209}Bi	$^9/_2$	16.0693	13.7	100	+	−37
^{235}U	$^7/_2$	1.8410	0.012	0.72	+	455
^0e	$^1/_2$	65.823×10^3	2.84×10^{10}	—	−	

[a]Resonance frequency in MHz in a magnetic field (approximately 2.35 T) in which the ^{1}H resonance of tetramethylsilane in CDCl$_3$ is exactly 100 MHz.
[b]Sensitivity S at constant field relative to $S(^1H) = 100$.
[c]Natural abundance in percent.
[d]Nuclear electric quadrupole moment in units of 10^{-26} cm^2.

Data adapted from an unpublished compilation by K. Lee and W. A. Anderson (Varian Associates, 1967) and from the article by R. K. Harris.[56]

ABX and AA'XX' Spectra

We present here a further discussion of the structure and analysis of two simple spin systems that are often encountered, even at high magnetic fields. These features are also indicative of considerations that are applicable to other, more complex, spin systems.

B.1 THE ABX SYSTEM

We discussed the ABX system briefly in Section 6.13. Here we provide additional details on the form of the matrix elements, the factoring of the secular equation, and the expressions for transition frequencies and intensities. In addition, we describe in some detail the use of the resultant algebraic expressions to analyze an experimental ABX spectrum. Although such analysis for a specific case can be carried out by computer spectral simulation, it is instructive to see the steps used in the general algebraic procedure, which is analogous to that used in Section 6.8 but is more tedious.

Structure of the Spectrum

The basis functions for the ABX system are given in Table 6.5 and are classified according to the operator F_z. As we saw for the ABC system in Section 6.13, the secular equation has two 3×3 factors. However, because $(\nu_A - \nu_B)$ and $(\nu_B - \nu_X)$ are much larger than J_{AX} and J_{BX}, the diagonal element $\mathcal{H}_{44}$ is much larger than the off-diagonal elements $\mathcal{H}_{24}$ and $\mathcal{H}_{34}$ that connect state ϕ_4 to states ϕ_2 and ϕ_3, and these off-diagonal elements can be set to zero. Hence ϕ_4 does not mix with other states, and the 3×3 determinant factors into 2×2 and a 1×1 determinants that are readily solved algebraically. As indicated in Section 6.13, this is tantamount to defining the approximate quantum numbers $f_z(AB)$ and $f_z(X)$. Analogous factoring occurs for the 3×3 block arising from the three functions with $f_z = -1$.

The computation of the matrix elements and the calculation of the transition frequencies and intensities are carried out in a manner analogous to that for the AB system in Section 6.8. The results are summarized in Table B.1 in terms of the quantities D_+, D_-, θ_+, θ_-, and ν_{AB}, defined by the five equations:

$$2D_+ \cos 2\theta_+ = (\nu_A - \nu_B) + \tfrac{1}{2}(J_{AX} - J_{BX}) \tag{B.1}$$

$$2D_- \cos 2\theta_- = (\nu_A - \nu_B) - \tfrac{1}{2}(J_{AX} - J_{BX}) \tag{B.2}$$

$$2D_+ \sin 2\theta_+ = J_{AB} \tag{B.3}$$

$$2D_- \sin 2\theta_- = J_{AB} \tag{B.4}$$

$$\nu_{AB} = \tfrac{1}{2}(\nu_A + \nu_B) \tag{B.5}$$

It is easier to use these relations in the analysis of an experimentally obtained ABX spectrum if certain limitations are imposed on these quantities. D_+ and D_- are defined as *positive* quantities, with $D_+ > D_-$. The nuclei are labeled so that $\nu_A \geq \nu_B$. The values of $2\theta_+$ varies from 0 to $\pi/2$, while $2\theta_-$ ranges from 0 to π. It can be shown that these limitations do not in any way impose any restrictions on the measurable physical quantities.

TABLE B.1 ABX Spectrum: Frequencies and Relative Intensities

Line	Origin	Energy	Relative intensity
1	B	$\nu_{AB} + \tfrac{1}{4}(-2J_{AB} - J_{AX} - J_{BX}) - D_-$	$1 - \sin 2\theta_-$
2	B	$\nu_{AB} + \tfrac{1}{4}(-2J_{AB} + J_{AX} + J_{BX}) - D_+$	$1 - \sin 2\theta_+$
3	B	$\nu_{AB} + \tfrac{1}{4}(2J_{AB} - J_{AX} - J_{BX}) - D_-$	$1 + \sin 2\theta_-$
4	B	$\nu_{AB} + \tfrac{1}{4}(2J_{AB} + J_{AX} + J_{BX}) - D_+$	$1 + \sin 2\theta_+$
5	A	$\nu_{AB} + \tfrac{1}{4}(-2J_{AB} - J_{AX} - J_{BX}) + D_-$	$1 + \sin 2\theta_-$
6	A	$\nu_{AB} + \tfrac{1}{4}(-2J_{AB} + J_{AX} + J_{BX}) + D_+$	$1 + \sin 2\theta_+$
7	A	$\nu_{AB} + \tfrac{1}{4}(2J_{AB} - J_{AX} - J_{BX}) + D_-$	$1 - \sin 2\theta_-$
8	A	$\nu_{AB} + \tfrac{1}{4}(2J_{AB} + J_{AX} + J_{BX}) + D_+$	$1 - \sin 2\theta_+$
9	X	$\nu_X - \tfrac{1}{2}(J_{AX} + J_{BX})$	1
10	X	$\nu_X + D_+ - D_-$	$\cos^2(\theta_+ - \theta_-)$
11	X	$\nu_X - D_+ + D_-$	$\cos^2(\theta_+ - \theta_-)$
12	X	$\nu_X + \tfrac{1}{2}(J_{AX} + J_{BX})$	1
13	Comb.	$2\nu_{AB} - \nu_X$	0
14	Comb.(X)	$\nu_X - D_+ - D_-$	$\sin^2(\theta_+ - \theta_-)$
15	Comb.(X)	$\nu_X + D_+ + D_-$	$\sin^2(\theta_+ - \theta_-)$

By combining Eq. B.1 with B.3 (and B.2 with B.4) and carrying out suitable algebraic operations, we obtain

$$2D_+ = \{[(\nu_A - \nu_B) + \tfrac{1}{2}(J_{AX} - J_{BX})]^2 + J_{AB}^2\}^{1/2} \qquad (B.6)$$

$$2D_- = \{[(\nu_A - \nu_B) - \tfrac{1}{2}(J_{AX} - J_{BX})]^2 + J_{AB}^2\}^{1/2} \qquad (B.7)$$

Equations B.6 and B.7 are analogous to Eq. 6.35 for the AB system, except that $(\nu_A - \nu_B)$ is replaced by the "effective" chemical shift difference $[(\nu_A - \nu_B) \pm \tfrac{1}{2}(J_{AX} - J_{BX})]$.

The wave functions are mixtures of A and B functions, and the transitions cannot strictly be called A or B transitions. The labeling "A" and "B" in the second column of Table B.1 is convenient but is strictly applicable only in the limiting case of large $(\nu_A - \nu_B)$. The "X" transitions are pure, and the combination transitions 14 and 15 also appear in the X region of the spectrum and may have intensities comparable with those of X lines. The combination transition 13 involves simultaneous flipping of all spins and is forbidden.

An ABX spectrum is completely described by nine quantities: the three chemical shifts ν_A, ν_B, and ν_X; the magnitudes of the three coupling constants $|J_{AB}|$, $|J_{AX}|$, and $|J_{BX}|$; and the signs of the three coupling constants. From Table B.1 and Fig. 6.8, it is apparent that the X lines are symmetrically arranged about ν_X, so that the value of ν_X can be immediately determined from an experimentally observed spectrum. From Table B.1, the average of the frequencies of all eight AB lines gives $\nu_{AB} = \tfrac{1}{2}(\nu_A + \nu_B)$, so that ν_A and ν_B can be determined separately if $(\nu_A - \nu_B)$ is also found, as indicated in the following.

Examination of the expressions in Table B.1 shows that a change in sign of J_{AB} has no effect on either the observed frequencies or intensities. The AB lines would have different labels, but no observable change would take place. Likewise, the absolute signs of J_{AX} and J_{BX} cannot be determined, because a change in sign of *both* J_{AX} and J_{BX} gives a spectrum that is unchanged in both frequency and intensity. A change in sign of only one of these two coupling constants may change some features in the spectrum, so that it is sometimes possible (depending on the parameters involved) to determine the *relative* signs of J_{AX} and J_{BX}.

Figure 6.8 shows that the AB portion of the spectrum may be decomposed into two AB-type quartets, or subspectra, which we designate $(ab)_+$ and $(ab)_-$. The values of J_{AB}, D_+, and D_- are readily determined, as indicated in Fig. 6.8, and the centers of the quartets are separated by $\tfrac{1}{2}|J_{AX} + J_{BX}|$. The absolute value symbol is used because we cannot know the labeling of the observed lines.

The X portion of the spectrum consists of three pairs of lines symmetrically placed around ν_X. The two strongest lines (9 and 12) are separated by $|J_{AX} + J_{BX}|$, which is just twice the separation of the centers of the $(ab)_+$ and $(ab)_-$ quartets. A pair of lines (10 and 11) is separated by $2(D_+ - D_-)$, while another pair (14 and 15) is separated by $2(D_+ + D_-)$. Lines 14 and 15 must lie outside lines 10 and 11, but the relation to lines 9 and 12 is variable. Frequently lines 14 and 15 (or alternatively lines 10 and 11) have so little intensity that they are not observed, and the X spectrum has four lines of virtually equal intensity.

Analysis of an ABX Spectrum

The analysis of an ABX spectrum to extract the magnitudes of the three chemical shifts and the three coupling constants is in principle straightforward, but in practice some care must be exercised to avoid ambiguities. As we have indicated, the values of J_{AB}, D_+, D_-, and $|J_{AX} + J_{BX}|$ are readily determined from the spectrum. To simplify notation, we define positive quantities M and N by the relations

$$\pm 2M \equiv \pm [4D_+{}^2 - J_{AB}{}^2]^{1/2} = (\nu_A - \nu_B) + \tfrac{1}{2}(J_{AX} - J_{BX}) \tag{B.8}$$

$$\pm 2N \equiv \pm [4D_-{}^2 - J_{AB}{}^2]^{1/2} = (\nu_A - \nu_B) - \tfrac{1}{2}(J_{AX} - J_{BX}) \tag{B.9}$$

Because D_+, D_-, and J_{AB} have been determined, both M and N are known experimentally. Mathematically, the following four solutions result from the possible choices of sign in Eq. B.8 and B.9:

	①	②	③	④
$\nu_A - \nu_B$	$M + N$	$M - N$	$-M + N$	$-M - N$
$\tfrac{1}{2}(J_{AX} - J_{BX})$	$M - N$	$M + N$	$-M - N$	$-M + N$

From the restrictions given previously, both $(\nu_A - \nu_B)$ and $(J_{AX} - J_{BX})$ must be positive; hence solutions ③ and ④ can be discarded. But because $M > N$, as a result of $D_+ > D_-$, ① and ② are both mathematically valid. Thus one of the two quantities $(M + N)$ or $(M - N)$ gives $(\nu_A - \nu_B)$, but we cannot tell at this point which one. In some cases we have enough information from prior studies of similar molecules to be able to reject one of the two solutions as including physically unreasonable parameters for either chemical shifts or coupling constants. In general, however, the resolution of this ambiguity requires a consideration of the intensity distribution in the X region, which depends on the values of θ_+ and θ_-.

Because M and N are both positive, solution ① gives $(\nu_A - \nu_B) > \tfrac{1}{2}(J_{AX} - J_{BX})$. From Eq. B.3, this means that $\cos 2\theta_- > 0$, so $0 < \theta_- < \pi/2$. Solution ② gives $(\nu_A - \nu_B) < \tfrac{1}{2}(J_{AX} - J_{BX})$ and $\pi/2 < \theta_- < \pi$. The two solutions usually give significantly different ratios for the intensities of the X lines, only one of which is compatible with the observed spectrum.

Table B.2 summarizes the procedure suggested in the foregoing paragraphs for a manual analysis of an ABX spectrum. Often, such analyses are carried out by computer simulation, but it is important to recognize that the ambiguities discussed are present in such a calculation, which generally relies only on frequencies, not intensities, of lines.

TABLE B.2   Procedure for the Analysis of an ABX Spectrum

1. Identify the two ab quartets on the basis of frequency and intensity relations. Note the value of J_{AB}.

2. Find the value of $\frac{1}{2}|J_{AX} + J_{BX}|$ from the separation of the centers of the two ab quartets.

3. Check the value of $|J_{AX} + J_{BX}|$ from the separation of the two strongest X lines, and identify lines 9 and 12 (see Fig. 7.8).

4. Find $2D_+$ and $2D_-$ from the separations of the first and third lines in the $(ab)_+$ and $(ab)_-$ quartets. Choose $2D_+$ as the larger. Check the values of $2D_+$ and $2D_-$ from the separations of lines in the X region and identify lines 10, 11, 14, and 15 (see Fig. 7.8).

5. Calculate M and N, where

$$2M = (4D_+{}^2 - J_{AB}{}^2)^{1/2}; \qquad 2N = (4D_-{}^2 - J_{AB}{}^2)^{1/2}$$

The two solutions for $(\nu_A - \nu_B)$ and $\frac{1}{2}(J_{AX} - J_{BX})$ are

	①	②
$\nu_A - \nu_B$	$M + N$	$M - N$
$\frac{1}{2}(J_{AX} - J_{BX})$	$M - N$	$M + N$

6. Find the value of $2\theta_+$, where $0 \le 2\theta_+ \le 90°$, from the relation

$$\sin 2\theta_+ = J_{AB}/2D_+$$

Find the *two* possible values of $2\theta_-$, where $0 \le 2\theta_- \le 180°$, from the relation

$$\sin 2\theta_- = J_{AB}/2D_-$$

Calculate the two possible values of $\sin(\theta_+ - \theta_-)$ and $\cos(\theta_+ - \theta_-)$ and from Table 6.5 compute the intensities of the X lines for each solution. If the smaller value of $\theta_-(0° - 45°)$ gives X intensities consistent with the observed spectrum, while the larger value $(45° - 90°)$ does not, then choose solution ① as the correct solution. If the converse is true, choose solution ②.

7. Find $\frac{1}{2}(\nu_A + \nu_B)$, which is the average of the centers of the $(ab)_+$ and $(ab)_-$ quartets, or equivalently, the average of the frequencies of all eight AB lines. From this value and the correct value of $(\nu_A - \nu_B)$ determined in steps 5 and 6, calculate ν_A and ν_B.

8. Assign to the sum $\frac{1}{2}(J_{AX} + J_{BX})$, for which the absolute value was found in step 2, a positive sign if the $(ab)_+$ quartet is centered at a higher frequency than the $(ab)_-$ quartet, or a negative sign if the reverse order is true. From this value and the correct value of $\frac{1}{2}(J_{AX} - J_{BX})$ determined in steps 5 and 6, calculate J_{AX} and J_{BX}.

B.2   THE AA'XX' SYSTEM

Structure of the Spectrum

As indicated in Section 6.17, an AA'XX' system has four separate coupling constants, $J_{AA'}$, $J_{XX'}$, J_{AX}, and $J_{AX'}$. In setting up the Hamiltonian in terms of 2^4 basis functions, we can simplify the resultant calculation by taking symmetry into account. Although the two A nuclei are not magnetically equivalent in the

TABLE B.3 Basis Functions for an AA'XX' Spin System

Basis functions	f_z	Symmetry	$f_z(A)$	$f_z(X)$
$\alpha_A\alpha_A\alpha_X\alpha_X$	2	s	1	1
$\alpha_A\alpha_A(\alpha_X\beta_X + \beta_X\alpha_X)$	1	s	1	0
$\alpha_A\alpha_A(\alpha_X\beta_X - \beta_X\alpha_X)$	1	a	1	0
$(\alpha_A\beta_A + \beta_A\alpha_A)\alpha_X\alpha_X$	1	s	0	1
$(\alpha_A\beta_A - \beta_A\alpha_A)\alpha_X\alpha_X$	1	a	0	1
$\alpha_A\alpha_A\beta_X\beta_X$	0	s	1	-1
$\beta_A\beta_A\alpha_X\alpha_X$	0	s	-1	1
$(\alpha_A\beta_A + \beta_A\alpha_A)(\alpha_X\beta_X + \beta_X\alpha_X)$	0	s	0	0
$(\alpha_A\beta_A - \beta_A\alpha_A)(\alpha_X\beta_X - \beta_X\alpha_X)$	0	s	0	0
$(\alpha_A\beta_A + \beta_A\alpha_A)(\alpha_X\beta_X - \beta_X\alpha_X)$	0	a	0	0
$(\alpha_A\beta_A - \beta_A\alpha_A)(\alpha_X\beta_X + \beta_X\alpha_X)$	0	a	0	0
$\beta_A\beta_A(\alpha_X\beta_X + \beta_X\alpha_X)$	-1	s	-1	0
$\beta_A\beta_A(\alpha_X\beta_X - \beta_X\alpha_X)$	-1	a	-1	0
$(\alpha_A\beta_A + \beta_A\alpha_A)\beta_X\beta_X$	-1	s	0	-1
$(\alpha_A\beta_A - \beta_A\alpha_A)\beta_X\beta_X$	-1	a	0	-1
$\beta_A\beta_A\beta_X\beta_X$	-2	s	-1	-1

molecule as a whole because of the generally unequal values of J_{AX} and $J_{AX'}$, for the basis functions it is valid to consider the two A nuclei to be symmetrically equivalent and to form symmetric and antisymmetric wave functions, as we did in Section 6.11 for the A_2 system, and of course, we can do the same for the two X nuclei. Thus we develop the basis functions given in Table B.3. These functions are classified by overall symmetry, by overall value of f_z, and by separate values of $f_z(A)$ and $f_z(X)$. As a result, the Hamiltonian factors into twelve 1×1 and two 2×2 blocks. There are 24 allowed transitions of significant intensity, half centered at ν_A and half at ν_X. Table B.4 gives the frequencies and intensities of the A transitions, and Fig. 6.15 depicts the half-spectrum schematically. (The X spectrum would be identical and furnishes no additional information on the coupling constants. It does give the X chemical shift.) It is apparent that the A spectrum is symmetric about its midpoint, ν_A. In Table B.4 the frequencies are given in terms of the parameters K, L, M, N, P, and R defined in the table. Lines 1 and 2 always coincide, as do lines 3 and 4. Thus the half-spectrum has only 10 lines.

Analysis of an AA'XX' Spectrum

The analysis of an AA'XX' spectrum is straightforward, but several ambiguities occur because of the symmetry of the spectrum. The chemical shifts ν_A and ν_X are

TABLE B.4 AA'XX' Spectrum: Frequencies and Relative
Intensities of the A Portion[a]

Line	Frequency relative to ν_A	Relative intensity
1	$\frac{1}{2}N$	1
2	$\frac{1}{2}N$	1
3	$-\frac{1}{2}N$	1
4	$-\frac{1}{2}N$	1
5	$P + \frac{1}{2}K$	$1 - K/2P$
6	$P - \frac{1}{2}K$	$1 + K/2P$
7	$-P + \frac{1}{2}K$	$1 + K/2P$
8	$-P - \frac{1}{2}K$	$1 - K/2P$
9	$R + \frac{1}{2}M$	$1 - M/2R$
10	$R - \frac{1}{2}M$	$1 + M/2R$
11	$-R + \frac{1}{2}M$	$1 + M/2R$
12	$-R - \frac{1}{2}M$	$1 - M/2R$

[a] $K = J_{AA'} + J_{XX'}$, $L = J_{AX} - J_{AX'}$, $M = J_{AA'} - J_{XX'}$, $N = J_{AX} + J_{AX'}$, $2P = (K^2 + L^2)^{1/2}$, $2R = (M^2 + L^2)^{1/2}$.

of course easily determined as the midpoints of the respective half-spectra. The
two strongest lines in the half-spectrum are separated by $|N|$. The remaining lines
can be shown to arise from two ab subspectra characterized by "effective coupling
constants" of K and M, respectively. P and R are defined by analogy to C in Eq.
6.35. Table B.4 and Fig. 6.15 show that the frequencies and intensity ratios con-
form to the AB pattern. The values of $|K|$ and $|M|$ are easily found from the sub-
spectra but cannot be distinguished from each other; $|L|$ is easily calculated from
the spectral line separations and the value of either $|K|$ or $|M|$, in a manner similar
to that illustrated in Fig. 6.3. The relative signs of J_{AX} and $J_{AX'}$ can be determined
by noting whether $|N|$ is larger or smaller than $|L|$. Because K and M cannot be
distinguished, we cannot ascertain the relative signs of $J_{AA'}$ and $J_{XX'}$. Finally, there
is no way from the spectrum alone that we can decide which of the calculated
pair is $J_{AA'}$ and which is $J_{XX'}$. The same ambiguity exists with J_{AX} and $J_{AX'}$. Usually,
however, the coupling constants can be assigned to the proper nuclei on the basis
of   analogy   to   other   systems.   For   example,   the   spectrum   of
1,1-difluoroethylene in Fig. 6.1 is an AA'XX' spectrum, which has been analyzed
to give the absolute values J_{HH} = 4.8 Hz, J_{FF} = 36.4 Hz, $J_{HF}(cis)$ = 0.7 Hz,
and $J_{HF}(trans)$ = 33.9 Hz. The assignments were readily made by analogy (see
Tables 5.1–5.3).

Review of Relevant Mathematics

This appendix includes some mathematical expressions that are used in this text. It is not intended to be comprehensive or to substitute for a mathematics textbook.

C.1 COMPLEX NUMBERS

It is frequently helpful to deal with a number z in the complex plane, where the x axis represents the "real" part of the number and the y axis represents the "imaginary" part of the number, designated by the coefficient $i = \sqrt{-1}$. A point in the complex plane may be represented by a pair of numbers, x and y, or by absolute value (modulus) r and an angle θ. The quantities are related as follows:

$$z = x + iy \qquad z = re^{i\theta} \qquad r^2 = x^2 + y^2 \qquad x = r\cos\theta \qquad y = r\sin\theta$$

$$(C.1)$$

The important Euler identity is

$$e^{i\theta} = \cos\theta + i\sin\theta$$

$$(C.2)$$

$$e^{-i\theta} = \cos\theta - i\sin\theta$$

The relations in Eq. C.2 may be added or subtracted to generate expressions for $\sin\theta$ and $\cos\theta$. Also, from Eq. C.2, we have the following relations:

$$e^{\pi i/2} = i \qquad e^{\pi i} = -1 \qquad e^{3\pi i/2} = -i \qquad e^{2\pi i} = 1 \qquad (C.3)$$

393

C.2 TRIGONOMETRIC IDENTITIES

The following trigonometric identities are very useful:

$$\sin(\theta \pm \psi) = \sin\theta\cos\psi \pm \cos\theta\sin\psi$$
$$\cos(\theta \pm \psi) = \cos\theta\cos\psi \mp \sin\theta\sin\psi \qquad\text{(C.4)}$$

The relations in Eq. C.4 may be added or subtracted to generate other useful expressions:

$$\cos\theta\cos\psi = \tfrac{1}{2}\left[\cos(\theta - \psi) + \cos(\theta + \psi)\right]$$
$$\sin\theta\sin\psi = \tfrac{1}{2}\left[\cos(\theta - \psi) - \cos(\theta + \psi)\right] \qquad\text{(C.5)}$$
$$\sin\theta\cos\psi = \tfrac{1}{2}\left[\sin(\theta - \psi) + \sin(\theta + \psi)\right]$$

C.3 VECTORS

A vector $\mathbf{P}$ is a quantity with both magnitude and direction, usually designated in boldface type (e.g., $\mathbf{B}$, ω), and represented pictorially as an arrow ↗. A vector in a space of N dimensions may be expressed in terms of N orthogonal unit basis vectors as a linear ordered array of numbers that describe the contributions from each basis vector:

$$\mathbf{P} = (p_1, p_2, p_3, \dots p_N) \qquad\text{(C.6)}$$

The scalar (dot) product of two vectors $\mathbf{P}$ and $\mathbf{R}$ at an angle θ to each other is given by

$$\mathbf{P}\cdot\mathbf{R} = |\mathbf{P}||\mathbf{R}|\cos\theta \qquad\text{(C.7)}$$

and may also be expressed in terms of the components in each dimension as

$$\mathbf{P}\cdot\mathbf{R} = p_1 r_1 + p_2 r_2 + \cdots + p_N r_N \qquad\text{(C.8)}$$

The vector (cross) product of vectors $\mathbf{P}$ and $\mathbf{R}$ is a vector $\mathbf{Q}$ orthogonal to both $\mathbf{P}$ and $\mathbf{R}$ of magnitude given by

$$\mathbf{Q} = \mathbf{P} \times \mathbf{R} = |\mathbf{P}||\mathbf{R}|\sin\theta \qquad\text{(C.9)}$$

In Cartesian coordinates, the components of $\mathbf{Q}$ are given by

$$Q_x = P_y R_z - P_z R_y$$
$$Q_y = P_z R_x - P_x R_z \qquad\text{(C.10)}$$
$$Q_z = P_x R_y - P_y R_x$$

C.4 MATRICES

A matrix M is an ordered array of numbers, usually describing a linear transformation of one vector $\mathbf{P}$ to another $\mathbf{R}$. The components of M relate the components of $\mathbf{R}$ to those of $\mathbf{P}$. If the dimensionalities of the vector spaces of $\mathbf{P}$ and $\mathbf{R}$ are equal, then the matrix is "square" with equal numbers of rows and columns. We deal only with square matrices except for the $N \times 1$ matrices that describe vectors as row or column matrices, for example:

$$\begin{bmatrix} p_1 \\ p_2 \\ p_3 \end{bmatrix} \qquad [p_1 \ p_2 \ p_3] \tag{C.11}$$

Addition of two square matrices $P + R$ is performed by taking the sums of corresponding elements, $P_{mn} + R_{mn}$. *Multiplication* of two square matrices P and R to give a square matrix Q is defined by

$$Q = PR = \begin{bmatrix} P_{11} & P_{12} & \cdots \\ P_{21} & P_{22} & \cdots \\ \cdots & \cdots & \cdots \end{bmatrix} \begin{bmatrix} R_{11} & R_{12} & \cdots \\ R_{21} & R_{22} & \cdots \\ \cdots & \cdots & \cdots \end{bmatrix} \tag{C.12}$$

where

$$Q_{ij} = \sum_{k=1}^{N} P_{ik} R_{kj} \tag{C.13}$$

For example,

$$\begin{bmatrix} aj + bm + cp & ak + bn + cq & al + bo + cr \\ dj + em + fp & dk + en + fq & dl + eo + fr \\ gj + hm + ip & gk + hn + iq & gl + ho + ir \end{bmatrix} = \begin{bmatrix} a & b & c \\ d & e & f \\ g & h & i \end{bmatrix} \begin{bmatrix} j & k & l \\ m & n & o \\ p & q & r \end{bmatrix} \tag{C.14}$$

Matrix multiplication is associative but often not commutative.

The *direct product* of two matrices $Q = P \times R$ is a matrix whose dimensionality is the product of the dimensionalities of the two matrices. The components of Q consist of all products of the components of the separate matrices, $P_{jk}R_{mn}$, with a convention as to ordering of the resultant components in Q. A specific example is given in Appendix D.

The *unit matrix*, usually given the symbol $\mathbf{1}$ or E, is a square matrix with all 1's along the principal diagonal and 0's elsewhere. For a square matrix P, the *inverse* of P is a square matrix P^{-1} that multiplies P to give the unit matrix: $PP^{-1} = P^{-1}P = \mathbf{1}$.

Many matrix manipulations involve a *similarity transformation* $R = P^{-1}QP$. In the situations that we consider, such as changes to the density matrix in

Chapter 11, matrix P describes a rotation, such as the result of a 90° pulse. In such instances matrix P is found to be *unitary* and to have a simple inverse, the complex conjugate transpose—$(P^{-1})_{mn} = P_{nm}{}^{*}$—in which rows and columns are interchanged and each i changes to $-i$. For example,

$$P = \frac{\sqrt{2}}{2}\begin{bmatrix} 1 & 0 & i & 0 \\ 0 & 1 & 0 & i \\ i & 0 & 1 & 0 \\ 0 & i & 0 & 1 \end{bmatrix} \qquad P^{-1} = \frac{\sqrt{2}}{2}\begin{bmatrix} 1 & 0 & -i & 0 \\ 0 & 1 & 0 & -i \\ -i & 0 & 1 & 0 \\ 0 & -i & 0 & 1 \end{bmatrix}$$

$$R = \frac{\sqrt{2}}{2}\begin{bmatrix} 1 & 0 & 1 & 0 \\ 0 & 1 & 0 & 1 \\ -1 & 0 & 1 & 0 \\ 0 & -1 & 0 & 1 \end{bmatrix} \qquad R^{-1} = \frac{\sqrt{2}}{2}\begin{bmatrix} 1 & 0 & -1 & 0 \\ 0 & 1 & 0 & -1 \\ 1 & 0 & 1 & 0 \\ 0 & 1 & 0 & 1 \end{bmatrix}$$

Spin Matrices

We summarize here the principal spin and rotation matrices for $I = \frac{1}{2}$, first the matrices for a single spin, then those for a two-spin system $I-S$ in which the eigenfunctions are products of the basis vectors α and β.

D.1 ONE SPIN

$$I_x = \frac{1}{2}\begin{bmatrix} 0 & 1 \\ 1 & 0 \end{bmatrix} \quad I_y = \frac{1}{2}\begin{bmatrix} 0 & -i \\ i & 0 \end{bmatrix} \quad I_z = \frac{1}{2}\begin{bmatrix} 1 & 0 \\ 0 & -1 \end{bmatrix}$$

$$I^+ = \begin{bmatrix} 0 & 1 \\ 0 & 0 \end{bmatrix} \quad I^- = \begin{bmatrix} 0 & 0 \\ 1 & 0 \end{bmatrix} \quad I^2 = \frac{3}{4}\begin{bmatrix} 1 & 0 \\ 0 & 1 \end{bmatrix}$$

$$R(90_x) = \frac{\sqrt{2}}{2}\begin{bmatrix} 1 & i \\ i & 1 \end{bmatrix} \quad R(90_y) = \frac{\sqrt{2}}{2}\begin{bmatrix} 1 & 1 \\ -1 & 1 \end{bmatrix}$$

$$R(180_x) = \begin{bmatrix} 0 & i \\ i & 0 \end{bmatrix} \quad R(180_y) = \begin{bmatrix} 0 & 1 \\ -1 & 0 \end{bmatrix}$$

D.2 TWO-SPIN SYSTEM

The basic 2×2 spin matrices for one spin can be expanded to 4×4 matrices for an $I-S$ spin system by taking the direct product of each with the 2×2 unit matrix, keeping the order I before S. For example,

$$I_x = I_x \times 1 = \frac{1}{2}\begin{bmatrix} 0 & 1 \\ 1 & 0 \end{bmatrix} \times \begin{bmatrix} 1 & 0 \\ 0 & 1 \end{bmatrix} = \frac{1}{2}\begin{bmatrix} 0 & 0 & 1 & 0 \\ 0 & 0 & 0 & 1 \\ 1 & 0 & 0 & 0 \\ 0 & 1 & 0 & 0 \end{bmatrix}$$

$$S_x = 1 \times I_x = \begin{bmatrix} 1 & 0 \\ 0 & 1 \end{bmatrix} \times \frac{1}{2}\begin{bmatrix} 0 & 1 \\ 1 & 0 \end{bmatrix} = \frac{1}{2}\left[\begin{array}{cc:cc} 0 & 1 & 0 & 0 \\ 1 & 0 & 0 & 0 \\ \hdashline 0 & 0 & 0 & 1 \\ 0 & 0 & 1 & 0 \end{array}\right]$$

If we designate the first 2×2 matrix P, the second 2×2 matrix R, and the 4×4 matrix Q, then the convention used here is to multiply matrix R by element P_{11} and place the resultant 2×2 submatrix in the upper left corner of the final matrix, as indicated, to give elements Q_{11}, Q_{12}, Q_{21}, and Q_{22}. Matrix R is then multiplied by element P_{12} and placed in the upper right corner to give elements Q_{13}, Q_{14}, Q_{23}, and Q_{24}, and the other multiplications are similarly carried out. This arrangement of elements ensures matrices with rows and columns that are in the order we customarily use—$\alpha_I\alpha_S$, $\alpha_I\beta_S$, $\beta_I\alpha_S$, $\beta_I\beta_S$.

For convenient reference, we give here the 15 matrices corresponding to the product operators for an I–S spin system.

$$I_x = \frac{1}{2}\begin{bmatrix} 0 & 0 & 1 & 0 \\ 0 & 0 & 0 & 1 \\ 1 & 0 & 0 & 0 \\ 0 & 1 & 0 & 0 \end{bmatrix} \qquad I_y = \frac{1}{2}\begin{bmatrix} 0 & 0 & -i & 0 \\ 0 & 0 & 0 & -i \\ i & 0 & 0 & 0 \\ 0 & i & 0 & 0 \end{bmatrix}$$

$$I_z = \frac{1}{2}\begin{bmatrix} 1 & 0 & 0 & 0 \\ 0 & 1 & 0 & 0 \\ 0 & 0 & -1 & 0 \\ 0 & 0 & 0 & -1 \end{bmatrix}$$

$$S_x = \frac{1}{2}\begin{bmatrix} 0 & 1 & 0 & 0 \\ 1 & 0 & 0 & 0 \\ 0 & 0 & 0 & 1 \\ 0 & 0 & 1 & 0 \end{bmatrix} \qquad S_y = \frac{1}{2}\begin{bmatrix} 0 & -i & 0 & 0 \\ i & 0 & 0 & 0 \\ 0 & 0 & 0 & -i \\ 0 & 0 & i & 0 \end{bmatrix}$$

$$S_z = \frac{1}{2}\begin{bmatrix} 1 & 0 & 0 & 0 \\ 0 & -1 & 0 & 0 \\ 0 & 0 & 1 & 0 \\ 0 & 0 & 0 & -1 \end{bmatrix}$$

$$I_xS_x = \frac{1}{4}\begin{bmatrix} 0 & 0 & 0 & 1 \\ 0 & 0 & 1 & 0 \\ 0 & 1 & 0 & 0 \\ 1 & 0 & 0 & 0 \end{bmatrix} \qquad I_yS_y = \frac{1}{4}\begin{bmatrix} 0 & 0 & 0 & -1 \\ 0 & 0 & 1 & 0 \\ 0 & 1 & 0 & 0 \\ -1 & 0 & 0 & 0 \end{bmatrix}$$

$$I_zS_z = \frac{1}{4}\begin{bmatrix} 1 & 0 & 0 & 0 \\ 0 & -1 & 0 & 0 \\ 0 & 0 & -1 & 0 \\ 0 & 0 & 0 & 1 \end{bmatrix}$$

$$I_xS_y = \frac{1}{4}\begin{bmatrix} 0 & 0 & 0 & -i \\ 0 & 0 & i & 0 \\ 0 & -i & 0 & 0 \\ i & 0 & 0 & 0 \end{bmatrix} \qquad I_yS_x = \frac{1}{4}\begin{bmatrix} 0 & 0 & 0 & -i \\ 0 & 0 & -i & 0 \\ 0 & i & 0 & 0 \\ i & 0 & 0 & 0 \end{bmatrix}$$

$$I_xS_z = \frac{1}{4}\begin{bmatrix} 0 & 0 & 1 & 0 \\ 0 & 0 & 0 & -1 \\ 1 & 0 & 0 & 0 \\ 0 & -1 & 0 & 0 \end{bmatrix}$$

$$I_yS_z = \frac{1}{4}\begin{bmatrix} 0 & 0 & -i & 0 \\ 0 & 0 & 0 & i \\ i & 0 & 0 & 0 \\ 0 & -i & 0 & 0 \end{bmatrix} \qquad I_zS_y = \frac{1}{4}\begin{bmatrix} 0 & -i & 0 & 0 \\ i & 0 & 0 & 0 \\ 0 & 0 & 0 & i \\ 0 & 0 & -i & 0 \end{bmatrix}$$

$$I_zS_x = \frac{1}{4}\begin{bmatrix} 0 & 1 & 0 & 0 \\ 1 & 0 & 0 & 0 \\ 0 & 0 & 0 & -1 \\ 0 & 0 & -1 & 0 \end{bmatrix}$$

The rotation matrices for the commonly used 90° and 180° pulses, both selective and nonselective, are given here:

$$R(90_x) = \frac{1}{2}\begin{bmatrix} 1 & i & i & -1 \\ i & 1 & -1 & i \\ i & -1 & 1 & i \\ -1 & i & i & 1 \end{bmatrix} \qquad R^I(90_x) = \frac{\sqrt{2}}{2}\begin{bmatrix} 1 & 0 & i & 0 \\ 0 & 1 & 0 & i \\ i & 0 & 1 & 0 \\ 0 & i & 0 & 1 \end{bmatrix}$$

$$R^S(90_x) = \frac{\sqrt{2}}{2}\begin{bmatrix} 1 & i & 0 & 0 \\ i & 1 & 0 & 0 \\ 0 & 0 & 1 & i \\ 0 & 0 & i & 1 \end{bmatrix}$$

$$R(90_y) = \frac{1}{2} \begin{bmatrix} 1 & 1 & 1 & 1 \\ -1 & 1 & -1 & 1 \\ -1 & -1 & 1 & 1 \\ 1 & -1 & -1 & 1 \end{bmatrix} \qquad R^I(90_y) = \frac{\sqrt{2}}{2} \begin{bmatrix} 1 & 0 & 1 & 0 \\ 0 & 1 & 0 & 1 \\ -1 & 0 & 1 & 0 \\ 0 & -1 & 0 & 1 \end{bmatrix}$$

$$R^S(90_y) = \frac{\sqrt{2}}{2} \begin{bmatrix} 1 & 1 & 0 & 0 \\ -1 & 1 & 0 & 0 \\ 0 & 0 & 1 & 1 \\ 0 & 0 & -1 & 1 \end{bmatrix}$$

$$R(180_x) = \begin{bmatrix} 0 & 0 & 0 & -1 \\ 0 & 0 & -1 & 0 \\ 0 & -1 & 0 & 0 \\ -1 & 0 & 0 & 0 \end{bmatrix} \qquad R^I(180_x) = \begin{bmatrix} 0 & 0 & i & 0 \\ 0 & 0 & 0 & i \\ i & 0 & 0 & 0 \\ 0 & i & 0 & 0 \end{bmatrix}$$

$$R^S(180_x) = \begin{bmatrix} 0 & i & 0 & 0 \\ i & 0 & 0 & 0 \\ 0 & 0 & 0 & i \\ 0 & 0 & i & 0 \end{bmatrix}$$

$$R(180_y) = \begin{bmatrix} 0 & 0 & 0 & 1 \\ 0 & 0 & -1 & 0 \\ 0 & -1 & 0 & 0 \\ 1 & 0 & 0 & 0 \end{bmatrix} \qquad R^I(180_y) = \begin{bmatrix} 0 & 0 & 1 & 0 \\ 0 & 0 & 0 & 1 \\ -1 & 0 & 0 & 0 \\ 0 & -1 & 0 & 0 \end{bmatrix}$$

$$R^S(180_y) = \begin{bmatrix} 0 & 1 & 0 & 0 \\ -1 & 0 & 0 & 0 \\ 0 & 0 & 0 & 1 \\ 0 & 0 & -1 & 0 \end{bmatrix}$$

Selected Answers to Problems

From Chapter 2

2.1   *a.* 298 MHz. *b.* 9.35 tesla. *c.* ^{1}H, 100; ^{31}P, 41; ^{17}O, 68; ^{59}Co, 165.

2.2 ^{1}H, 10,000; ^{13}C, 1.76; ^{15}N, 0.04; ^{57}Fe, 0.0066.

2.4 $[I_x, I_z] = -iI_y$; $[I_x, I_y] = iI_z$.

2.5 $P \propto \gamma^3$.

2.6   For ^{19}F: (*a*) 9×10^{-6}; (*b*) 5400×10^{-6}; (*c*) 32×10^{-6}.

2.7   11.6 gauss

2.8 $\nu_{\frac{1}{2}} = \left(\dfrac{\ln 2}{\pi}\right)^{\frac{1}{2}} \dfrac{1}{T_2} = \dfrac{1}{2.14 T_2}$

2.9   *a.* $\ln[(n_{eq} - n)/2\, n_{eq}] = -R_1 t$.

     b. $\ln[(n_{eq} - n)/(1 + f)n_{eq}] = -R_1 t$, where f is the fraction of M_0 that has been inverted.

     c. From (*a*) at t_{null}, $\ln \frac{1}{2} = -R_1 t_{null}$, but from (*b*) no such simple relation occurs.

2.10 $B_{eff}/B_1 = 1.10$. The angle between $\mathbf{B}_1$ and $\mathbf{B}_{eff}$ is 25°.

2.11 $\gamma B_1/2\pi = 12.5$ kHz, so a range of ± 6 kHz from the pulse frequency gives $> 90\%$ intensity. For a 40 μs pulse with the same value of B_1 there would be no signal.

From Chapter 3

3.1   The sampling rate must be at least 10,000 points/second. For a digital resolution of 0.5 Hz, 20,000 data points are required.

3.2 $S/N \propto (1 - e^{-\pi t})/t^{1/2}$

     S/N: $t = 0.05$ s, 110; 1 s, 95; 2 s, 71; 5 s, 45

3.3   For ^{17}O, $S/N \propto (1 - e^{-20 t})/t^{1/2}$.

     At $t = 1$ s, $S/N = 1$. The optimum S/N comes at $t \approx 1.2\, T_2^* = 0.06$ s, where $S/N = 2.85$.

3.4 $S/N \propto (1 - e^{-40 t})/(1 - e^{-20 t})^{1/2}$.

     At $t = 1$ s, $S/N = 1$. At 0.06 s, $S/N = 1.09$.

3.5 $\ln[(n_{eq} - n)/n_{eq}] = -R_1 t$

3.6   At $t = 1.27\ T_1$, $(n_{eq} - n)/n_{eq} = 0.72$, or 72% recovery. In 1000 T_1 there are 787 repetitions, and $S/N \propto (787)^{1/2}(0.72) = 20$. At $t = 5\ T_1$, $(n_{eq} - n)/n_{eq} = 0.993$. In 1000 T_1 there are 200 repetitions, and $S/N \propto (200)^{1/2}(0.0.993) = 14$.

3.7   Since the folded lines are actually at higher frequencies than shown in the spectrum, they should have a larger phase correction than is applied at the "observed" frequency.

From Chapter 4

4.1   60 MHz: 215 Hz, 3.58 ppm; 500 MHz: 3.58 ppm, 1791 Hz; 7.0 T: 3.58 ppm, 1068 Hz

4.2   The β protons are more shielded. The α and β protons experience nearly the same ring current effect from the ring in which they reside, but the α protons also experience a deshielding ring current from the other ring.

4.3 $\delta_H(CHCl_3) - \delta_H(C_6H_6) = 0.825$ ppm. $\delta_C(CHCl_3) - \delta_C(C_6H_6) = -52$ ppm. $\kappa(CHCl_3) = -0.740 \times 10^{-6}$ and $\kappa(C_6H_6) = -0.617 \times 10^{-6}$ (both in cgs units). Therefore, $\delta_{measured} - \delta_{true} = 0.257$ ppm, which is 31% of the observed ^1H chemical shift difference but only 0.49% of the ^{13}C chemical shift difference.

4.4   The nucleus experiences a ring current shielding of approximately 0.85 ppm.

4.5   Fig. 4.15: $(CH_3)_2C(OCH_3)_2$. Fig. 4.16: CH_3OCH_2COOH. The COOH line is more intense than it should be because the hygroscopic sample has absorbed water, which is in rapid exchange with the carboxyl proton.

4.6   110 ppm. Probably an amide nitrogen.

From Chapter 5

5.1 $\delta_A = 0.38$ ppm, $\delta_X = 1.77$ ppm, and $J = 8$ Hz. At 300 MHz the lines would occur at 110, 118, 527, and 535 Hz from TMS.

5.2 $^2J_{HD}$ is measured in CH_3D, and the value of $^2J_{HH}$ is found from Eq. 5.2.

5.3 $J(^{15}NH) = -56$ Hz. $J(^{15}ND) = -8.6$ Hz.

5.4 $K(^{14}NH)/K(^{13}CH) = 1.53$.

5.5   The observed splitting is the weighted average of $^1J(^{15}NH) \approx 90$ Hz in **VII** and the ^{15}N . . . H coupling through the hydrogen bond in **VI**, which has not been measured but is probably $\ll 90$ Hz. Change in the tautomeric equilibrium with temperature alters the weighting factors. The coupling persists in this rapid intramolecular exchange process involving a single proton, whereas it is averaged to zero in an intermolecular process where a given proton is replaced by another proton with random spin orientation.

5.6   *a.* The splitting arises from ^{13}C coupling to ^2H, which is not decoupled.

b. $(CH_3)_2CHCCH\!\!=\!\!CH_2$
$\qquad\quad |$
$\qquad\ \, OH$

5.7   Fig. 5.5: CH_3CHCl_2

Fig. 5.6: $CH_2COOC_2H_5$
$\qquad\quad\ |$
$\qquad\quad CH_2COOC_2H_5$

Fig. 5.7: $(C_2H_5O)_2PH$ $\qquad$ $^1J_{PH} = 688\ Hz;\,^3J_{PH} = 9\ Hz.$
$\qquad\qquad\ \ \|$
$\qquad\qquad\ \ O$

Fig. 5.8:

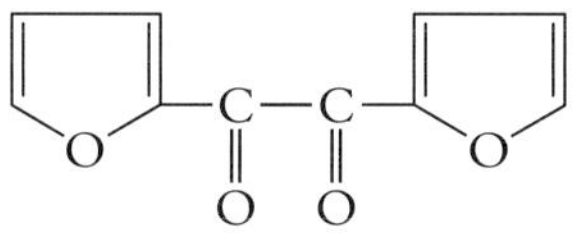

Fig. 5.9:

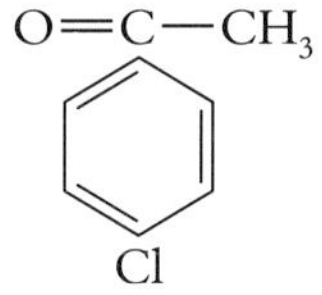

From Chapter 6

6.1 $CH_2\!\!=\!\!CHF$, ABMX; PF_3, A_3X; cubane, A_8; $CH_3CHOHCH_3$, A_6MX if OH is not exchanging rapidly or A_6M with exchange; chlorobenzene, AA'BB'C or AA'MM'X, depending on field strength; $CH_3CH_2CH_3$, A_6B_2 or $A_3A'_3B_2$ if long-range coupling is taken into account.

6.4   (*a*) and (*d*) are AB spectra.

6.5   *a.* Alternative 1: Two magnetically equivalent protons are coupled to three magnetically equivalent ^{19}F nuclei, giving an A_2X_3 system. Alternative 2: Two coupled chemically nonequivalent protons constitute an AB system.

b. Alternative 1: $\delta_A = 4.51$ ppm. Alternative 2: $\delta_A = 4.53$ ppm; $\delta_B = 4.49$ ppm.

c. Alternative 1: 2693 (1), 2702 (3), 2711 (3) 2720 (1). Alternative 2: 2690 (1), 2699 (2.1), 2715 (2.1), 2724 (1).

6.6   *a.* $\nu_A = 144.6$; $\nu_B = 120.4$ Hz. *b.* $\nu_A = 219.2$; $\nu_B = 212.8$ Hz.

6.8   (*b*), where $^4J_{HH} \approx 2$ Hz for the ring protons. In (*c*) the corresponding $^5J_{HH}$ for the ring protons is probably too small to permit observable effects under normal experimental conditions.

6.9

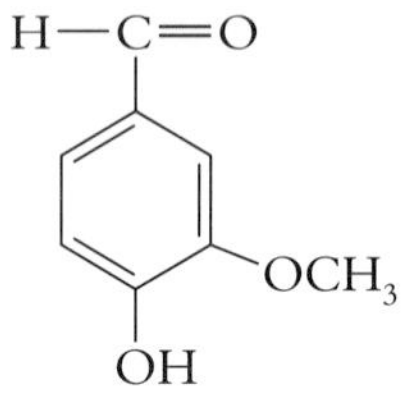

$\nu_A = 748.2$ Hz, $\nu_B = 747.3$ Hz, $\nu_X = 703.7$ Hz, $J_{AB} = 1.8$ Hz, $J_{AX} = 8.2$ Hz, $J_{BX} = 0.4$ Hz.

6.10 From the molecular formulas, these spectra must arise from *cis* and *trans* 1,2-difluoroethylene. The spectrum of the third isomer, 1,1-difluoroethylene, was shown in Fig. 6.1 and is analyzed in Appendix B.

Fig. 6.17: From analysis of the AA'XX' spectrum (see Appendix B), $J_{AA'} = 19.1$ Hz, $J_{XX'} = 2.1$ Hz, $J_{AX} = 72.0$ Hz, and $J_{AX'} = 20.5$ Hz.

Fig. 6.18: Analysis of the AA'XX' spectrum gives $J_{AA'} = 131.9$ Hz, $J_{XX'} = 9.8$ Hz, $J_{AX} = 75.7$ Hz, and $J_{AX'} = 2.7$ Hz.

From relevant entries in Tables 5.1 to 5.3, it is clear that $J_{AX} = {}^2J_{HF}$, $J_{AX'} = {}^3J_{HF}$, $J_{AA'} = {}^3J_{FF}$, and $J_{XX'} = {}^3J_{HH}$. Although the specific values of J in the tables are only illustrative and can vary significantly with chemical substituents, it is generally true that ${}^3J(trans) > {}^3J(cis)$ for comparable couplings. A consistent pattern of coupling constants is obtained by assigning the spectrum of Fig. 6.17 to *cis* CHF=CHF and that of Fig. 6.18 to *trans* CHF=CHF.

6.11 *a*. A_3X_2; *b*. AA'XX'; *c*. A_3; *d*. ABX; *e*. AX.

6.12 *a*. ABB'XX'; *b*. AA'XX' and ABXY, with intensity ratio of $1:2$ if the three conformations are equally populated; *c*. ABC; *d*. three different ABX systems; *e*. three different AX systems.

6.13

n	Relative Intensity						
0				1			
1			1	1	1		
2		1	2	3	2	1	
3	1	3	6	7	6	3	1

6.14 The spectrum consists of a single strong line from several isotopomers containing Hg isotopes of $I = 0$ (collectively accounting for about 70% natural abundance) and a pair of satellites from the isotopomer containing ^{199}Hg (17%), $I = {}^1/_2$. A second set of satellites from ^{201}Hg (13%), $I = 3/2$, occurs also, but quadrupole broadening (see Chapter 8) would probably render them unobservable.

6.15 *a*. The spectrum of H_a is a quartet (1:1:1:1) with spacing 137 Hz. The spectrum of H_b is a septet (1:2:3:4:3:2:1) with spacing 48 Hz.

b. Unresolved two-bond and three-bond couplings, as well as strong coupling effects, lead to additional unresolved lines. Also, quadrupolar relaxation of ^{11}B can broaden ^{1}H lines, as shown in Chapter 8.

c. $^{1}J(^{10}BH_a) = 46$ Hz.

From Chapter 7

7.1   110 kHz

7.2   Energy levels:

$E_1 = -\frac{1}{2}(\nu_A + \nu_X) - \frac{1}{4}D(3 \cos^2\theta - 1)$

$E_1 = -\frac{1}{2}(\nu_A + \nu_X) + \frac{1}{4}D(3 \cos^2\theta - 1)$

$E_1 = \frac{1}{2}(\nu_A + \nu_X) + \frac{1}{4}D(3 \cos^2\theta - 1)$

$E_1 = \frac{1}{2}(\nu_A + \nu_X) - \frac{1}{4}D(3 \cos^2\theta - 1)$

$D = \gamma_A \gamma_X \hbar^2 / r^3$

Spectral lines for A:

$\nu_1 = E_4 - E_2 = \nu_A - \frac{1}{2}D(3 \cos^2\theta - 1)$

$\nu_2 = E_3 - E_1 = \nu_A + \frac{1}{2}D(3 \cos^2\theta - 1)$

7.3   Energy levels:

$E_1 = -\nu_H - \frac{1}{4}D(3 \cos^2\theta - 1)$ *symmetric*

$E_2 = 0 + \frac{1}{2}D(3 \cos^2\theta - 1)$ *symmetric*

$E_3 = 0 + 0$ *antisymmetric*

$E_4 = +\nu_H - \frac{1}{4}D(3 \cos^2\theta - 1)$ *symmetric*

$D = \gamma_H^2 \hbar^2 / r^3$

Spectrum:

$\nu_1 = E_4 - E_2 = \nu_H - \frac{3}{4}D(3 \cos^2\theta - 1)$

$\nu_2 = E_2 - E_1 = \nu_H + \frac{3}{4}D(3 \cos^2\theta - 1)$

7.4 C_2H_4: Powder pattern like that of Fig. 7.8a, with $\sigma_{11} = -81.1$, $\sigma_{22} = 84.3$, and $\sigma_{33} = 177.9$ ppm. C_2H_2: Powder pattern like Fig. 7.8b, with $\sigma_\perp = 39.0$ and $\sigma_\| = 279.4$ ppm.

7.5   a.

	f_z	$\alpha\alpha$	$\alpha\beta$	$\beta\alpha$	$\beta\beta$
	f_z	+1	0	0	−1
$\alpha\alpha$	+1	A	C	C	E
$\alpha\beta$	0	D	A	B	C
$\beta\alpha$	0	D	B	A	C
$\beta\beta$	−1	F	D	D	A

The operators step f_z up or down; for example, $|\alpha\beta> = D|\alpha\alpha>$, $\Delta f_z = -1$.

b. $I_{1x}I_{2x} + I_{1y}I_{2y} = \frac{1}{2}[I_1^{+}I_2^{-} + I_1^{-}I_2^{+}]$

The product of the raising and lowering operators increases f_z for one nucleus and simultaneously decreases f_z for the second nucleus, each by one unit.

From Chapter 8

8.1   a. $R_1^D(CH) = 1.12 \times 10^9 \tau_c$ s^{-1}

b. $R_1^D(CN) = 0.227 \times 10^9 \tau_c$ s^{-1}

$c.\ R_1^{CSA} = 2.19 \times 10^9\ \tau_c\ \mathrm{s}^{-1}$
$R_1 = 3.54 \times 10^9\ \tau_c\ \mathrm{s}^{-1}$

8.2 $R_2^{SC} = 6.9\ \mathrm{s}^{-1}$
$T_2^{SC} = 0.14\ \mathrm{s}$
$\nu_{1/_2} = 2.3\ \mathrm{Hz}$

8.3 T_1/T_2: $0.2\ \nu_0$, 1.08; ν_0, 2.27; $1.2\ \nu_0$, 2.70; $1.8\ \nu_0$, 4.00

8.4 $R_1^{Dipolar} = 0.036\ \mathrm{s}^{-1}$; $R_1^{Other} = 0.021\ \mathrm{s}^{-1}$

8.5 $\eta(^{29}\mathrm{Si}) = -2.52$

8.6   The derivation is tedious but straightforward. Note that at large values of τ_c, higher powers of τ_c dominate and smaller terms can be ignored.

8.7   The dipolar interaction predominates at lower temperatures, and spin-rotation interaction is important at high temperatures.

8.8 $a.\ \tau_c(\mathrm{ring}) = 1.93 \times 10^{-12}\ \mathrm{s}$
$\tau_c(\mathrm{methyl}) = 5.1 \times 10^{-13}\ \mathrm{s}$
$b.\ T_1^D(^{13}\mathrm{C}) = 30\ \mathrm{s}$

8.9 $T_1 = 2.2\ \mathrm{s}$

From Chapter 9

9.3   5th echo: CP, signal reduced by 6%; MG, reduction of 0.2%. 20th echo: CP, reduction of 83%; MG, no reduction.

9.4 $\omega - \omega_o = 3.87\ \gamma B_1$.

9.5   A vector twice as long as those shown in Fig. 9.2 would precess at ω and be refocused at each echo. The modulation of the other half of the echo would be at frequency $2\pi J$.

9.6 $a.$ From Table 6.4 and Fig. 6.7, the antisymmetric transition between ϕ_3 and ϕ_7 comes at frequency ν_C. The six symmetric functions give three ^{13}C transitions ($\Delta f_C = \pm 1$) and four ^{1}H transitions ($\Delta f_H = \pm 1$).
$b.$ Excess populations: ϕ_1, $+17$; ϕ_5, $+15$; ϕ_2, $+1$; ϕ_6, -1; ϕ_4, -15; ϕ_8, -17; ϕ_3, $+1$; ϕ_7, -1.
$c.$ Excess populations: ϕ_1, -15; ϕ_5, $+15$; ϕ_2, $+1$; ϕ_6, -1; ϕ_4, $+17$; ϕ_8, -17; ϕ_3, $+1$; ϕ_7, -1.
^{13}C transitions:

		Relative Intensity		
	Frequency	Equil.	Perturbed	Δ
$\phi_1 \rightarrow \phi_5$	$\nu_C - J$	2	-30	-32
$\phi_2 \rightarrow \phi_6$	ν_C	2	2	0
$\phi_4 \rightarrow \phi_8$	$\nu_C + J$	2	$+34$	$+32$
$\phi_3 \rightarrow \phi_7$	ν_C	2	2	0

From Chapter 10

10.2 CH_3, 14 ppm; $CH_2=$, 117 ppm; $CH=$, 135 ppm; and $CDCl_3$, 78 ppm are assigned definitively. CH_2, 20 ppm and 32 ppm; and CH_2O, 70 ppm and 71 ppm show ambiguities.

10.3 CH_3, 0.9 ppm; $CH=$, 5.9 ppm; and $CHCl_3$, 7.25 ppm are assigned definitively. $CH_2=$, 5.15 ppm and 5.25 ppm; CH_2, 1.4 ppm and 1.55 ppm; and CH_2O, 3.4 ppm and 3.9 ppm show ambiguities. However, on the basis of chemical shift correlations, CH_2 adjacent to both $C=C$ and O has a lower shielding ($\delta = 3.9$ ppm).

10.4 CH_3, 0.9 ppm; CH_2, 1.4 ppm; CH_2, 1.55 ppm; CH_2O, 3.4 ppm; CH_2O, 3.9 ppm; CH, 5.9 ppm; and $CHCl_3$, 7.25 ppm are assigned definitively. $CH_2=$, 5.15 ppm and 5.25 ppm remain ambiguous.

10.5   All ambiguities regarding ^{13}C assignments are readily resolved. The only remaining ambiguity is the ^1H chemical shifts for $CH_2=$, which can best be resolved by examining the coupling patterns on an expanded display as an AMX spectrum.

From Chapter 11

11.7   With the numbering used in Fig. 11.1, at 2τ $\rho(4)$ is:

a. $\rho(4) = \rho(1)$ for a 180_y pulse or $-\rho(1)$ for a 180_x pulse, where $\rho(1)$ is given in Eq. 11.53.

b.

$$\rho(4) = \begin{bmatrix} -\epsilon_S & 0 & -i\epsilon_I e^C & 0 \\ 0 & \epsilon_S & 0 & -i\epsilon_I e^C \\ i\epsilon_I e^{-C} & 0 & -\epsilon_S & 0 \\ 0 & i\epsilon_I e^{-C} & 0 & \epsilon_S \end{bmatrix}$$

$$C = 2\pi i v_I(2\tau) = 4\pi i v_I \tau$$

Note that the S populations are inverted and the coherences show no dependence on J but continue to evolve at frequency v_I.

c.

$$\rho(4) = \begin{bmatrix} -\epsilon & 0 & i\epsilon_I e^{-D} & 0 \\ 0 & \epsilon_S & 0 & i\epsilon_I e^D \\ -i\epsilon_I e^D & 0 & -\epsilon_S & 0 \\ 0 & -i\epsilon_I e^{-D} & 0 & \epsilon_S \end{bmatrix}$$

$$D = 2\pi i J(2\tau) = 4\pi i J \tau$$

The coherences evolve with J but the chemical shift has been eliminated at the echo.

11.8   The process may be summarized in terms of the following six operations:

$$I_z + S_z \xrightarrow{90°I_x} \xrightarrow{\Omega_I \tau} \xrightarrow{\pi J I_z S_z \tau} \xrightarrow{180°(a, b, c)} \xrightarrow{\Omega_I \tau} \xrightarrow{\pi J I_z S_z \tau},$$

where the characteristics of the 180° pulse differ for *a*, *b*, and *c*. Each of the four evolution steps gives two terms in *I*, so there are a total of 16 terms in *I*. *S* does not evolve and remains either as S_z or $-S_z$. For example, at time τ we obtain

$$-I_y \cos \Omega_I \tau \cos \pi J \tau + I_x S_z \cos \Omega_I \tau \sin \pi J \tau + I_x \sin \Omega_I \tau \cos \pi J \tau$$
$$+ I_y S_z \sin \Omega_I \tau \sin \pi J \tau + S_z.$$

Without the "shortcuts" from recognizing the refocusing effects of the echoes, a great deal of tedious algebra would be required to obtain the results, which are (for a 180_x pulse):

a. $I_y + S_z$

b. $-I_y \cos 2\Omega_I \tau + I_x \sin 2\Omega_I \tau - S_z$

c. $I_y \cos 4\pi J \tau - I_x S_z \sin 4\pi J \tau - S_z$

11.9 $\mathrm{Tr}\,[M_{xy}\rho''(5)] \propto \mathrm{Tr}\,[I^+\rho''(5)] = 0$

From Chapter 12

12.1 $I_z + S_z \xrightarrow{90°I_x, \ 90°S_x} -(I_y + S_y) \xrightarrow{180°I_x, \ 180°S_x} I_y + S_y \xrightarrow{90°I_x, \ 90°S_x} I_z + S_z$

Ideally an uncoupled ^{13}C shows only longitudinal magnetization, hence gives no signal. In practice, pulse imperfections give rise to a component M_{xy}, which is cancelled by the phase cycle shown for INADEQUATE.

12.3   For convenience, the coherence pathway can be divided into evolutions due to chemical shifts and coupling, as follows:

$$I_z + S_z \xrightarrow{90°I_x, \ 90°S_x} -(I_y + S_y) \xrightarrow{\Omega_I t_1/2, \ \Omega_S t_1/2} \xrightarrow{180°I_y, \ 180°S_y} \xrightarrow{\Omega_I t_1/2, \ \Omega_S t_1/2}$$

$$-(I_y + S_y) \xrightarrow{\Omega_I(\Delta - t_1), \ \Omega_S(\Delta - t_1)} -I_y \cos\Omega_I(\Delta - t_1) + I_x \sin\Omega_I(\Delta - t_1)$$
$$- S_y \cos\Omega_S(\Delta - t_1) + S_x \sin\Omega_S(\Delta - t_1)$$

$$I_z + S_z \xrightarrow{90°I_x, \ 90°S_x} -(I_y + S_y) \xrightarrow{\pi J I_z S_z \Delta} -I_y \cos\pi\Delta + I_x S_z \sin\pi\Delta$$
$$- S_y \cos\pi\Delta + I_z S_x \sin\pi\Delta$$

The first pathway shows modulation with t_1, whereas the second pathway does not.

12.4   The initial steps (up to time 5) are given in Eq. 11.84. The steps after time 9 involve refocusing of chemical shifts and evolution of the coupling during the period $1/2J$ as follows:

$$I_x S_x \cos\Omega_S t_1 + I_x S_z \sin\Omega_S t_1 \xrightarrow{\pi J I_z S_z \frac{1}{2J}} I_x S_x \cos\Omega_S t_1 + I_y \sin\Omega_S t_1$$

The multiple quantum term $I_x S_x$ produces no signal, and the term $I_y \sin\Omega_S t_1$ evolves during t_2 to produce the signal.

12.5   Because of the three evolution periods, this pathway is best handled in a manner similar to that in Table 12.1. Using I and S for the proton spins and T for ^{13}C and taking into account the refocusing effect of the 180° pulses, the initial portion (up to time 5) is:

$$I_z + S_z \xrightarrow{90°I_x, S_x} -I_y - S_y \xrightarrow{\pi J_{CH} \frac{1}{2J}} I_x T_z + S_x T_z$$

$$\xrightarrow{90°T_x} -I_x T_y - S_x T_y \xrightarrow{\Omega_T t_1} -(I_x T_y + S_x T_y)\cos\Omega_T t_1$$

$$+ (I_x T_x + S_x T_x)\sin\Omega_T t_1$$

$$\xrightarrow{90°T_x} -(I_x T_z + S_x T_z)\cos\Omega_T t_1 + (I_x T_x + S_x T_x)\sin\Omega_T t_1$$

$$\xrightarrow{\pi J_{CH} \frac{1}{2J}} -(I_y + S_y)\cos\Omega_T t_1 + (I_x T_x + S_x T_x)\sin\Omega_T t_1$$

$$\xrightarrow{\Omega_I, S t_2} -I_y \cos\Omega_T t_1 \cos\Omega_I t_2 - S_y \cos\Omega_T t_1 \cos\Omega_I t_2$$

$$+ I_x \cos\Omega_T t_1 \sin\Omega_I t_2 + S_x \cos\Omega_T t_1 \sin\Omega_I t_2$$

$$+ I_x T_x \sin\Omega_T t_1 \cos\Omega_I t_2 + I_y T_x \sin\Omega_T t_1 \sin\Omega_I t_2$$

$$+ S_x T_x \sin\Omega_T t_1 \cos\Omega_S t_2 + S_y T_x \sin\Omega_T t_1 \sin\Omega_S t_2$$

(This pathway ignores the initial T_z, which cycles back to T_z at time 5.) The 90° I,S pulse at time 5 generates longitudinal I and S polarization for the NOESY relaxation period τ, and an HMQC transfer to ^{15}N and back to ^1H follows. Phase cycling removes many unwanted terms.

References

1. *Encyclopedia of Nuclear Magnetic Resonance*, D. M. Grant and R. K. Harris, Eds. (John Wiley & Sons, Chichester, England, 1996).
2. W. Pauli, *Naturwissenschaften* **12**, 741 (1924).
3. D. M. Dennison, *Proc. Roy. Soc. London* **A115**, 483 (1927).
4. O. Stern, *Z. Phys.* **7**, 249 (1921).
5. W. Gerlach, O. Stern, *Z. Phys.* **8**, 110 (1921).
6. J. M. B. Kellogg, I. I. Rabi, J. R. Zacharias, *Phys. Rev.* **50**, 472 (1936).
7. I. I. Rabi, J. R. Zacharias, S. Millman, P. Kusch, *Phys. Rev.* **53**, 318 (1938).
8. C. J. Gorter, *Physica* **3**, 995 (1936).
9. H. C. Torrey, *Encyclopedia of NMR*, p. 666.
10. E. M. Purcell, H. C. Torrey, R. V. Pound, *Phys. Rev.* **69**, 37 (1946).
11. F. Bloch, W. W. Hansen, M. Packard, *Phys. Rev.* **69**, 127 (1946).
12. W. G. Proctor, F. C. Yu, *Phys. Rev.* **77**, 717 (1950).
13. W. C. Dickenson, *Phys. Rev.* **77**, 736 (1950).
14. G. Lindstrom, *Phys. Rev.* **78**, 817 (1950).
15. H. A. Thomas, *Phys. Rev.* **80**, 901 (1950).
16. J. T. Arnold, S. S. Dharmatti, M. E. Packard, *J. Chem Phys.* **19**, 507 (1951).
17. R. R. Ernst, W. A. Anderson, *Rev. Sci. Instrum.* **37**, 93 (1966).
18. S. W. Homans, *A Dictionary of Concepts in NMR* (Oxford University Press, Oxford, 1992).
19. R. Freeman, *A Handbook of Nuclear Magnetic Resonance* (Longman Scientific and Technical, Harlow, Essex, England, 1987).
20. A. R. Lepley, G. L. Closs, *Chemically Induced Magnetic Polarization* (John Wiley & Sons, New York, 1973).
21. C. R. Bowers *et al.*, *Adv. Magn. Reson.* **14**, 269 (1990).
22. W. Happer, *Encyclopedia of NMR*, p. 3640.
23. F. Bloch, *Phys. Rev.* **70**, 460 (1946).
24. E. L. Hahn, *Phys. Rev.* **80**, 580 (1950).
25. F. A. Bovey, *Chem. Eng. News* **43**, 98 (1965).
26. R. K. Harris, J. Kowalewski, S. Cabral de Menezes, *Pure Appl. Chem.* **69**, 2489 (1997).

27. R. R. Ernst, G. Bodenhausen, A. Wokaun, *Principles of Nuclear Magnetic Resonance in One and Two Dimensions* (Clarendon Press, Oxford, 1987).

28. P. J. Hore, *Nuclear Magnetic Resonance* (Oxford University Press, Oxford, 1995).

29. R. J. Abraham, J. Fisher, P. Loftus, *Introduction to NMR Spectroscopy* (John Wiley & Sons, Chichester, England, 1988).

30. C. P. Slichter, *Principles of Magnetic Resonance* (Springer-Verlag, Berlin, 1990).

31. M. Goldman, *Quantum Description of High-Resolution NMR in Liquids* (Oxford University Press, Oxford, 1988).

32. R. K. Harris, *Nuclear Magnetic Resonance Spectroscopy: A Physicochemical View* (Longman Scientific and Technical/John Wiley & Sons, Harlow, Essex, England/New York, 1983).

33. A. Abragam, *The Principles of Nuclear Magnetism* (Oxford University Press, London, 1961).

34. J. A. Pople, W. G. Schneider, H. J. Bernstein, *High Resolution Nuclear Magnetic Resonance* (McGraw-Hill, New York, 1959).

35. J. W. Emsley, J. Feeney, L. H. Sutcliffe, *High Resolution Nuclear Magnetic Resonance Spectroscopy* (Pergamon Press, London, 1965).

36. L. M. Jackman, F. A. Cotton, *Dynamic NMR Spectroscopy* (Academic Press, New York, 1975).

37. C.-N. Chen, D. I. Hoult, *Biomedical Magnetic Resonance Technology* (Adam Hilger, Bristol, 1989).

38. A. G. Redfield, S. D. Kunz, *J. Magn. Reson.* **19**, 250 (1975).

39. J. J. Led, H. Gesmar, *Encyclopedia of NMR*, p. 2119.

40. O. W. Howarth, *Encyclopedia of NMR*, p. 3967.

41. S. Braun, H.-O. Kalinowski, S. Berger, *150 and More Basic NMR Experiments—A Practical Course* (Wiley-VCH, New York, 1998).

42. D. Canet, *Nuclear Magnetic Resonance: Concepts and Methods* (John Wiley & Sons, Chichester, England, 1996).

43. W. A. Anderson, *Encyclopedia of NMR*, p. 2126.

44. J. C. Hoch, A. S. Stern, *Encyclopedia of NMR*, p. 2980.

45. H. D. W. Hill, *Encyclopedia of NMR*, p. 4505.

46. H. D. W. Hill, *Encyclopedia of NMR*, p. 3762.

47. V. W. Miner, W. W. Conover, *Encyclopedia of NMR*, p. 4340.

48. W. S. Warren, W. Richter, *Encyclopedia of NMR*, p. 1417.

49. W. E. Lamb, Jr., *Phys. Rev.* **60**, 817 (1941).

50. N. F. Ramsey, *Phys. Rev.* **78**, 699 (1950).

51. G. A. Webb, *Encyclopedia of NMR*, p. 4307.

52. J. C. Facelli, *Encyclopedia of NMR*, p. 4327.

53. C. J. Jameson, *Encyclopedia of NMR*, p. 1273.

54. *Pure Appl. Chem.* **29**, 627 (1972).

55. J. L. Markley, A. Bax, Y. Arata, C. W. Hilbers, R. Kaptein, B. D. Sykes, P. E. Wright, K. Wüthrich, *Pure Appl. Chem.* **70**, 130 (1998).

56. R. K. Harris, *Encyclopedia of NMR*, p. 3301.

57. H. Spiesecke, W. G. Schneider, *J. Chem. Phys.* **35**, 722 (1961).

58. H. Spiesecke, W. G. Schneider, *J. Chem. Phys.* **35**, 731 (1961).

59. C. E. Johnson, F. A. Bovey, *J. Chem. Phys.* **29**, 1012 (1958).

60. K. Wüthrich, *NMR of Proteins and Nucleic Acids* (John Wiley & Sons, New York, 1986).

61. G. N. La Mar, W. D. Horrocks, Jr., R. H. Holm, *NMR of Paramagnetic Molecules* (Academic Press, New York, 1973).

62. I. Bertini, C. Luchinat, *NMR of Paramagnetic Molecules in Biological Systems* (Benjamin/ Cummings, Menlo Park, CA, 1986).

63. R. E. Sievers, *Nuclear Magnetic Resonance Shift Reagents* (Academic Press, New York, 1973).

64. H. Günther, *NMR Spectroscopy* (John Wiley & Sons, Chichester, England, 1995).

65. F. A. Bovey, *NMR Spectroscopy* (Academic Press, New York, 1988).

66. F. W. Wehrli, T. Wirthlin, *Interpretation of Carbon-13 NMR Spectra* (Heyden, London, 1976).

67. G. C. Levy, R. L. Lichter, *Nitrogen-15 NMR Spectroscopy* (John Wiley & Sons, New York, 1979).

68. P. Laszlo, *NMR of Newly Accessible Nuclei* (Academic Press, New York, 1983).

69. B. R. Seavey, E. A. Farr, W. M. Westler, J. L. Markley, *J. Biomol. NMR* **1**, 217 (1991).

70. N. F. Ramsey, E. M. Purcell, *Phys. Rev.* **85**, 143 (1952).

71. M. Karplus, *J. Chem. Phys.* **30**, 11 (1959).

72. A. C. Wang, A. Bax, *J. Am. Chem. Soc.* **118**, 2483 (1996).

73. J. C. Facelli, *Encyclopedia of NMR*, p. 2516.

74. M. Barfield, *Encyclopedia of NMR*, p. 2520.

75. I. Ando, *Encyclopedia of NMR*, p. 2512.

76. T. Schaefer, *Encyclopedia of NMR*, p. 4571.

77. M. J. Minch, *Concepts Magn. Reson.* **6**, 41 (1994).

78. C. W. Haigh, *J. Chem. Soc. A* 1682 (1970).

79. P. L. Corio, *Structure of High-Resolution NMR Spectra* (Academic Press, New York, 1966).

80. P. L. Corio, S. L. Smith, *Encyclopedia of NMR*, p. 797.

81. D. S. Stephenson, *Encyclopedia of NMR*, p. 816.

82. J. Kaski, J. Vaara, J. Jokisaari, *J. Am. Chem. Soc.* **118**, 8879 (1996).

83. S. R. Hartmann, E. L. Hahn, *Phys. Rev.* **128**, 2042 (1962).

84. G. E. Pake, *J. Chem. Phys.* **16** 327 (1948).

85. J. S. Waugh, L. M. Huber, U. Haeberlen, *Phys. Rev. Lett.* **20**, 180 (1968).

86. E. O. Stejskal, J. D. Memory, *High Resolution NMR in the Solid State* (Oxford University Press, Oxford, 1994).

87. C. A. Fyfe, *Solid State NMR for Chemists* (C.F.C. Press, Guelph, Ontario, Canada, 1983).

88. J. W. Emsley, J. C. Lindon, *NMR Spectroscopy Using Liquid Crystal Solvents* (Pergamon Press, Oxford, 1975).

89. N. Bloembergen, E. M. Purcell, R. V. Pound, *Phys. Rev.* **73**, 679 (1948).

90. I. Solomon, *Phys. Rev.* **99**, 559 (1955).

91. A. D. Bain, E. P. Mazzola, S. W. Page, *Magn. Reson. Chem.* **36**, 403 (1998).

92. C. M. Venkatachalam, D. W. Urry, *J. Magn. Reson.* **41**, 313 (1980).

93. P. S. Hubbard, *Phys. Rev.* **131**, 1155 (1963).

94. T. C. Farrar, T. C. Stringfellow, *Encyclopedia of NMR*, p. 4101.

95. K. Pervushin, R. Riek, G. Wider, K. Wüthrich, *J. Am. Chem. Soc.* **120**, 6394 (1998).

96. G. Lipari, A. Szabo, *J. Am. Chem. Soc.* **104**, 4546 (1982).

97. T. C. Farrar, E. D. Becker, *Pulse and Fourier Transform NMR* (Academic Press, New York, 1971).

98. D. Neuhaus, M. P. Williamson, *The Nuclear Overhauser Effect in Structural and Conformational Analysis* (VCH, New York, 1989).

99. D. Neuhaus, *Encyclopedia of NMR*, p. 3290.

100. G. C. Levy, D. J. Kerwood, *Encyclopedia of NMR*, p. 1147.

101. J. Kowalewski, *Encyclopedia of NMR*, p. 3456.

102. D. E. Woessner, *Encyclopedia of NMR*, p. 1068.

103. R. E. D. McClung, *Encyclopedia of NMR*, p. 4530.

104. H.Y. Carr, E. M. Purcell, *Phys. Rev.* **94**, 630 (1954).

105. S. Meiboom, D. Gill, *Rev. Sci. Instrum.* **29**, 688 (1958).

106. R. Freeman, *Spin Choreography* (Spektrum Academic Publishers, Oxford, 1997).

107. D. J. States, R. A. Haberkorn, D. J. Ruben, *J. Magn. Reson.* **48**, 286 (1982).

108. D. Marion, K. Wüthrich, *Biochem. Biophys. Res. Commun.* **113**, 967 (1983).

109. F. J. M. van de Ven, *Multidimensional NMR in Liquids* (VCH Publishers, New York, 1995).

110. R. S. Macomber, *A Complete Introduction to Modern NMR Spectroscopy* (John Wiley & Sons, New York, 1998).

111. W. R. Croasmun, R. M. K. Carlson (Eds.), *Two-Dimensional NMR Spectroscopy: Applications for Chemists and Biochemists* (VCH Publishers, New York, 1994).

112. A. Bax, *Two-Dimensional Nuclear Magnetic Resonance in Liquids* (Delft University Press, Delft, Holland, 1982).

113. T. C. Farrar, J. E. Harriman, *Density Matrix Theory and Its Applications in NMR Spectroscopy* (Farragut Press, Madison, WI, 1998).

114. G. D. Mateescu, A. Valeriu, *2D NMR Density Matrix and Product Operator Treatment* (PTR Prentice-Hall, Englewood Cliffs, NJ, 1993).

115. M. Munowitz, *Coherence and NMR* (John Wiley & Sons, New York, 1988).

116. W. S. Brey (Ed.), *Pulse Methods in 1D and 2D Liquid-Phase NMR* (Academic Press, San Diego, 1988).

117. R. Burger, P. Bigler, *J. Magn. Reson.* **135**, 529 (1998).

118. D. Marion, L. E. Kay, S. W. Sparks, D. A. Torchia, A. Bax, *J. Am. Chem. Soc.* **111**, 1515 (1989).

119. N. Chandrakumar, S. Subramanian, *Modern Techniques in High-Resolution FT-NMR* (Springer-Verlag, New York, 1987).

120. J. Cavanagh, W. J. Fairbrother, A. G. Palmer, III, N. J. Skelton, *Protein NMR Spectroscopy* (Academic Press, San Diego, 1996).

121. H. Kessler, M. Gehrke, C. Griesinger, *Angew. Chem. Int. Ed.* **27**, 490 (1988).

122. F. A. Bovey, *High Resolution NMR of Macromolecules* (Academic Press, New York, 1972).

123. M. Ikura, L. E. Kay, A. Bax, *Biochemistry* **29**, 4659 (1990).

124. C. A. Bewley, K. R. Gustafson, M. R. Boyd, D. G. Covell, A. Bax, G. M. Clore, A. M. Gronenborn, *Nature Struct. Biol.* **5**, 571 (1998).

125. J. K. M. Sanders, B. K. Hunter, *Modern NMR Spectroscopy—A Guide for Chemists* (Oxford University Press, Oxford, 1993).

126. A. E. Derome, *Modern NMR Techniques for Chemistry Research* (Pergamon Press, Oxford, 1987).

127. H. Friebolin, *One and Two-Dimensional NMR Spectroscopy* (VCH Publishers, New York, 1993).

128. E. Breitmaier, *Structure Elucidation by NMR in Organic Chemistry: A Practical Guide* (John Wiley & Sons, Chichester, England, 1993).

129. J. K. M. Sanders, E. C. Constable, B. K. Hunter, *Modern NMR Spectroscopy—A Workbook of Chemical Problems* (Oxford University Press, Oxford, UK, 1989).

130. G. E. Martin, A. S. Zektzer, *Two-Dimensional NMR Methods for Establishing Molecular Connectivity* (VCH Publishers, New York, 1988).

131. G. M. Clore, A. M. Gronenborn, *NMR of Proteins* (Macmillan, London, 1993).

132. J. N. S. Evans, *Biomolecular NMR Spectroscopy* (Oxford University Press, Oxford, 1995).

133. C. T. W. Moonen, P. C. M. van Zijl, J. A. Frank, D. Le Bihan, E. D. Becker, *Science* **250**, 53 (1990).

134. R. J. Gillies (Ed.), *NMR in Physiology and Biomedicine* (Academic Press, San Diego, 1998).

135. P. T. Callaghan, *Principles of Nuclear Magnetic Resonance Microscopy* (Oxford University Press, Oxford, 1991).

Index